선박 내연기관
운 용 실 무

MAN ES 캠 구동 엔진(MC-C)

정현우 편저

해광출판사

"인생은 한 권의 책과 같다.
어리석은 자는 마구 넘겨 버리지만, 현명한 자는 열심히 읽는다.
인생은 단 한 번만 읽을 수 있다는 것을 알기 때문이다."

평생을 자식들 걱정과 뒷바라지를 위해 살아오시고,
책을 내는데 있어서 아낌없는 조언과 격려를 해주신
아버지와 어머니께
그리고
항상 나를 믿고 응원해주는 사랑하는 아내 나경이와
나의 햇살, 딸 하율이에게 사랑의 마음을 담아 이 책을 바칩니다.

머리말

이 책은 Main Engine중 MAN ES 캠 구동 엔진(MC-C)에 대한 내용으로, MAN ES는 선박용 Engine에서 압도적인 점유율을 차지하고 있어 많은 기관사들이 이미 익숙하게 경험한 엔진입니다.

그러나 이 책은 그러한 경험 많은 분들을 위해 작성된 것이 아닙니다.

진급이 빠른 요즘, 예상치 못한 상황이 발생했을 때 상대적으로 경험이 부족한 초임 1기사 분들 또는, MAN ES 캠 구동 엔진이 낯선 후배들을 위해 이 책을 준비했습니다.

이 책이 그분들에게 좋은 길잡이가 되기를 진심으로 바랍니다.

승선 중 발생할 수 있는 다양한 문제들에 대해 기본적인 조치 방법과 정비 과정을 초임 1기사의 시각에서 이해하기 쉽게 설명하고, 작업의 순서에 따라 사진 자료를 첨부해 실용성을 높였습니다. 앞서 출간한 선박용 전자제어 엔진 책과는 기본적인 엔진의 구동 원리는 다르지만, 운용 관점에서 전자제어 시스템의 유무만 다를 뿐, 유지보수 측면에서는 큰 차이가 없습니다.

오히려 캠 구동 엔진은 더 단순하고 수월한 작업이 많습니다.

비록 이 책이 선박 엔진에서 발생할 수 있는 모든 문제에 대한 정답을 제시하지는 못하겠지만, 캠 구동 엔진을 처음 접하는 분들께 반드시 도움이 될 것이라 확신합니다.

현재 더 이상 승선하지는 않지만, 이 책을 읽으시는 모든 분들이 더 나은 책을 집필하여 후배들에게 도움을 줄 수 있는 선배 해기사가 되기를 바랍니다.

'한 사람의 열 걸음보다 열 사람의 한 걸음이 더 큰 걸음'이라는 말이 있습니다.

저 역시 후배들에게 든든한 선배가 되고자 현업에서 최선을 다하겠습니다.

끝으로 이 책의 교정과 출간을 위해서 물심양면으로 도움을 주신 해광출판사와 문우당서점 대표님께 깊은 감사의 마음을 전합니다.

오늘도 끝이 보이지 않는 먼 바다에서 인생의 파도에 맞서는 젊은 초임 사관들을 응원합니다.

함께 노력하며 성실한 승선 생활을 통해 안전하고 건강한 운항을 이루길 기원합니다.

1ST Engineer **정 현 우**

Slow and Steady Wins the Race!

목차

[1] MAN ES 캠 구동 엔진(MC-C) 운용

MAN ES 캠 구동 엔진(MC-C)에 대한 기본적인 운용에 대하여 설명하자면, Man ES 캠 구동 엔진의 경우 **각 Cylinder에 F.O.P(Fuel Oil Pump)가 설치**되어 분사를 담당하며 **F.O.P(Fuel Oil Pump) Cam Position** 문제가 자주 발생한다.

Cylinder마다 독립적으로 F.O.P(Fuel Oil Pump)가 존재하여 일부 F.O.P의 Cam Position 다르거나 제대로 변경되지 않아서(Ahead/Astern) Engine Start시 Cam Position Alarm 및 시동 불량 등이 발생할 수 있다.

Cam Position이 제대로 변경되지 않았다면 Engine Stop 후 Engine을 재 시동하거나 입·출항이 임박한 상황에서는 Half Ahead, Slow Astern까지는 사용해도 되며, 상기 사항이 반복된다면 Reversing Mechanism의 Air Cylinder를 분해하여 점검해야 한다.

Engine의 원리에 대한 이론적인 설명 대신 실무에 적합한 교재이므로 운용에 대하여 설명하고자 한다.

1) 캠 위치(CAM POSITION) ABNORMAL

Cam Position이 Ahead / Astern 방향으로 정확히 가지 않을 경우,

"WRONG WAY/FUEL CAM POSITION ABNR"와 같은 알람이 AMS(Alarm Monitoring System)에 발생하며 Indicator Panel의 **Ahead / Astern** Lamp에 불이 들어오지 않고, C/R Control Panel 내부 Safety System의 **Fuel Cam Position**에 **"빨간 불"**이 점등된다.

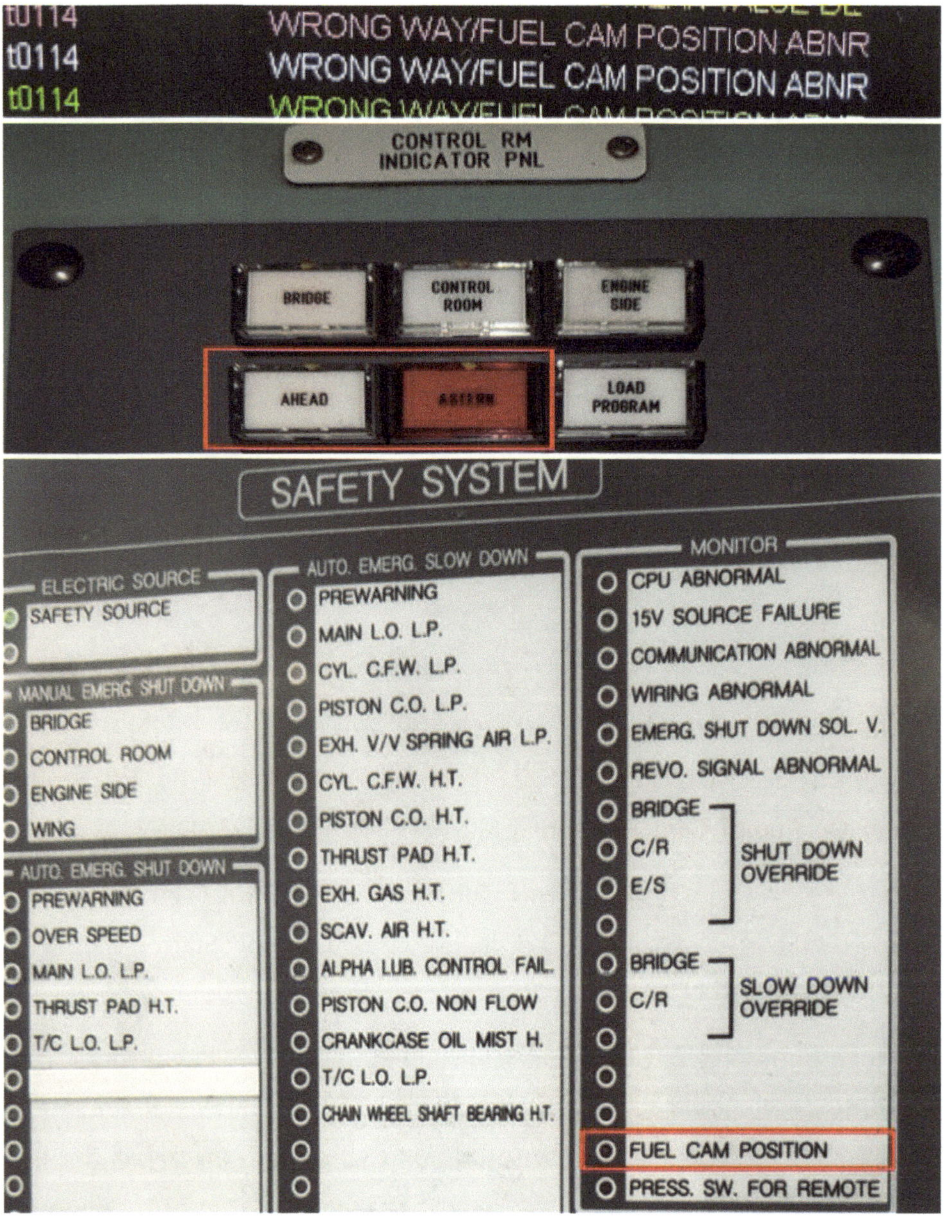

해당 Alarm이 발생할 경우 실제 Local에서 Cam Position이 변경되지 않은 Cylinder

를 확인하고 Engine Stop후 다시 재시동을 한다.

빈번하게 발생되는 Cylinder의 경우 Air Cylinder를 분해 점검한다.

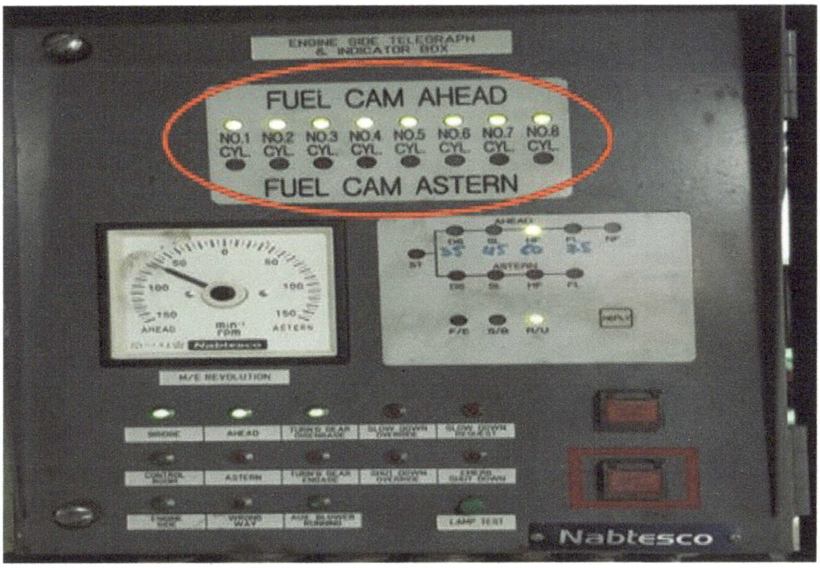

< Engine Side Telegraph & Indicator Box >

< F.O.P Cam Position >

상기 그림과 같이 Cam Position이 빠져나오면 'Ahead Position'
Cam Position이 들어가면 'Astern Position'

만약 Indicator Panel의 Ahead / Astern Lamp가 깜빡이거나 점등되지 않을 경우, Air Cylinder의 Clamping Piece의 Hexagon Screw 분해하여 취외 후 소제한 다음 조립한다.

또는 Cam Position Switch Connector에 연결된 Power Cable 불량이거나 Junction Box Power Cable 불량일 수도 있다.

오랫동안 교환하지 않았고 예비품 보유 중이라면 Clamping Piece, Position Switch, Connector 교환하는 것을 권한다.

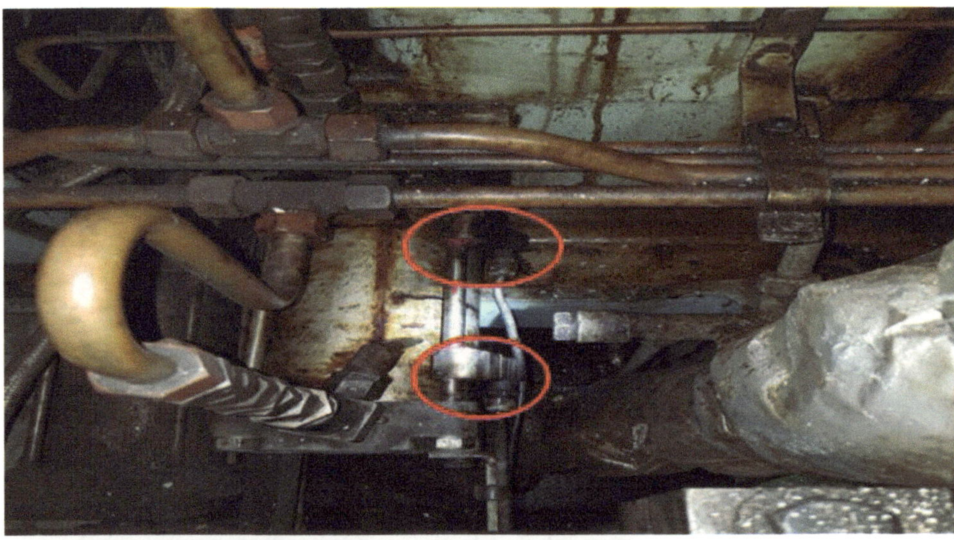

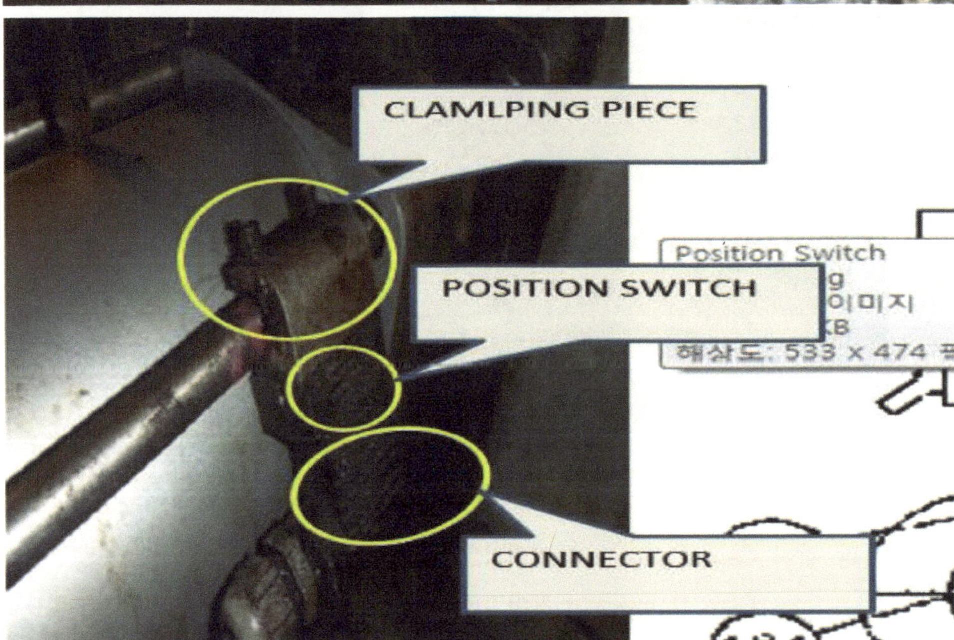

2) 기관 제어실(ENGINE CONTROL ROOM)

< Indicator Panel >

AUX. Blower Control 할 수 있으며 현재 Engine 운전 상태를 파악할 수 있다.

Manual Slow Turning Button을 이용하여 Slow Turning을 할 수 있다.

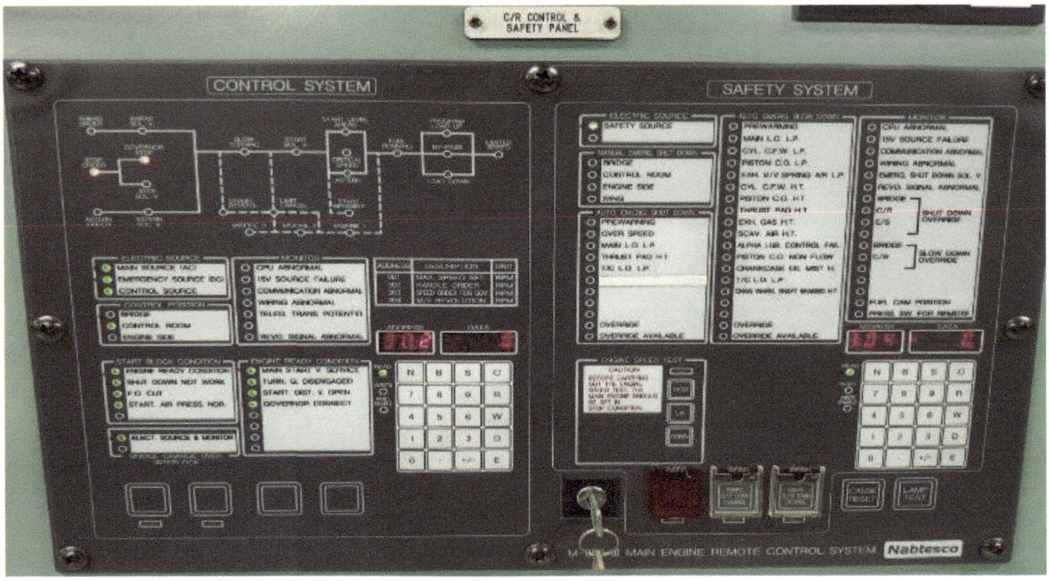

< C/R Control & Safety Panel >

각 종 Safety Alarm들을 확인 및 Setting Value 조절할 수 있다.
아래에 보면 Shut-down & Slow-down Cancel할 수도 있다.

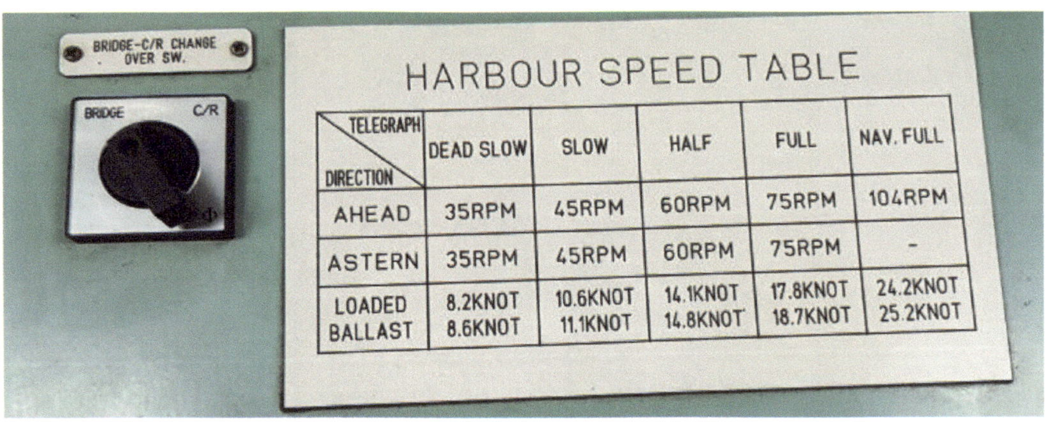

HARBOUR SPEED TABLE

TELEGRAPH / DIRECTION	DEAD SLOW	SLOW	HALF	FULL	NAV. FULL
AHEAD	35RPM	45RPM	60RPM	75RPM	104RPM
ASTERN	35RPM	45RPM	60RPM	75RPM	-
LOADED BALLAST	8.2KNOT 8.6KNOT	10.6KNOT 11.1KNOT	14.1KNOT 14.8KNOT	17.8KNOT 18.7KNOT	24.2KNOT 25.2KNOT

< Bridge / ECR Change Over SW & Speed Table >

엔진 사용 종료(Finished With Engine) 후 Air Blow를 위해 Change Over SW를 ECR로 변경한다.
출항 시 Engine 준비 완료되면 Bridge로 Change Over SW를 변경한다.
Speed Table로 Engine 사용 범위를 확인한다.

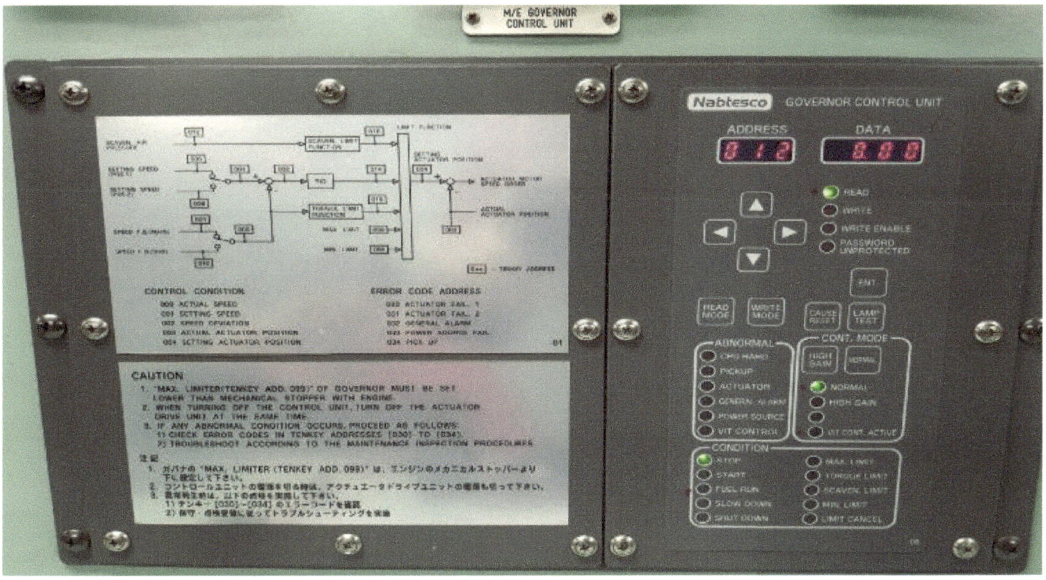

< M/E Governor Control Unit >

Control Mode에서 **'High Gain'**은 황천 시에 사용하는 Mode이며
사용 시 Main Engine의 Load가 Setting 된 값 이상으로 올라가지 않는다.

* HIGH GAIN MODE

7.9.2 HIGH GAIN mode

The actuator operating upper limit is reduced to a preset value and the control sensitivity is also increased.
This mode is effective to control the fluctuation in rotation during racing.
This mode is selected by pressing the HIGH GAIN push button on the front panel.

2.4 CONTROL MODE Indication Lamp／ Selector Switch

(Indication Lamps)		(Switch Location)
1) NORMAL	: Constant-speed control mode	Front Panel
2) HIGH GAIN	: High-sensitivity control mode	

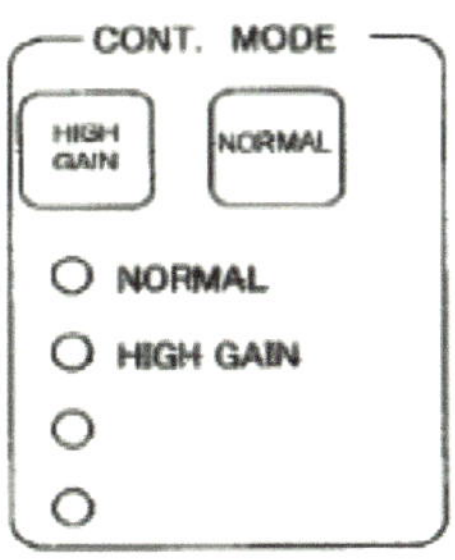

< CYL. Cut-out Operating Panel & F.O.P Roller Guide Lifting Valve >

3.

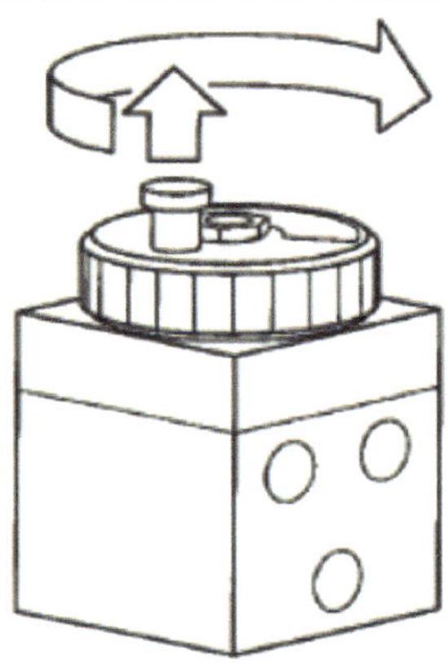

3. When the roller guide has been lifted, the dial of the 4/3 directional control valve can be returned to zero position. Lift the knob to release the dial from position 1.

The roller guide is now locked in a position free of the fuel cam, and the engine can be operated on the remaining cylinders, *see Volume I, Chapter 704.*

CYL Cut-out Operating Panel에서 홀수 또는 짝수기통 감통운전이 가능하며, 실제 Local M/E FOP Roller Guide Lifting Valve를 이용하여 개별 기통 감통 운전이 가능하다.

Engine운항 중에 F.O Pump 고착된 경우 Roller Guide Lifting Valve 사용하여 감통 운전한다.

Lifting Valve 사용 방법은 일반적으로 Engine 운전 중에 F.O Pump 고착된 경우에도 Camshaft에 의해 Cam이 지속적으로 회전하므로 운항 중 상기 Lifting Valve 도면의 Knob 들어 올린 상태에서 Dial의 Position을 '0' → '1'로 돌리면 '딸각' 소리가 나면서 더 이상 돌아가지 않는다.

그 때 Lifting Valve에서 Air Cylinder측으로 Compressor Air 공급되고, Lifting Pin이 Roller Guide 측으로 나와서 최종적으로 운전 중인 Roller Guide를 들어올려 F.O Pump 감통 운전된다.

그 후 들고 있는 Knob을 내리면 아래로 들어가 Lifting Valve가 고정된다.

아래 측 길게 빠져나와 있는 Air Cylinder Rod 끝단의 구멍에 Safety Pin을 집어넣어 고정한다.

감통 운전을 해제할 경우에는 반드시 Engine Stop중에 시행한다.

Engine운전 중에 Lifting Valve 원상복구 시 F.O Pump Roller Guide가 갑자기 아래로 떨어질 경우 Camshaft 의해 회전하는 Cam과 F.O Pump Roller 부딪혀 Cam이 손상될 수 있기 때문이다.

따라서 감통된 F.O Pump 원상복구할 때는 반드시 Engine Stop 하고, Turning Gear로 Engine을 Turning하면서 Cam의 위치를 확인하고 복구한다.

만약에 복구가 제대로 되지 않는다면 Lifting Valve가 불량한 것이므로 Air Cylinder Rod 복구시키는 Air Line을 취외하여 Lifting Pin이 빠져나올 때까지 Compressor Air를 강제로 공급하면 복구된다.

정상적인 방법으로는 Engine Stop후 Turning Gear로 Engine Turning하여 Cam의 볼록한 부분이 상부인 Position에서 Lifting Valve의 Knob 들어 올려 Dial의 Position을 '0' → '1'로 변경한다.

그 때 Lifting Valve에서 Air Cylinder측으로 Compressor Air가 공급되고, Lifting Pin이 Roller Guide 측으로 나와서 최종적으로 운전 중인 Roller Guide를 들어올려 F.O Pump 감통 운전된다.

그 후 들고 있는 Knob을 내리면 아래로 들어가면서 Lifting Valve가 고정된다.

마지막으로 Rod 끝단의 구멍에 Safety Pin을 집어넣어 고정시킨다.

감통 운전을 해제할 경우는 앞서 언급한 바와 같이 Engine Stop하고 시행한다.

예전에는 Lifting 공구에 Chain과 Chain Wheel을 설치하여 감통했다면 요즘은 Air Cylinder에서 Air를 공급하여 감통하는 형식이다.

승선했던 선박에서는 운항 중에 M/E NO.6 Exhaust Deviation Temp이 발생하여 현장 확인 시 실제 NO.6 Cylinder 배기가스 온도가 현저히 낮아서 F.O Pump확인하니 Pumping이 되지 않으며 Roller Guide가 상부로 올라간 상태에서 고착되었다.

F.O Pump가 고착되면 Plunger를 움직여주는 Regulation Rack도 고착되지만 손으로 움직였을 때 부드럽게 움직여서 Roller Guide쪽에 문제가 있는 것으로 판단했다.

Engine 운항 중 앞서 언급한 방법대로 감통 운전을 하고 정박 후 감통 복귀하여 Pump OV'HL 결과 Lifting Valve가 불량하여 타 기통과 Air Cylinder Rod 길이 차이가 없을 정도로 Air가 약하지만, 지속적으로 Shaft쪽으로 공급되었고, Shaft가 들어가면서 Lifting Pin이 Roller Guide와 맞물리고 Roller Guide를 들어 올려 감통 운전이 되었다.

따라서 주기적으로 Lifting Valve에서 Air Cylinder Rod로 공급되는 두 개의 Air Line(공급 & 복구) 취외하여 Air가 나오는지 반드시 확인하고, Air가 나온다면 Lifting Valve 교환한다.

정상적인 경우는 Lifting Valve에 연결된 3개 Air Line 중 맨 아래의 Air Line은 항상 Air가 공급 되고, Lifting Valve Position을 변경하면 Air 공급라인을 통해 Air Cylinder Rod 밀어주어 Roller Guide 최종적으로 들어 올려 감통 운전 되고, Lifting Valve Postion 원상복귀하면 Air 복구라인을 통해 Air Cylinder Rod 원상복귀하여 최종적으로 Roller Guide를 내려준다.

Lifting Pin	Roller Guide의 Lifting Pin조립되는 부분

< 1 & 2번 Air Line으로 Air Cylinder Rod 작동시키며 3번 Air Line은 항상 Air 공급 중 >

3번 Air Line과 VIT Air Line과 연결

Lifting Valve 작동불량 후 Ball Valve 설치

Lifting Gear Checking

The lifting gear for the fuel pump roller guide is checked and activated by manually operating the four-way valve of the lifting gear while the engine is running or being turned.

Note!
Whereas the roller guide may be lifted when the engine is running, the roller guide may only be lowered when the engine is stopped. Otherwise, damage may be caused to the roller guide.

1. Lift the knob on the dial of the 4/3 directional control valve, and turn the dial to position 1.

   Thereby compressed air is led to the bearing of the lifting device, causing the shaft, which functions as piston, to be pressed in against the roller guide.

   The air cylinder is constantly vented through a bore in the shaft until the shaft is in the correct position.

2. When the shaft is pressed fully in, its lifting pin is positioned under the reversing arm of the roller guide, and the duct (which is connected to the lifting cylinder) is uncovered, thus activating the air cylinder.

   The air cylinder lifts the coupling arm, which turns the shaft, and the lifting pin on the shaft thus lifts the roller guide clear of the arm.

   When the roller guide has been successfully lifted, it is no longer possible, with the hand on the fuel oil high-pressure pipe, to feel the beat of the oil pressure pulse.

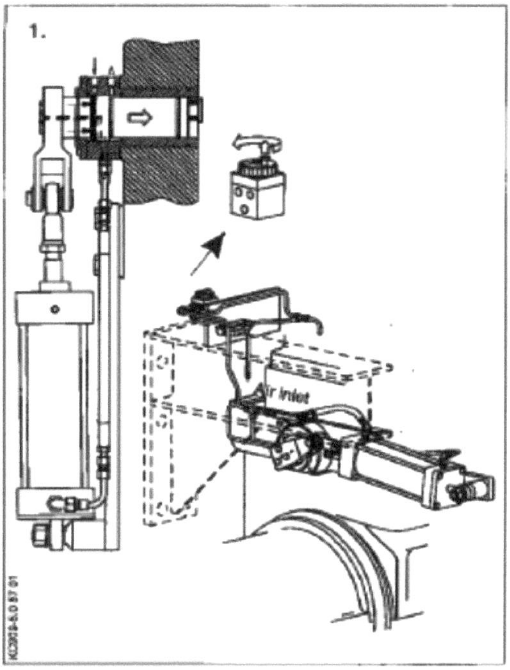

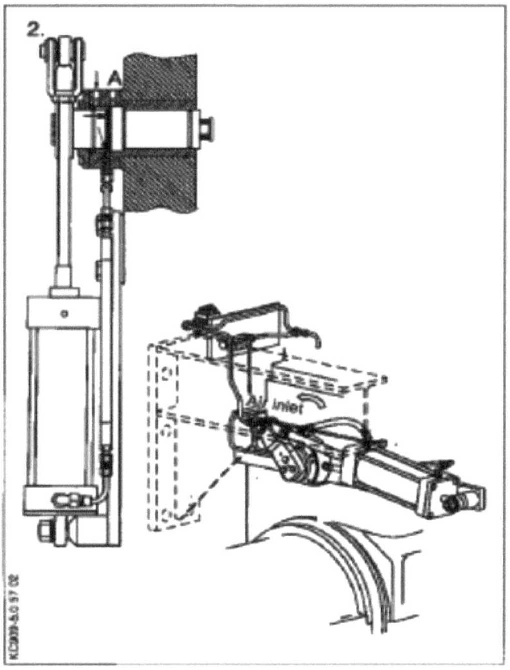

Fuel Pump Gear - Details

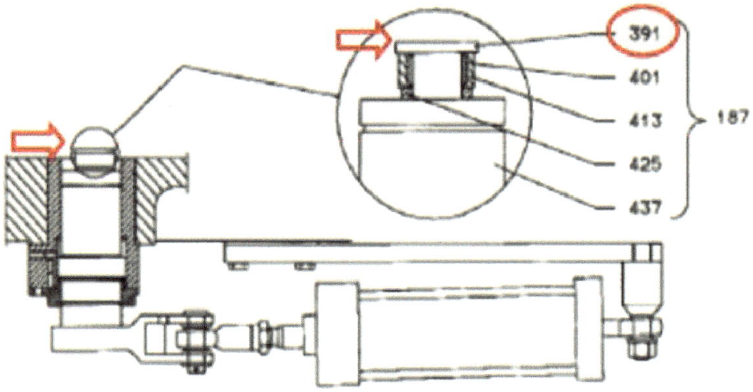

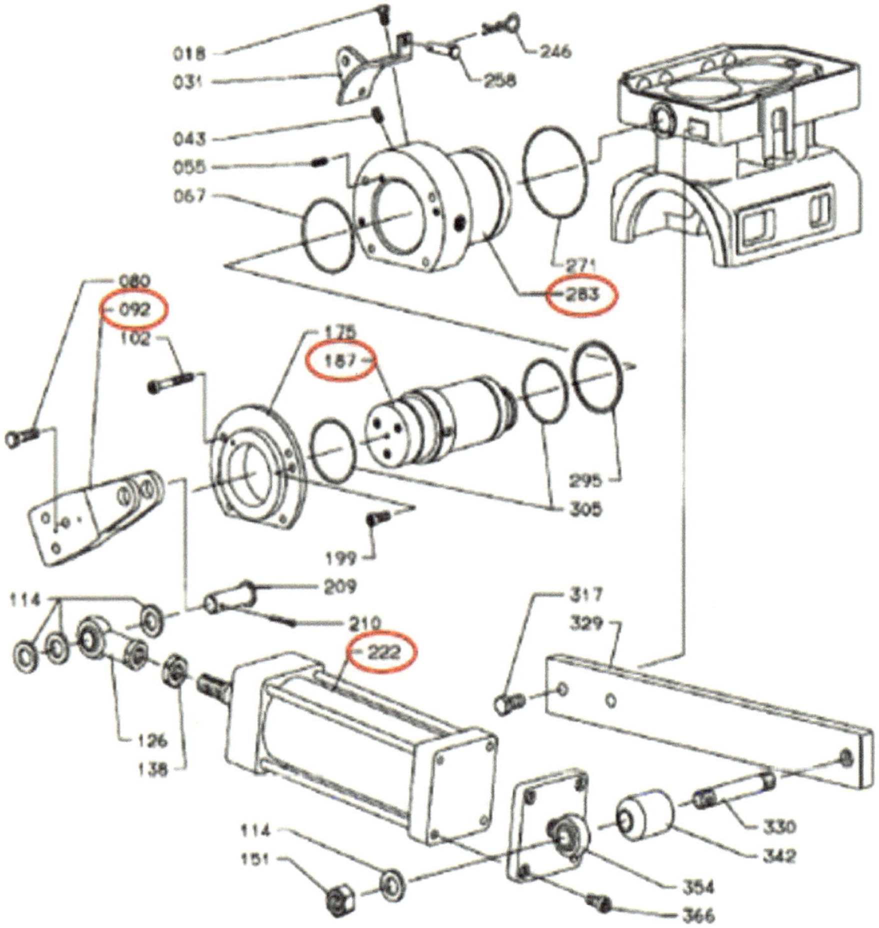

Fuel Pump Gear

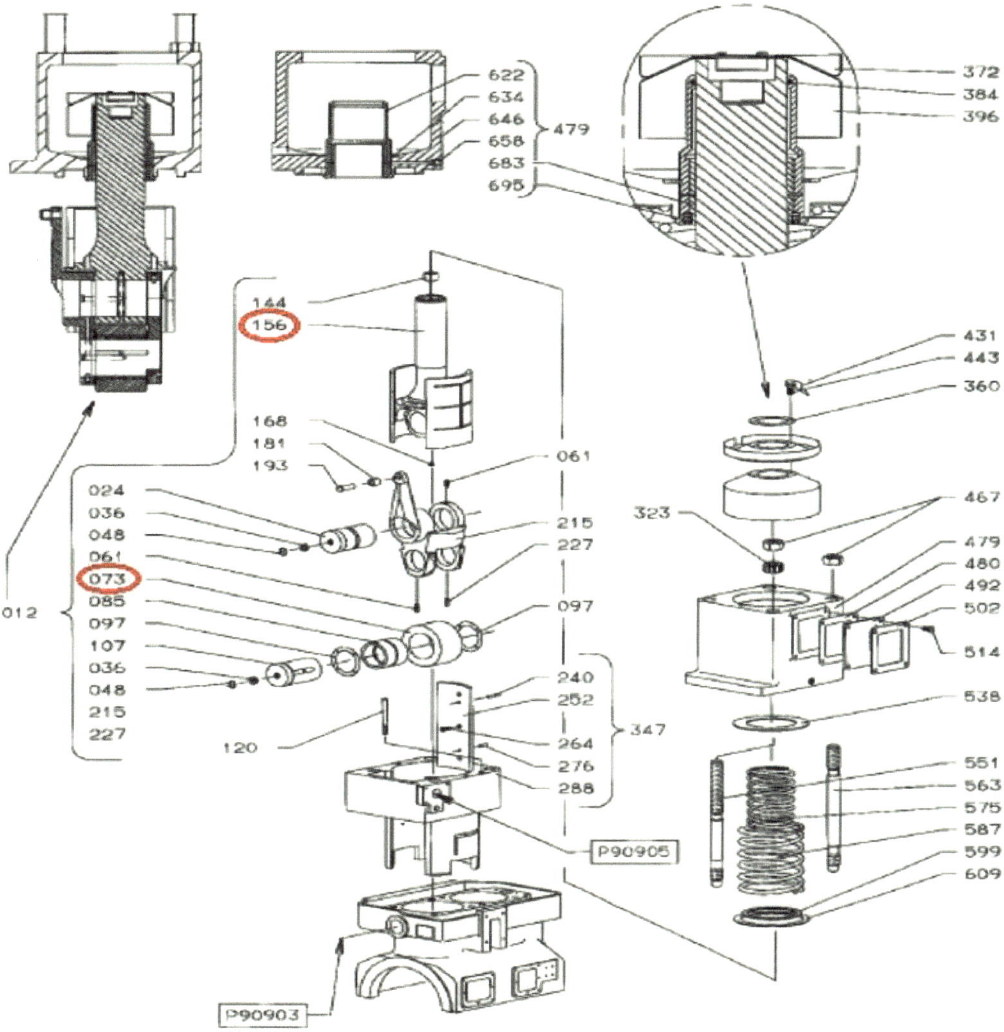

< 156 : Roller Guide / 073 : Roller >

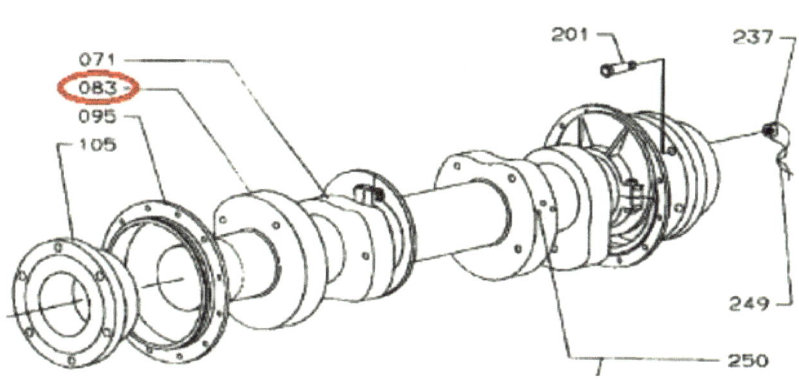

< 083 : Fuel Cam >

감통 전

감통 후(Dial & Knob Position 차이)

감통 전 Air Cylinder

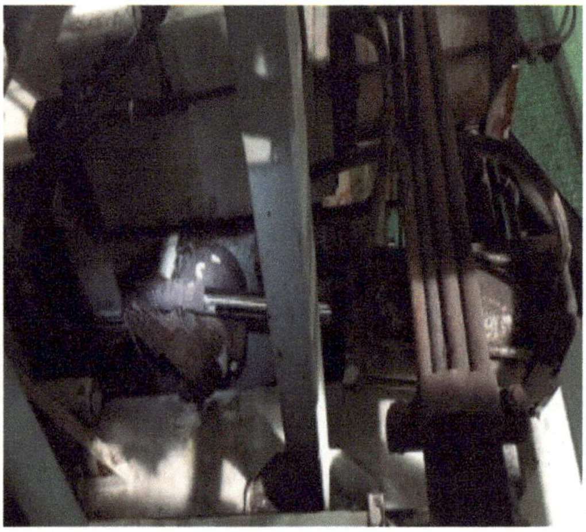

감통 후 Air Cylinder(Rod 길이 차이)

감통 전 Safety Pin

감통 후 Safety Pin 체결

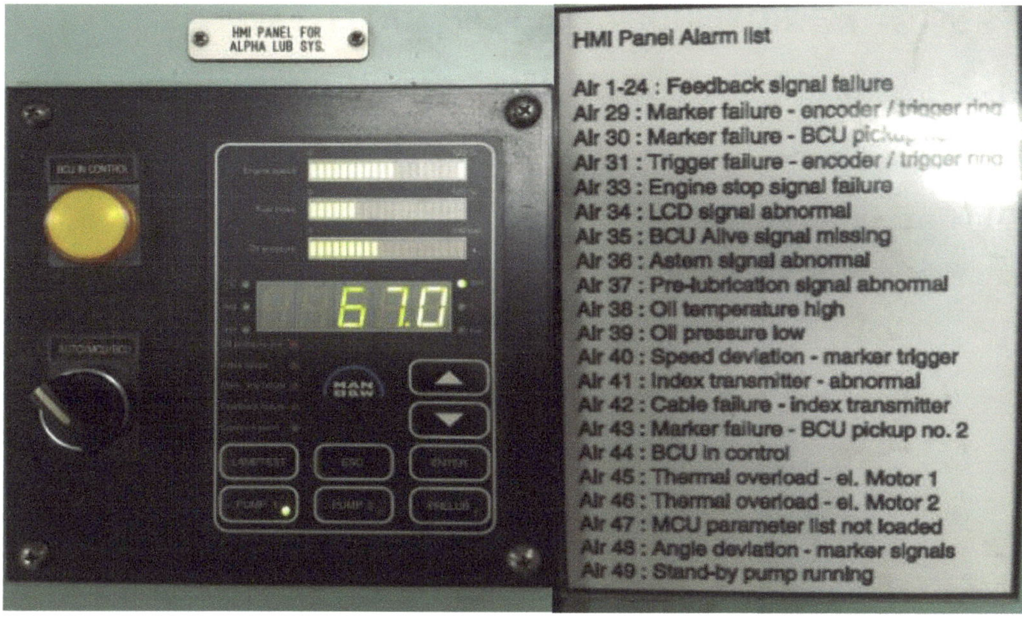

< Alpha LUB System Panel >

[PUMP1] = NO.1 PUMP START & STOP

[PUMP2] = NO.2 PUMP START & STOP

[ESC] + [PUMP1] = NO.1 PUMP로 교대 운전

[ESC] + [PUMP2] = NO.2 PUMP로 교대 운전

[ESC] + [PRELUB] = PRELUB TEST(Lubricators will lubricate 1000 times)되고, 'PRELUB' BUTTON을 누르면 STOP

[LAMPTEST] = Test Indicator Lamp

3) 실린더 유 공급량(CYLINDER OIL FEED RATE) 조절 방법

(1) 위,아래 Button을 눌러 Display에 'F.rAtE'가 나오면 [Enter]를 누른다.

(2) 위,아래 Button을 눌러 주유량 조절할 Cylinder가 나오면 [Enter]를 누른다.

(3) 'SEt.Al'를 이용하면 전 실린더 주유량을 조절할 수 있다.

(4) 주유량을 조절 후(주유율 표를 참조) [Enter]를 누르면 Save 된다.

(5) [ESC]를 세 번 눌러 MAIN MENU로 복구한다.

< Local Alpha LUB Pump >

4) 비상 주유(EM'CY LUBRICATION)

아래와 같이 Jumping하면 EM'CY Lubrication 된다.

5.5 Emergency running without external trigger signals

We strongly recommend that all efforts are made to repair the trigger system before switching to the emergency mode.

In the event that all trigger signals to the lubricator system fails (angle encoder and pickups for the BCU system), it is possible to maintain a certain level of lubrication by means of an internal stroke generator on the MCU or BCU card.

This stroke generator emits one pulse per second, i.e. approx. 60 rpm.

To be able to run in this "emergency mode", the following electrical connections must be changed:

Remove the wire from MCU-J22 terminal 1

Connect an extra wire from MCU-J52 terminal 4 to MCU-J22 terminal 1.

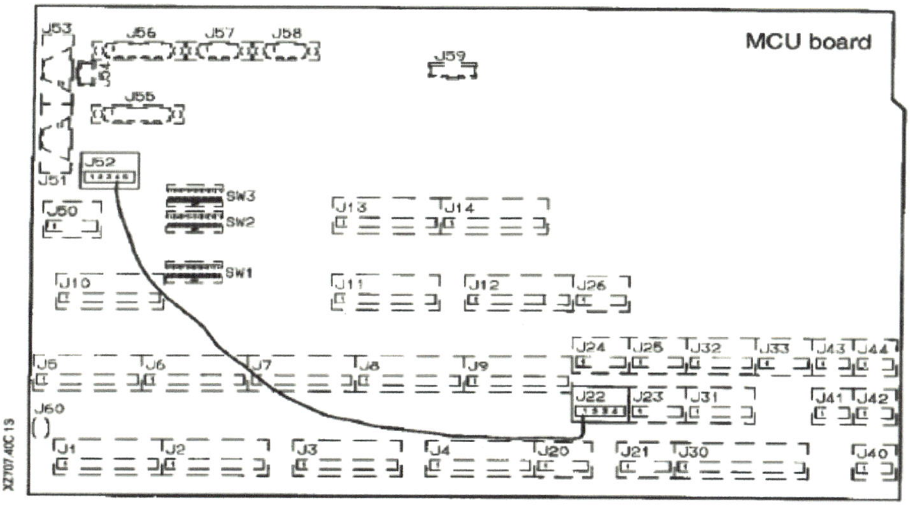

The system now starts lubricating regardless of whether the engine is running or stopped.

The system will also give various alarms.

However, feedback alarms will still function and indicate whether all cylinders are being lubricated. When running in this mode, the BCU system is not active.

The MCU slow-down function remains active.

Lubrication is effected as "random" and "rpm-dependent", however with a supplement of 25% of the amount set for the individual cylinders, but with a fixed number of revolutions of 60.

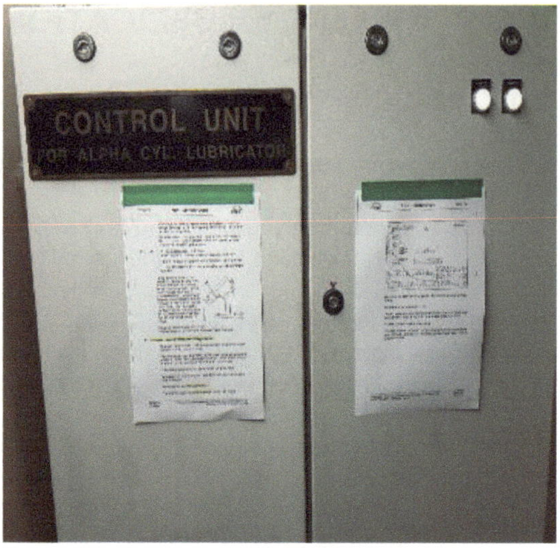

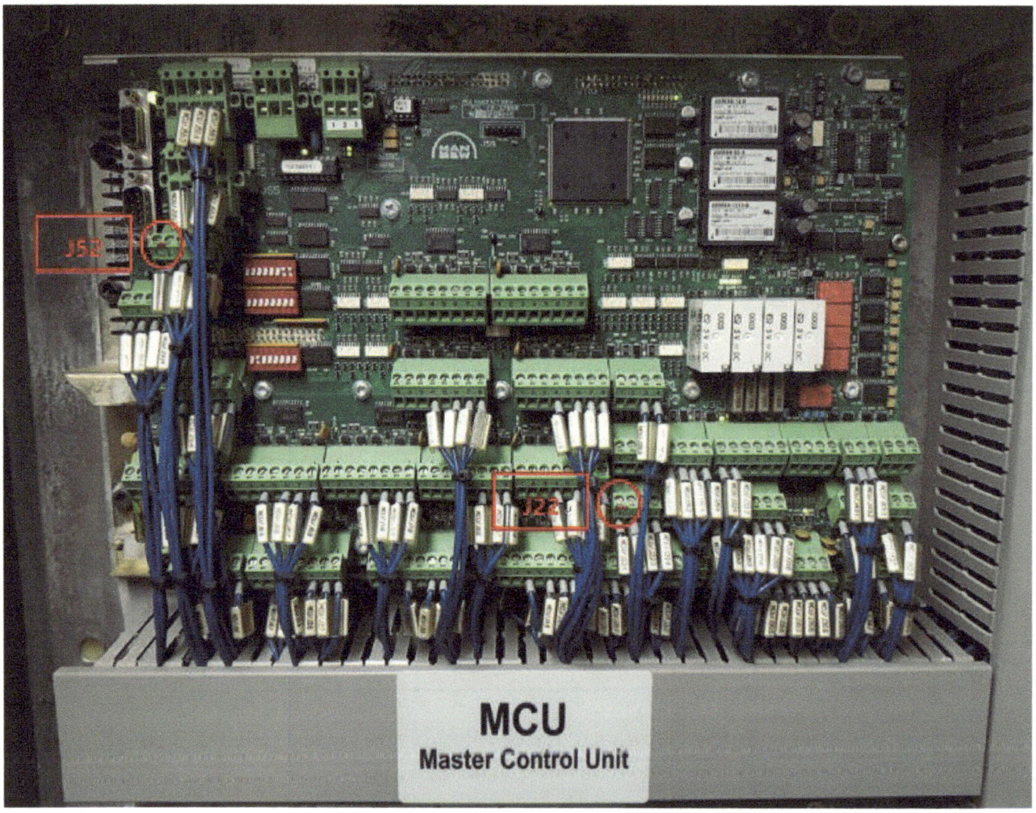

* 추가로 BCU Pick-up Sensor(No.1 or No.2) Fail Alarm 발생 시 전원공급 확인, Pick-up Sensor 소제, No.1 & 2 Pickup Sensor Power Cable 교환해보고 상태 동일하다면 문제 발생한 Pick-up Sensor 교환한다.

5) 엔진 슬로우 터닝(ENGINE SLOW TURNING)

< Main Starting Air & Slow Turning Air Line / Slow Turning Valve >

Slow Turning으로 윤활을 다시 시켜주므로 Main Engine 30분이상 Stop 후 Engine 재시동할 때는 Slow Turning을 하는 것이 좋다.

Bridge Indicator Panel상의 Manual Slow Turning Button 누른 상태에서 조절 핸들로 Start하거나, Bridge → Engine으로 Change Over Switch 변경하고 Engine Room Indicator Panel상의 Manual Slow Turning Button을 눌러서 하는 방법이 있다.

Slow Turning Button 누르면 Solenoid Valve 작동시켜서 Main Starting Air Valve 개방용 공기를 배출하여 Main Starting Air Valve는 계속 닫혀 있고 작은 Slow Turning Valve만 개방된다.

Manual Slow Turning Button 누른 상태에서 조절 핸들로 Start 명령을 내린다.

Crankshaft가 최소한 1회전 했을 때 Slow Turning Switch Stop한다.

6) 메인 엔진 에어 블로우(M/E AIR BLOW)

Engine Stop후에 연소실의 잔류 가스 및 이물질을 불어내기 위해 Air Blow를 한다.

Indicator Panel에서 AUX.Blower 기동 후 Air Blow 시행하며 아래 사진을 참조한다.

먼저 Telegraph Receiver를 Dead-slow Ahead / Astern Position에 놓고, Maneuvering Handle을 눌러 위쪽에 표시된 홈에 끼우면 Air Blow가 시행된다.

Rev. Indicator 상으로 **20RPM**되면 Maneuvering Handle 원상복귀하고 Telegraph Stop한다.

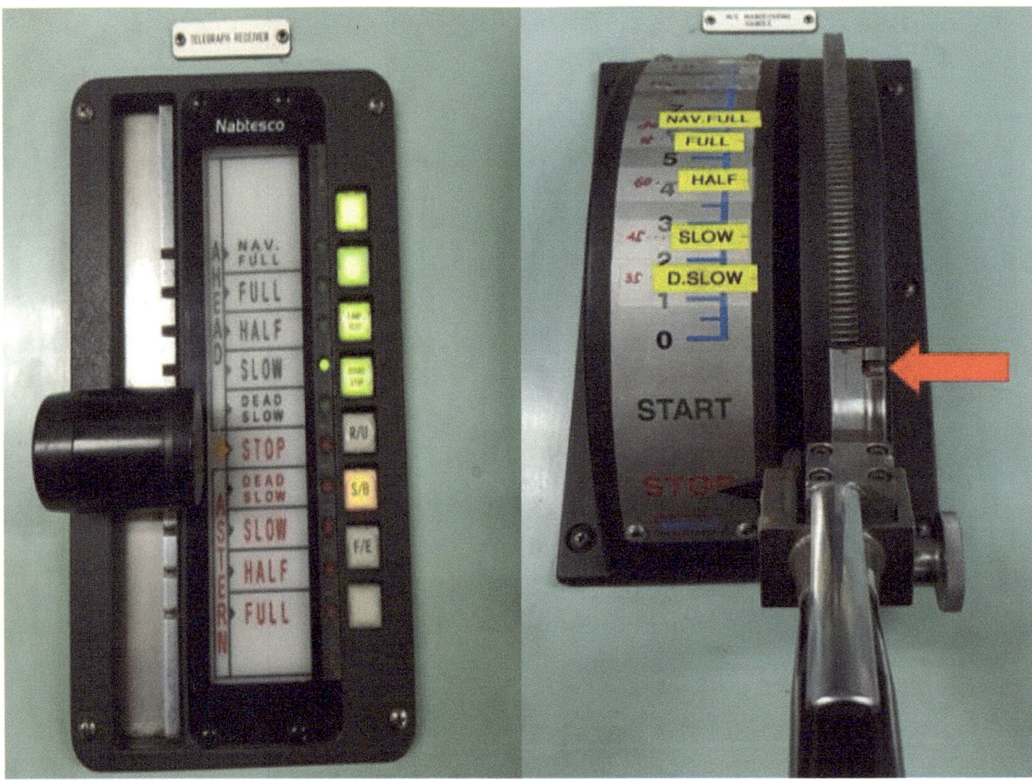

< Telegraph Receiver / Maneuvering Handle >

Telegraph Receiver Dead-slow Ahead(&Astern)

Maneuvering Handle 올려
위쪽 홈에 끼운다.

RPM 20미만에서 Maneuvering Handle 내려
아래 홈에 끼운다.

Telegraph Receiver Stop

REV. Indicator(20미만에서 Air Blow Stop)

7) CYL. OIL 계산법

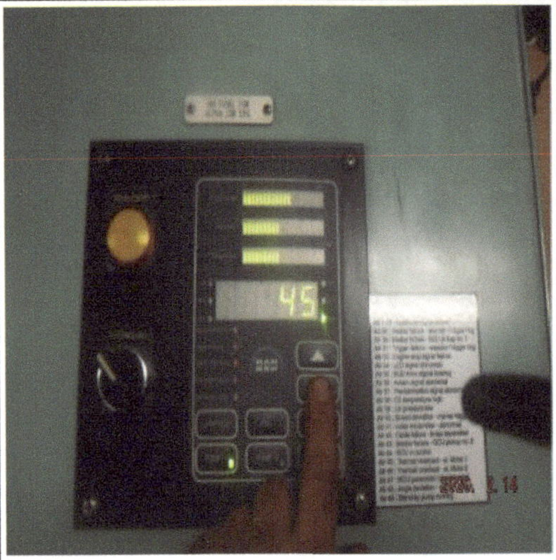

△▽ 이동

disP → ENTER

Str.hi → ENTER

앞다섯자리

△▽ 이동	disP → ENTER

Str.Lo → ENTER	뒤 다섯 자리

위와 같은 경우 총 주유 횟수는 402684560 이다.

Reading of total strokes [Str.hi] and [Str.lo]

1. Press [▲] or [▼] until diSP is shown in the display.
2. Press [ENTER] and [▲] or [▼] until Str.hi or Str.lo is shown in the display.
3. Press [ENTER] to read the stroke counter.

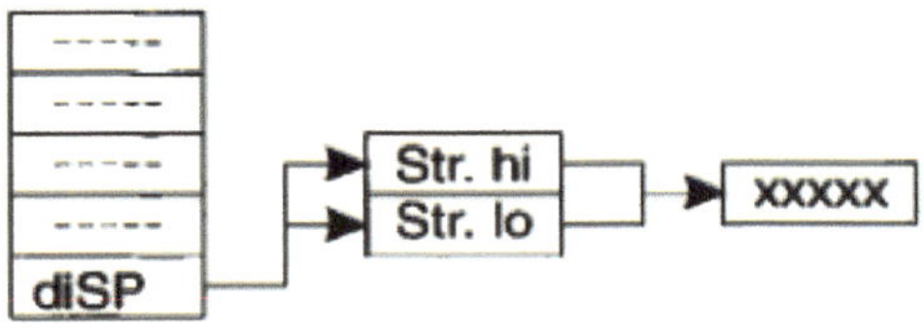

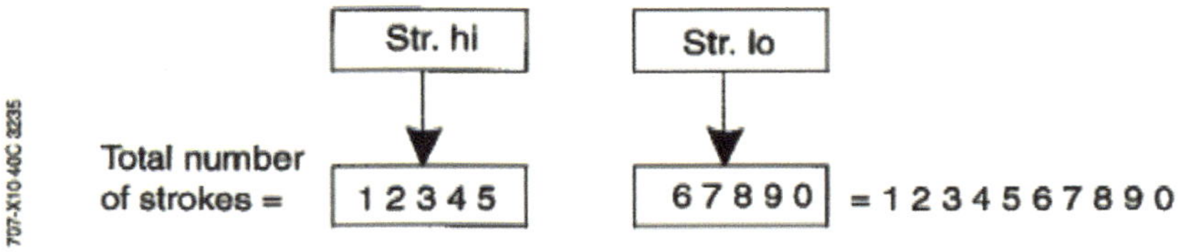

1회 주유량은 선박 마다 다르며 Lubricator 관련 문서에 나와있다.

Lubricator constants :

Lubricator piston diameter:	6	[mm]
Lubricator stroke:	13	[mm]
Number of pistons per lubricator	5	[pcs]
Number of lubricators per Cylinder	2	[pcs]
Lubricator volumetric efficiency:	0,97	
Lubricator injection volume:	1,78	[cm3/stroke]

Cylinder Oil 소모량 계산식은 다음과 같다.

주유 횟수(STROKE) x 1회 주유량(CM^3) ÷ 1000 = CYL.OIL 소모량 (Liter)

소모량(CM^3)=주유 횟수(STROKE) x 1회 주유량(CM^3)

소모량(Liter)= 주유량(CM^3) ÷ 1000

8) 입항 스탠바이 (STAND-BY)

(1) 기름 계산 후 Logbook에 작성하고 당직 항해사에게 유류 잔량을 알려준다.

(2) Main Air & Aux' Air Tank Drain, Safety & Control Air Drain 하여 수분 및 유분 확인한다.
(Main Air Tank & Aux' Air Tank 에 직접 가보면 표시가 되어있거나 Name Tag가 있다.)

< Main Air Tank >

< AUX' Air Tank >

< MAN ES Engine Starting Air Drain Valve >

< RT-FLEX Engine FWD & AFT Air Drain Valve >

* RT-FLEX Engine 참고용

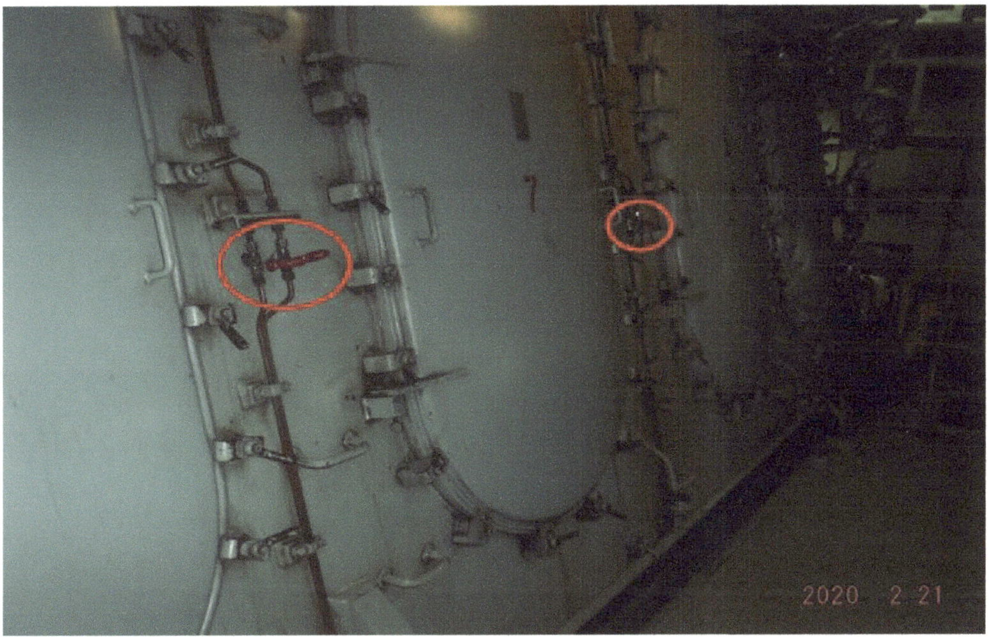

< Control & Safety Air, Control Air Filter Drain Valve >

(3) 연료유 SETT'& SERV' Tank Drain을 하여 수분을 제거한다.

(Drain Valve는 동일하므로 SERV' Tank 사진만 참고)

< SERV' Tank Drain Valve >

(4) 입항 S/BY 전날 혹은 당일에는 조수기는 미리 Stop한다.

9) 엔진 사용 종료 (FINISHED WITH ENGINE)

(1) 입항 스탠바이 중 트랜시버로 **"Almost in position"** 방송과 함께 기름 계산을 시작한다. 정박하면 기름 잔량(R.O.B)를 Deck Office의 당직 항해사에게 알려주어야 한다.

(2) 기름 계산이 끝나고 기름 잔량(R.O.B)을 알려준 다음 Main Engine Air Blow준비를 한다. **(ECR로 Change Over SW 변경해야 ECR에서 Air Blow 가능하다.)**

Test Cock을 Open하고 1기사에게 트랜시버로 보고한다.

〈 Main Engine Test Cock 〉

(3) Air Blow 상태 확인 후 이상유무를 보고하고 Main Starting Air & Distributor Valve Close한다.

〈 Main Starting Air & Distributor Valve 〉

〈 RT-FLEX Main Starting Air Valve 〉

표시한 Main Starting Air Valve Open & Close하면 Limit Switch에 의하여 Interlock이 걸리고, Console에서Open & Close Lamp 로 표시된다.

RT-FLEX Main Starting Air Valve 경우에 고정 Lever가 있으므로 밀어서 풀어준 상태에서 Valve를 조작한다.

Open & Close시 Valve Stem측에 홈이 있으며 그곳에 고정 Lever를 걸어준다.

(Air Blow시 Test Cock에서 Air가 아닌 다음의 색깔들이 확인된다면 1기사에게 보고한다. 흰색 : 수분, 청색 : 윤활유, 불꽃 : 잔류물, 검은색 : 연료)

(4) Engine Jacket Warming Valve를 서서히 Hammering이 발생하지 않도록 열어준다.

(5) Turning Gear Engage 및 Engine Turning하면서 Engine Control Room에서 실린더유 주유한다. Turning Gear를 Engine Flywheel에 넣고 빼는 것은 쉽다.

Lever 이용하여 Turning Gear를 움직이되 넣을 때 고정 핀을 꼽아두고 특히 뺄 때 고정 핀을 반드시 뺀 상태에서 Lever가 움직이니 명심하도록 한다.

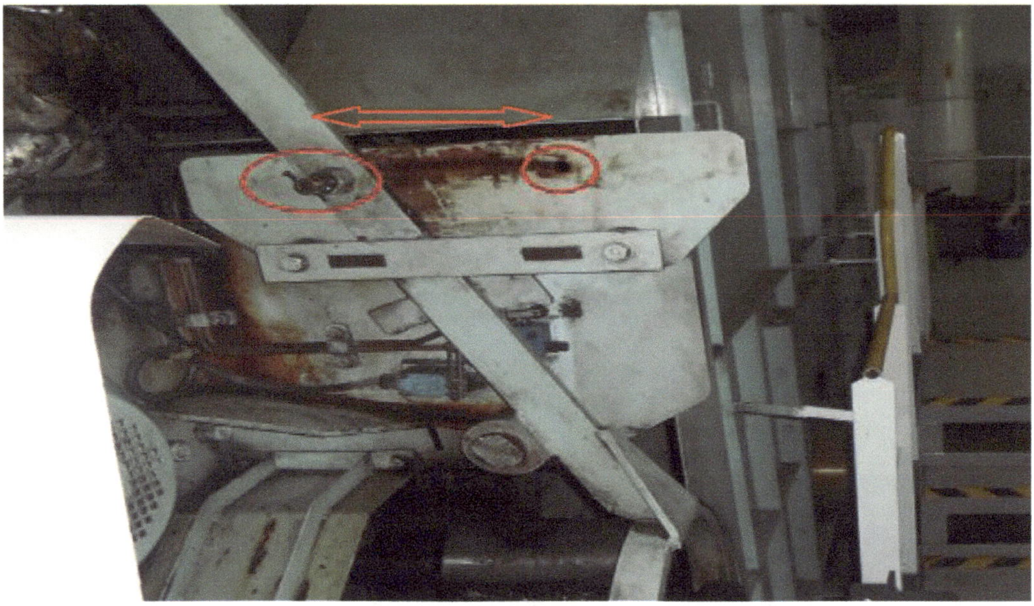

< 고정 핀 >

Turning Gear Engage

Turning Gear Disengage

* Main Engine Load가 낮을 때는 Cylinder Liner에 많은 양의 실린더유가 주유되므로 추가적으로 주유까지 할 필요는 없다.

다만, 입항 스탠바이 중에 Engine Stop 시간이 다소 길었으면 Cylinder Liner에 주유된 Cylinder Oil이 중력으로 아래로 흐르기 때문에, 이런 경우에는 정지 후 추가적인 주유를 하도록 한다.

반면, 출항 스탠바이 때는 Cylinder Liner및 Main Bearing에 Cylinder Oil, System Oil이 중력으로 아래 Crank chamber로 빠져 윤활이 되어있지 않다. 따라서 반드시 Engine Turning하여 주유한다.

※ Air Blow 목적: Engine연소실 잔류물 제거

※ Turning 목적: Engine열 응력 제거(입항 직후) / Cylinder & Main Bearing 윤활 (출항 때)

10) 출항 스탠바이 (STAND-BY)

출항 한 시간 전 선내 방송이 나오면 기관실로 내려가 출항 준비를 시작한다.

(1) Main Air & Aux′ Air Tank Drain, Safety & Control Air Drain 한다.

(2) 기름 계산 후 Deck Office 당직 항해사에게 유류 잔량을 알려준다.

(3) Main Engine Turning을 하고 Engine Control Room에서 Cylinder Liner주유를 한다.

(4) 주유가 끝나면 내려가서 Turning Gear를 Disengage한다.

(5) Main Starting Valve Open하고, Main Engine Side로 올라와서 Test Cock Open 재확인한다.

(6) 1기사에게 트랜시버로 Air Blow 준비 되었다고 보고를 한 다음, Air Blow 상태 보고한다.

(7) Main Engine Test Cock Close하고 Control Room으로 올라간다.
(Engine이 준비되었다면 Bridge로 Change Over SW 변경한다)

(8) Engine Control Room에서 있다가 Bridge에서 Engine 기동하면 Jacket Warming Valve Close 및 기관실 순찰을 돌며 이상유무를 확인 한다.

11) R/UP (RUNG UP)

(1) R/UP Engine 후 Departure Report를 작성한다.

(2) Departure Report 완료 되었으면 기관실 순찰 시행한다.

만약 대양 항해 시작이라면 기관장 또는 일기사에게 보고 후 조수기 기동을 한다.

12) 오일 미스트 디텍터(OIL MIST DETECTOR)

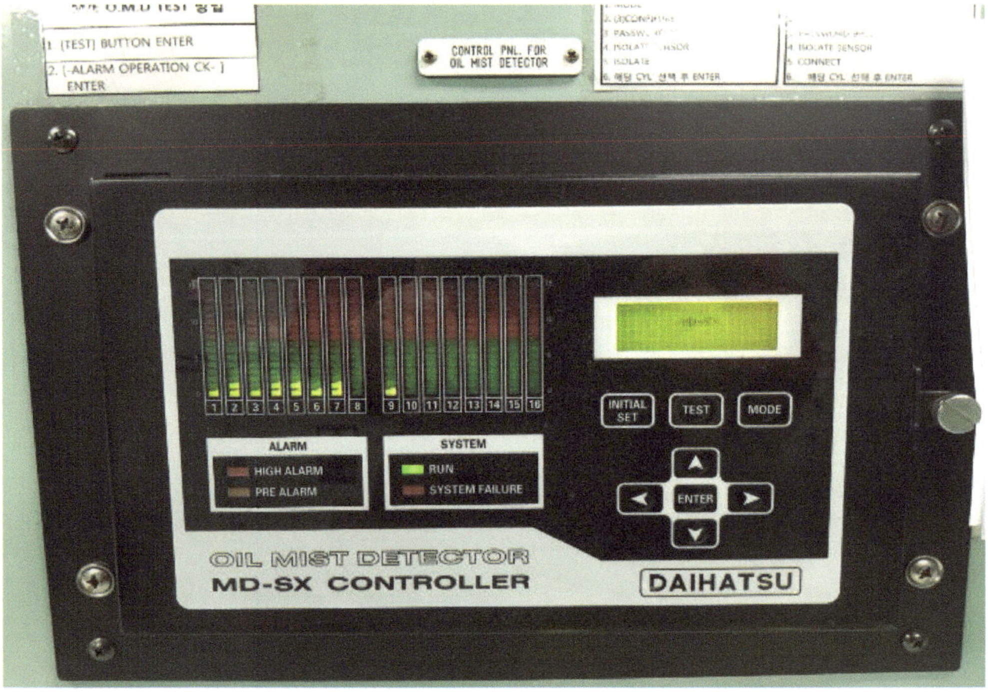

(1) ISOLATE 방법

① MODE

② (3)CONFIGURE

③ PASSWORD '9999' (선박 마다 다름)

④ ISOLATE SENSOR

⑤ ISOLATE

⑥ 해당 CYL'선택 후 ENTER

(2) ISOLATE 해제

① MODE

② (3)CONFIGURE

③ PASSWORD '9999' (선박 마다 다름)

④ ISOLATE SENSOR

⑤ CONNECT

⑥ 해당 CYL'선택 후 ENTER

(3) M/E O.M.D TEST

[TEST] Enter → [ALARM OPERATION CK] Enter

13) VIT(VARIABLE INJECTION TIMING) 조절

< 좌측 59번 / 우측 45번 >

VIT 조절은 Barrel 위치를 조절하면 Plunger 위치가 그만큼 달라지므로 Barrel이 위로 움직이면 분사 타이밍 늦어지고 아래로 움직이면 분사 타이밍 빨라진다.

따라서 Barrel 위치에 따라 분사타이밍이 조절되는데 그 조절을 Air로 해준다.

위 사진의 Air 압력 현재 2.2~5 상태에서 Air 압력 낮춰주면 BARREL 위치가 낮아져 분사 타이밍이 빨라지며, 분사 타이밍을 늦추려면 Air 값을 증가시키면 된다.

P-MAX, P-COMP을 보고 필요에 따라 조절할 때 사용하며 사진에서 59 = MANUAL ASTERN, 45 = MANUAL AHEAD 값 조절하는 것이다.

(단 , 반드시 Engine Stop한 상태에서 조절!)

* 각 개별 Cylinder VIT Rack Position 변경으로 분사타이밍을 조절할 수도 있다.

14) ENGINE LOCAL TEST

(1) 운용 (LOCAL)

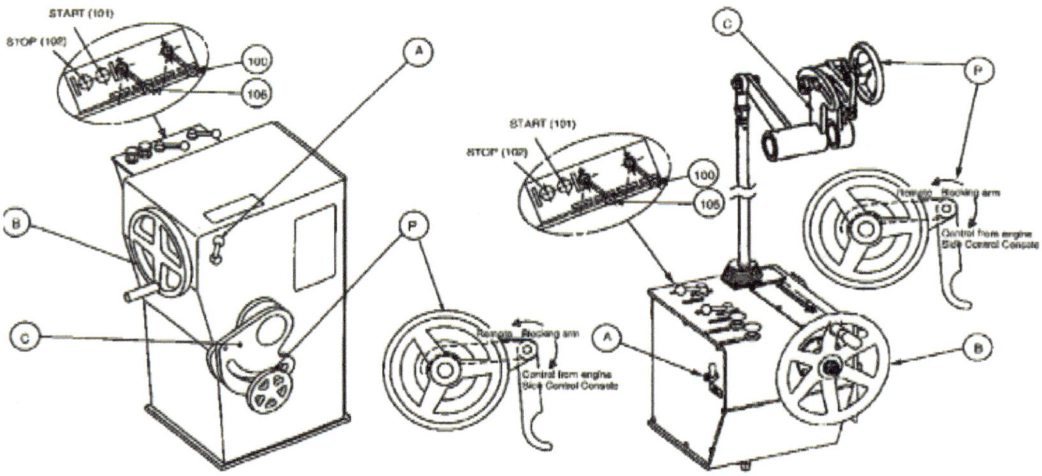

(MC Type Engine) (MC-C Type Engine)

For changing-over to 'Control from Engine Side Control Console' with **running engine,** *see Section 703-10 Item 4 'Control from Engine Side Control Console'.*

1. Check that valve (105), which is the "telegraph handle" of the engine side control system, is in the required position. Note that reversing to a new direction is only possible when stop valve (102) is activated.

2. Turn the handle 'A' anti-clockwise to free the regulating handwheel 'B'.

3. Put the blocking arm in position for control from engine side control console.

4. Turn handwheel 'B' to move the innermost lever of the change-over mechanism 'C' to a position where the impact handwheel 'P' is able to enter the tapered slots in both levers. Quickly, turn the impact handwheel 'P' anti-clockwise. This causes disconnection of the governor and connection of the regulating handwheel 'B' to the fuel pumps.

   The impact handwheel must be very firmly tightened so as to ensure that it does not work itself loose. This firm tightening must also be carried out when changing over from Manual Control to Normal running.

   It is vital that the impact handwheel is used to firmly secure the tapered pin and thus the cone. This will prevent the impact wheel from inadvertently causing the cone to loosen at certain vibration levels.

5. Change position of valve (100) from Normal to Control from Engine Side Control Console. Now air supply is led to the valves of the manoeuvring system for control from engine side control console.

6. Ready for start. Start is described in Section 703-10 Item 4 'Control from Engine Side Control Console'.

Note: Always keep the threads of the change-over mechanism well lubricated, to ensure a quick changing-over.

Bridge → Engine으로 Change Over Switch 변경한 다음 Local에서 상기와 그림과 같이 한다.

① Telegraph Position을 확인(Ahead/Astern)하고 'A'를 반 시계방향으로 변경한다.

② Blocking Arm을 Emergency 위치에 놓는다.

③ Hand wheel 'B'를 돌려서 Impact wheel 'P'가 두 레버의 경사진 홈에 들어갈 수 있는 위치인 절환장치 Lever 'C'의 가장 안쪽 위치에 오도록 한다.
Impact wheel 'P' 반 시계방향으로 돌리면 Governor와 분리되고 Hand wheel 'B'가 F.O Pump와 연결된다.

④ Valve(100)의 위치를 Remote에서 Emergency로 변경하면 비상운전용 조종 시스템 Valve로 공기가 공급된다.

⑤ Hand wheel 'B'로 엔진 회전수를 직접 설정한 다음 Start Button 시동될 때가지 누른다.

⑥ 처음 시동될 때 Air가 많이 들어가고 초기 기동 RPM이 높아 연료도 많이 들어간다. 그러므로 Start후 Hand wheel 'B' 반 시계방향으로 돌려서 최소 회전수(Min.RPM)로 내린다.

⑦ 그 후 Stop Button을 누르면 Engine Stop한다.

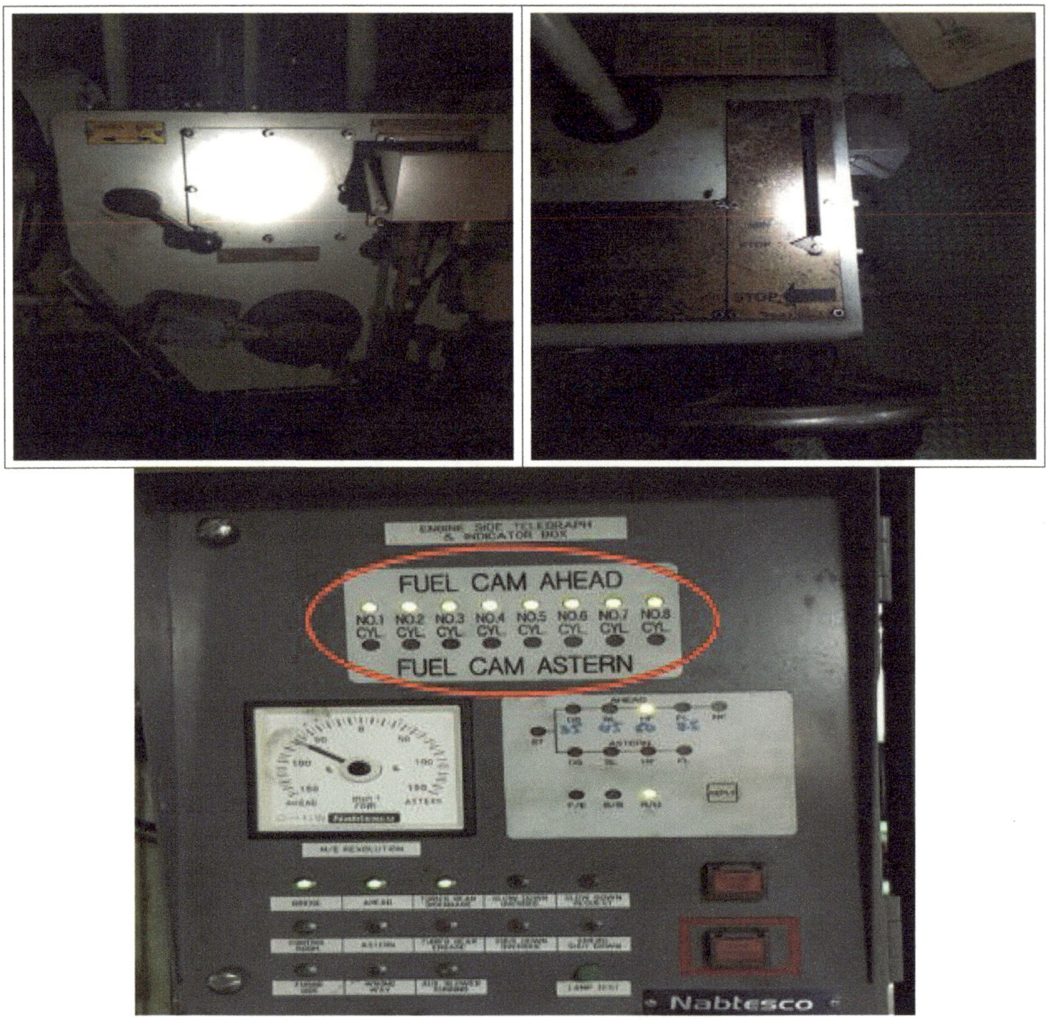

< Local(Emergency) Control Stand >

(2) 정지

- Push Stop Switch

(3) REMARK

Connection to regulating Shaft
(See also Plate 70303, 'Manoeuvring Gear')

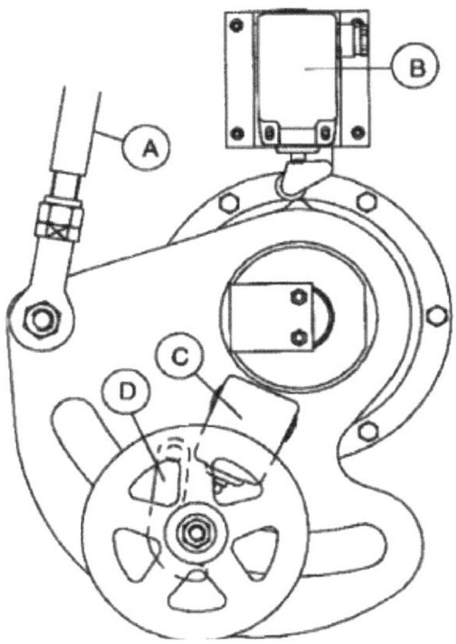

A) Pull rod connetion from governor.

B) Stop indicator.

C) Indicator from engine side control.

D) Blocking arm "Normal control"-position.

E) Hollow shaft connected to regulating handwheel on engine side control console.

F) Shaft connected to regulating arms on fuel pumps.

G) Keys and keyways.

H) Plate connected to regulating handwheel.

I) Plate connected to governor.

J) Plate connected to regulating shaft.

P) Impact handwheel. Changing from Normal to Control from Engine Side Control Console, *see Plate 70302*.

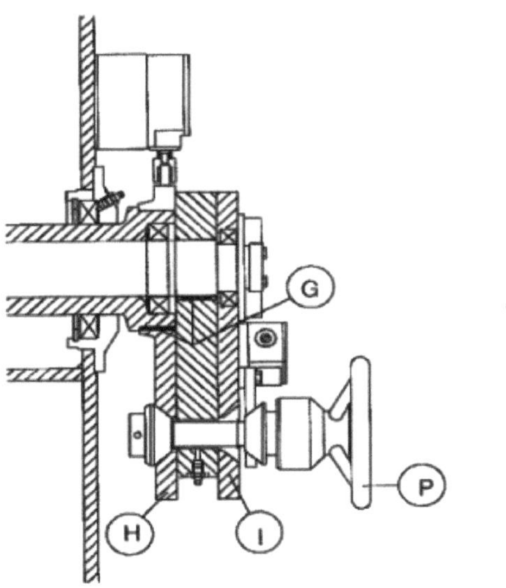

Control from Engine Side Control Console

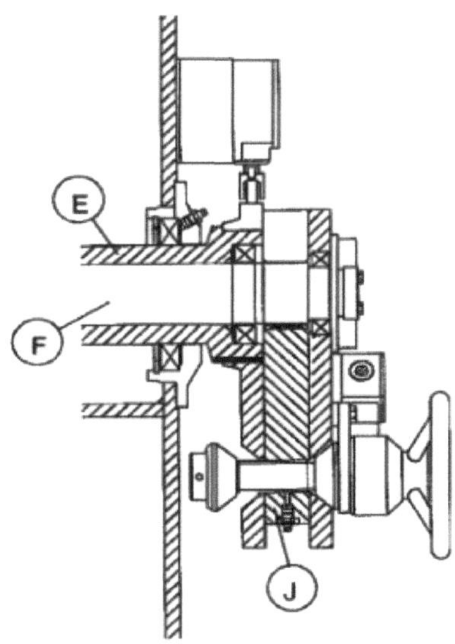

Normal Control

[2] MAN ES 캠 구동 엔진(MC-C) 부품 정비 및 운용

1) 연료 분사 밸브

정비 주기 : 8000H

Fuel Valve

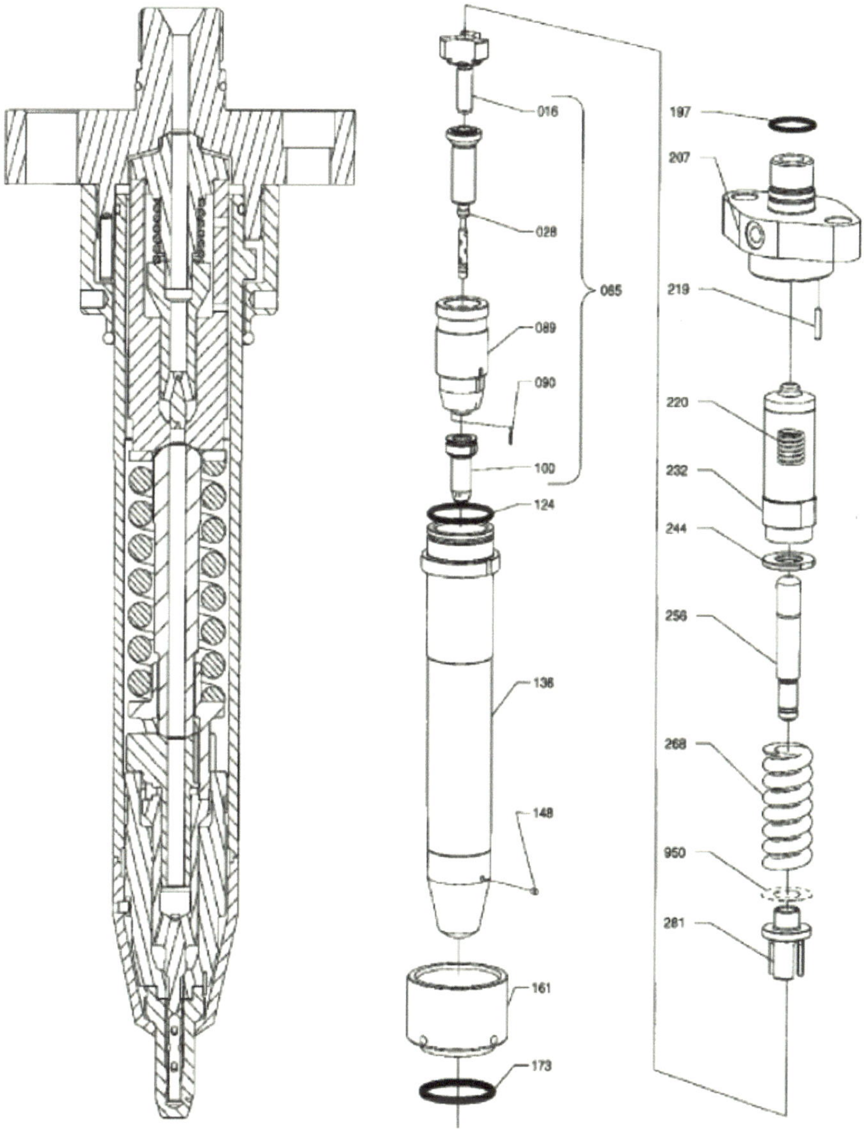

Fuel Valve and Fuel Pump Panel

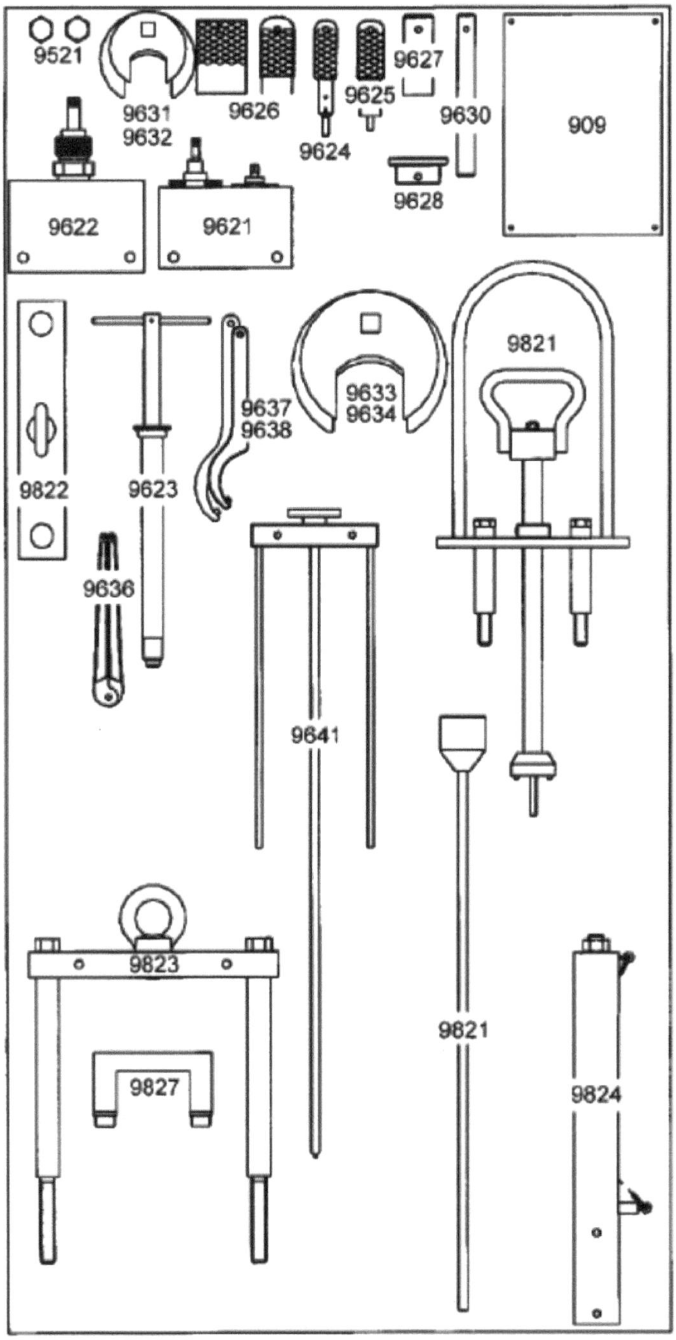

909 : Tool panel

9521 : Screw

9621 : Milling tool for fuel oil pipe seat

9622 : Milling tool for fuel oil pipe

9623 : Grinding mandrel for holder, inside

9624 : Grinding mandrel for valve head

9625 : Grinding mandrel for thrust spindle

9626 : Grinding mandrel for holder, outside

9627 : Drift for non-return valve

9628 : Flange for dismantling of non-return valve

9630 : Drift for spindle guide

9631, 9632, 9633, 9634 : Crowfoot wrench head

9636 : Pin Wrench, special

9637, 9638 : Hook wrench

9641 : Measuring tool for fuel pump lead

9821, 9823 : Lifting tool for fuel pump

9822 : Lifting tool for fuel pump housing

9824 : Tool for P.max rack

9827 : Assembling tool

Fuel Valve - Tools

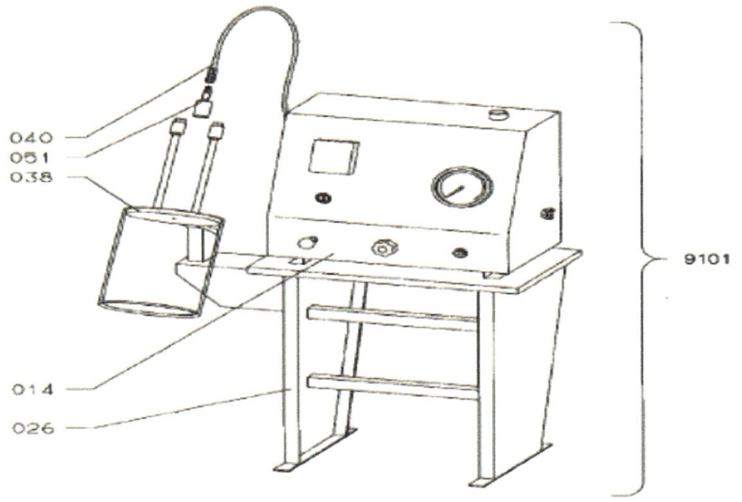

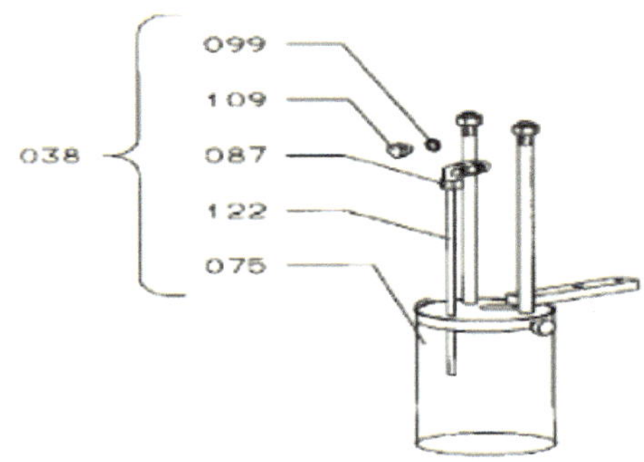

014 : High-pressure pump

026 : Table

038 : Rig fuel oil valve

040 : High-pressure hose

051 : Connection piece, M42X2

075 : Cover

087 : Angle union

099 : Gasket

109 : Plug screw

122 : Hose

9101 : Test rig, complete

Fuel Valve Nozzle - Tools

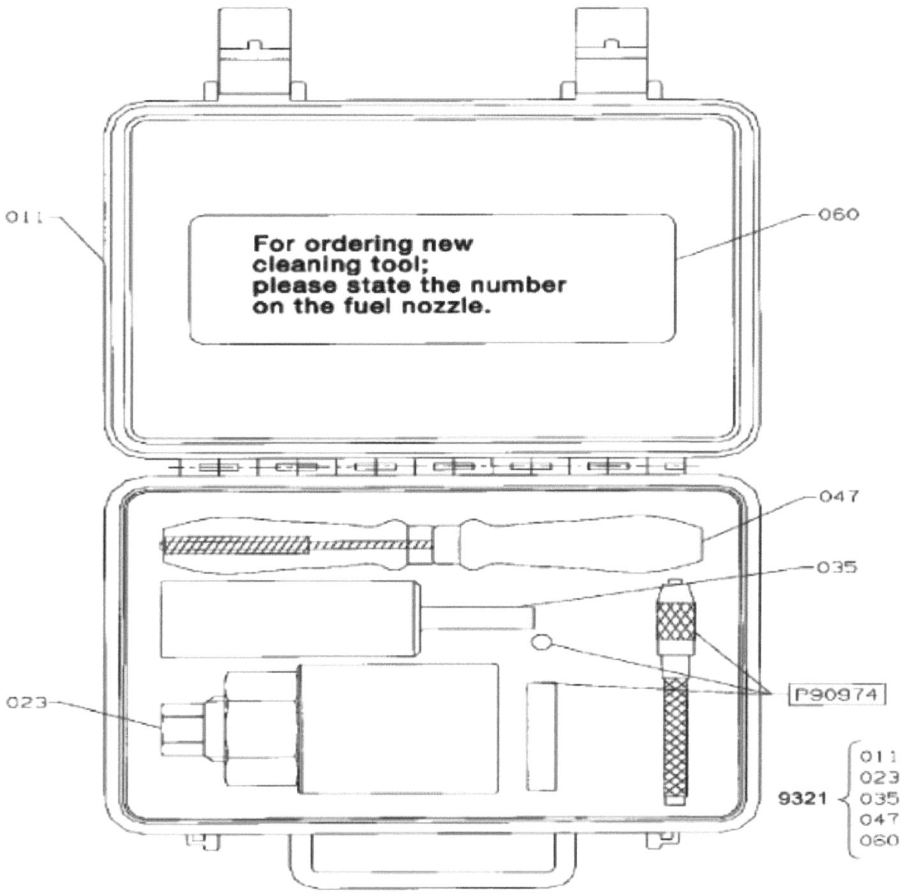

011 : Tool box

023 : Extractor for atomizer

035 : Drift for atomizer

047 : Cleaning brush

060 : Instruction plate

9321 : Toolset, complete

Fuel Valve - Inspection Tools

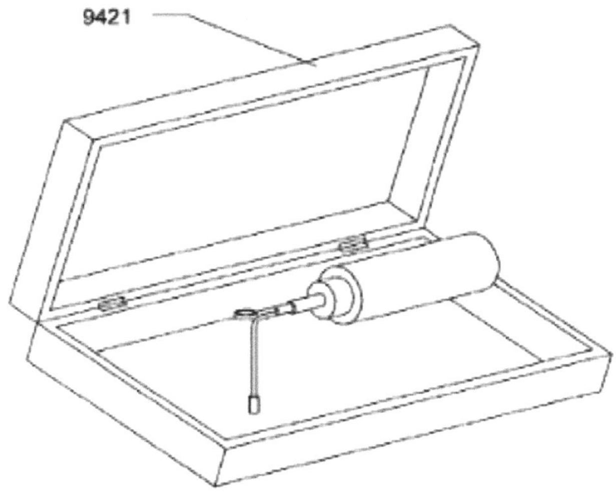

9421 : Probe light with magnifier, complete

Fuel Valve Nozzle - Cleaning Tools

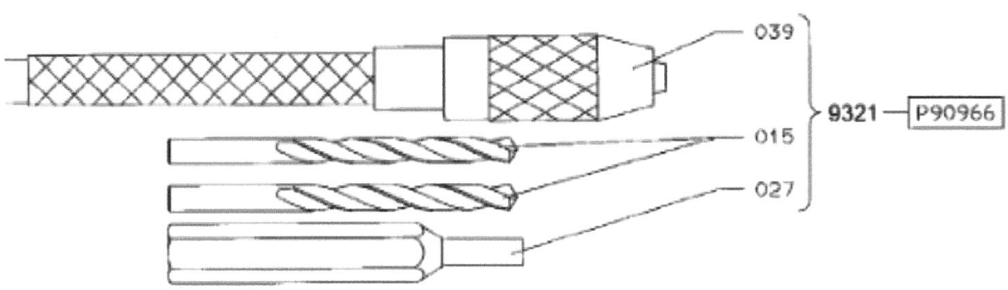

015 : Drill

027 : Test mandrel

039 : Pin vice

9321 : Cleaning tool set, complete

Fuel Valve

SAFETY PRECAUTIONS *For detailed sketch, see 900-2*

X	Stopped engine
X	Shut off starting air supply – *At starting air receiver*
X	Block the main starting valve
X	Shut off starting air distributor/distributing system supply
X	Shut off safety air supply – *Not ME-engines*
X	Shut off control air supply
X	Shut off air supply to exhaust valve – *Only with stopped lubricating oil pumps*
	Engage turning gear
	Shut off cooling water
X	Shut off fuel oil
	Stop lubricating oil supply
	Lock the turbocharger rotors

Data

Ref.	Description	Value	Unit
D09-40	Fuel valve opening pressure	300 - 380	bar
D09-42	Inlet seat, max. diameter	25	mm
D09-43	Fuel valve	16	kg

F.I.V(Fuel Injection Valve) 경우 빈번한 엔진 정박작업 중 하나이다.

Instruction Book에서 권고하는 교환 주기는 8000H 이다.

하지만 일반적으로 교환 주기 도래 전 O-ring 경화 또는 Nozzle 누설하여 교환 작업을 시행한다.

주기적으로 메인 엔진 피스톤 링 & 라이너 점검 시 피스톤 헤드 Fuel Oil 자국으로 F.I.V 누설을 확인 할 수 있다.

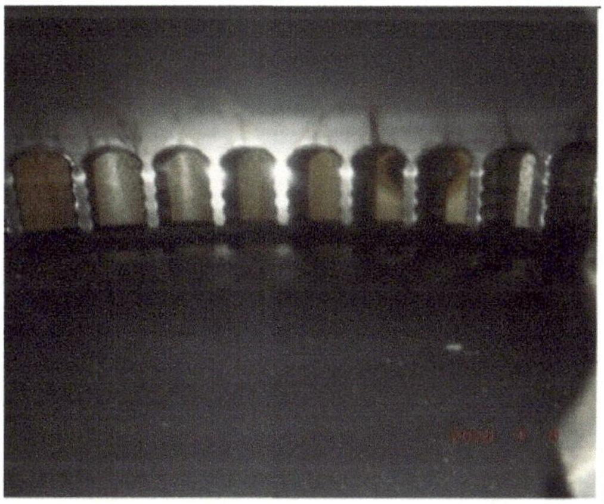

-> PISTON CROWN TOP SIDE & SCAV' AIR PORT & NOZZLE 누설 (교환 완료)

작업 설명에 있어 교환, 분해, 소제, 정비의 순서로 설명을 시작하겠다.

(1) 교환 방법

① 앞으로 설명할 모든 엔진 작업들은 기본적으로 엔진 사용 종료(Finished With Engine) 상태에서 작업을 진행하며 Engine Stop, Main Starting Air Valve Close, Starting Air Distributor Close, Turning Gear Engage 상태를 포함하여 엔진 사용 종료(Finished With Engine)라 지칭한다.

엔진 사용 종료(Finished With Engine) 후, F.O Pump Stop 후 F.O In & Out Valve Close한다.
(Control Air & Safety Air Close 할 필요 없다.
Alarm Monitoring System에서 M/E F.O Low Pressure Alarm 발생한다.)

F.O.Pump Inlet Valve Close, Drain Valve Open하여 F.O Pump 내부의 F.O Drain 한다.
(단, M/E & G/E F.O Line Common Line이라면 Pump Stop 하지 않고 F.O Line Inlet & Outlet Valve Close 후 F.O Pump 앞에 위치한 Inlet Valve Close 한다.)

*** 만약에 상기와 같은 방법으로 F.O Pressure Drop되지 않는다면 F.O Valve가 누설하는 것이므로 Pipe Line 확인한 다음 누설 부 Valve 교환한다.**

② F.O High-Pressure Pipe의 Union Nut**(90913-0130-048)**를 Hook Spanner 이용하여 분해한 다음 Union Nipple**(90913-0130-024 & 107)** 분해하여 F.O High-Pressure Pipe(3EA)를 취외한다.

③ F.I.V의 Return Pipe, Nut**(90101-0160-210)** 및 Spring Housing**(90101-0160-222)**를 분해한 다음 F.I.V Lifting Tool(아래사진 참조)을 이용하여 F.I.V 취외한다.

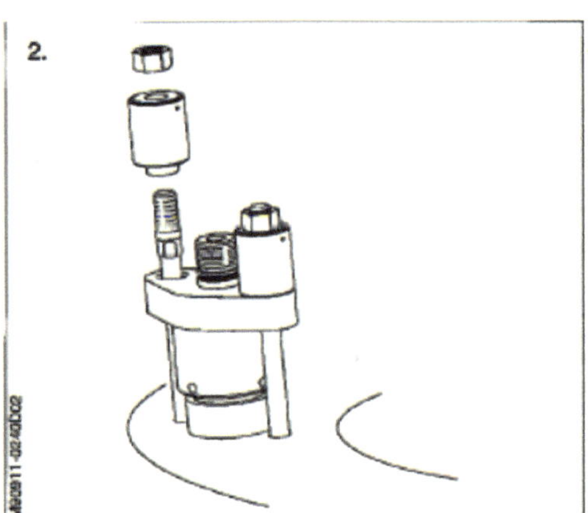

< Spring Housing & Nut >

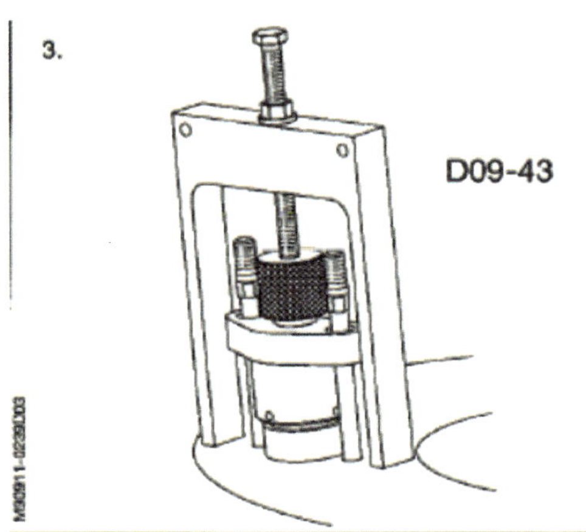

< F.I.V Lifting Tool : 90151-0204-9822 >

④ F.I.V를 취외한 다음 내부를 Brush를 이용하여 깨끗하게 소제한다.

⑤ F.I.V에 연결된 Return Line Coupling을 취외하고 Brush 소제한 다음 Size맞는 Copper(구리) Packing과 함께 정비한 F.I.V에 조립한다.
(Copper Packing 누락하여 조립하지 않도록 한다.
Copper Packing은 재고 없을 시 Gas Torch로 가열 후 상온 냉각하면 재사용 가능하다.)

< F.I.V Return Line Coupling 및 Copper Packing >

⑥ F.I.V의 O-ring**(90910-0143-173, 90910-0143-197)** 조립하고 F.I.V 하단 Cylinder Head 접촉 부 Molykote 도포한 다음 정비된 F.I.V를 Cylinder Head에 안착 시키고, Spring Housing 및 Nut 장착 후 Spring Housing 평평하게 되도록 Nut를 Tight한다.
(평평하게 Tight하는 것이 중요하고, 제대로 안착 되면 분사 시 설정한 압력에 맞춰 분사 된다. 그렇지 않으면 설정한 압력과 다르게 분사되거나 밖으로 누설한다.)

⑦ 정비된 F.O High Pressure Pipe Distance(Pipe to Thrust Bushing Lower Edge)가 **14mm** 맞는지 재확인하고 필요 시 조정한다.
O-ring**(90913-0130-119)** 조립하고 F.O High Pressure Pipe 연결한다.

⑧ F.O High Pressure Pipe의 Union Nipple Full Tight하고 Union Nut를 Hook Spanner 이용하여 Hand Tight한다.
Instruction Book상으로 Union Nipple **300Nm**이다.

⑨ F.O Line 복구하고 Line Venting을 위해 F.O Pump Drain Valve Open하여 Venting 한다.
Venting 후 Drain Valve Close 한다.

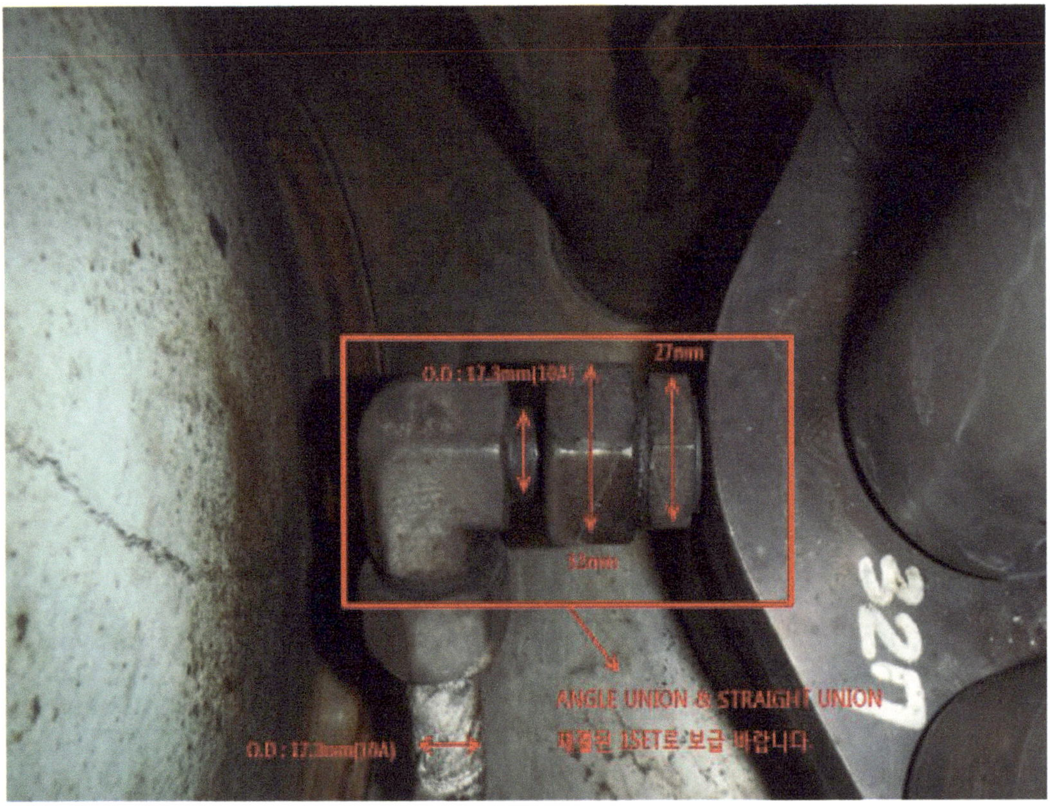

〈 F.I.V Return Line Coupling 〉

F.I.V Return Line Coupling의 **Copper Packing이 불량하여 누설하는 경우,**
Spring Housing을 평평하게 Tight하지 않아 Housing과 Cylinder Head사이에서 누설하는 경우,
Coupling자체가 불량하여 누설하는 경우가 있다.

상기 언급한 부분을 대비하여 Coupling 및 Copper Packing의 재고는 항상 넉넉하게 보유한다.

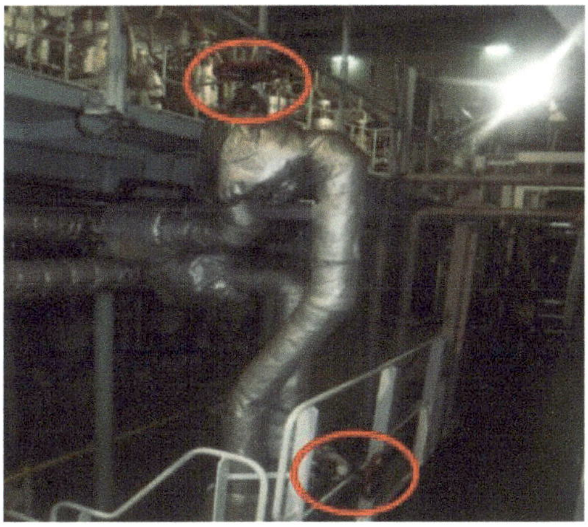

F.O Line In & Out Valve

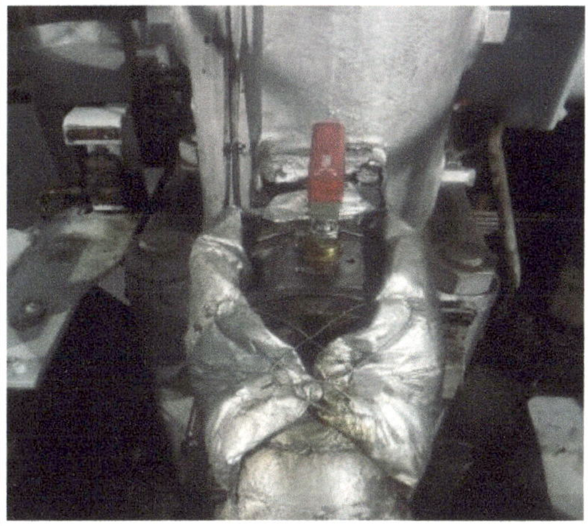

F.O.P Inlet Valve

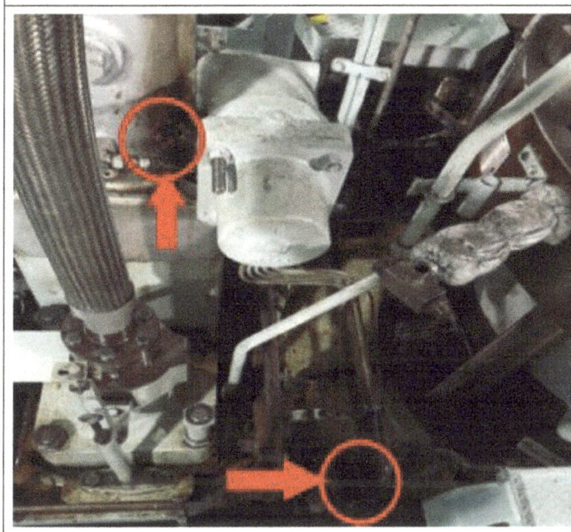

F.O.P Drain Valve 및 Drain Position

Return Line 분해

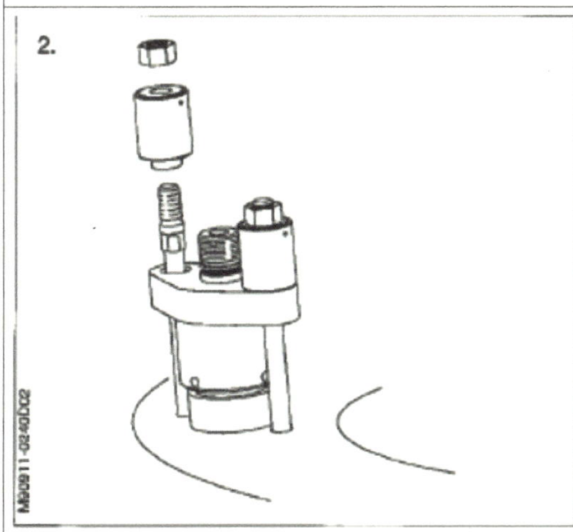

Spring Housing & Nut 분해

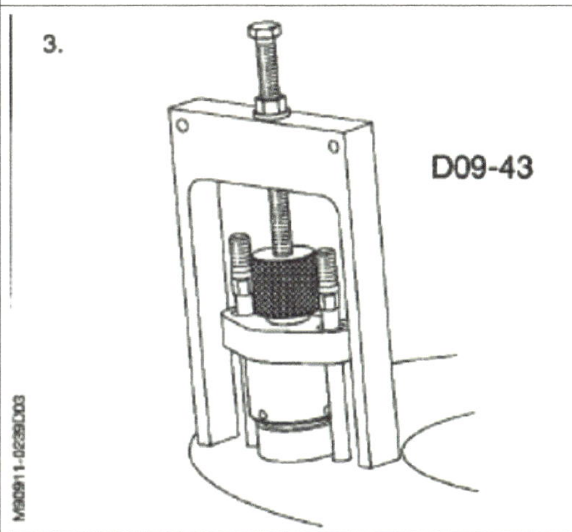

F.I.V Lifting Tool 이용하여 취외

Cylinder Head에서 F.I.V 취외

Return Line Coupling Cooper Packing

173 & 197 O-ring & F.I.V 하단부 Molykote 도포

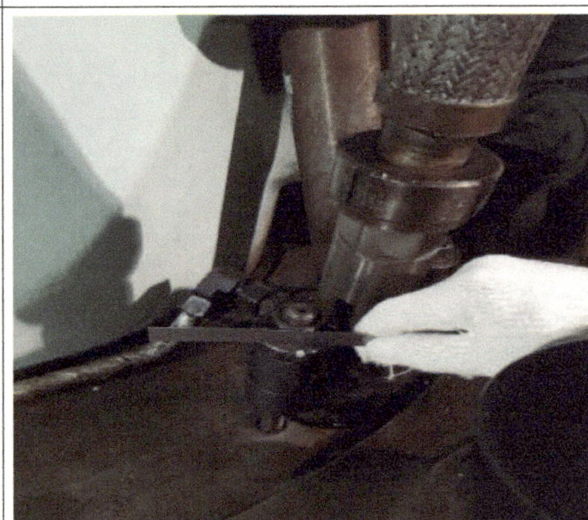

장착 및 Spring Housing Tight

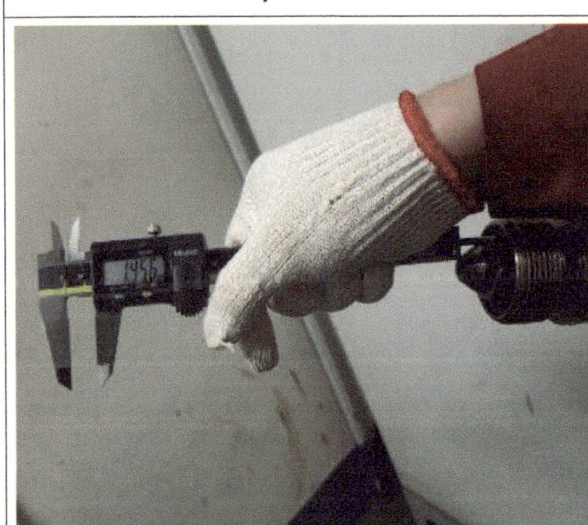

F.O High Pressure Pipe O-ring & 14MM

F.O High Pressure PIPE 장착

(2) 분해 방법

F.I.V 분해 및 조립 작업대

F.I.V 분해 및 조립 Tool

Tool 장착

Tool 장착 후 Tight

Hook Spanner와 Hammer로 Union Nut 분해

Union Nut 분해

F.I.V Guide Pin

Rubber Hammer로 Valve Head 분해

F.I.V 내부 부품

F.I.V Housing 내부 Guide Pin

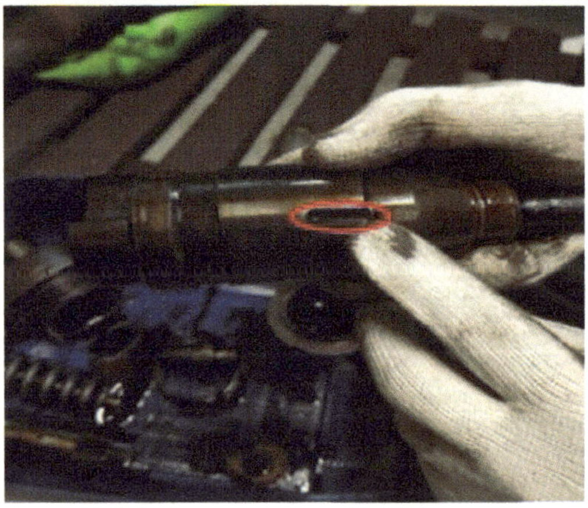

Housing Guide Pin에 연결되는 부분
(Spindle Guide Complete)

그림을 참조하여 순차적으로 분해하고
깨끗하게 소제한다
Spindle Guide 조립할 때는 Housing 내부 Pin과
Spindle Guide 홈에 제대로 조립 되도록 한다.

(3) 조립 방법

조립은 분해의 역순이므로 중요 Point만 집어본다.

① Nozzle Holder(**90910-0143-136**)의 O-ring(**90910-0143-124**)를 조립한다.

② 도면을 참조하여 **Spindle guide → Thrust foot → Disc(Optional extras) → Spring → Thrust spindle → Disc → Non-return valve → Valve head → Union Nut** 순으로 조립한다.

상기 부품 조립할 때 각 부품의 상태를 주의 깊게 확인하고 필요 시 교체한다.

Holder에 Spindle Guide 넣고 Spindle Guide의 Nozzle tip(Atomizer)를 돌리면서 당기면 더 빠져나오며, 이 상태가 Holder Guide Pin에 Spindle Guide가 정확히 조립된 것이다.

(단, RT-FLEX Engine의 경우 조정 너트(Adjusting Nut)로 분사 압력을 조정하는 것과 다르게 MAN ES 캠 구동 엔진은 Disc(90910-0143-950, 통상 Thim Plate라 부름)를 가감하여 분사 압력을 조정한다.

참고로 Disc의 두께가 다르면 분사 압력도 달라지니 보유 중인 예비품 Disc두께를 확인한다. 가급적이면 Disc 2개까지만 넣고 그 이상 필요한 경우에는 Spring을 신품으로 신환 한다.)

③ Valve head와 Union Nut의 Thread에 Molykote 도포한 다음 Tool을 설치 및 Tight 한다.

④ Tight하면 Spring이 아래로 눌러지고 Valve head와 Nozzle Holder를 연결하는 Guide Pin이 제대로 안착 되었는지 확인 한 다음 Union Nut를 Hook Spanner이용하여 Tight한다.

⑤ Tool을 풀어내고 F.I.V Tester에 올려서 분사Test (**300~380 Bar**) 한다.

Additional Disc(Thim Plate)로 분사 압력 조절

Spindle Guide, Thrust Foot, Additional Disc

Holder Guide Pin에 Spindle Guide가 제대로 조립되도록 한다.

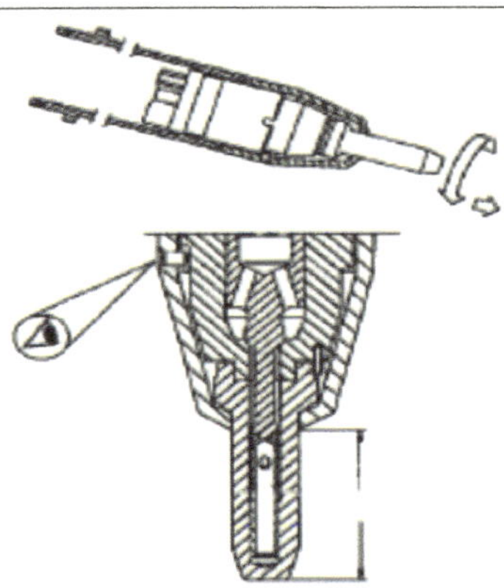

Spindle Guide를 Valve Housing에 조립하고 돌리면서 당기면 Housing Guide Pin에 Spindle Guide가 제대로 장착된다.

1.

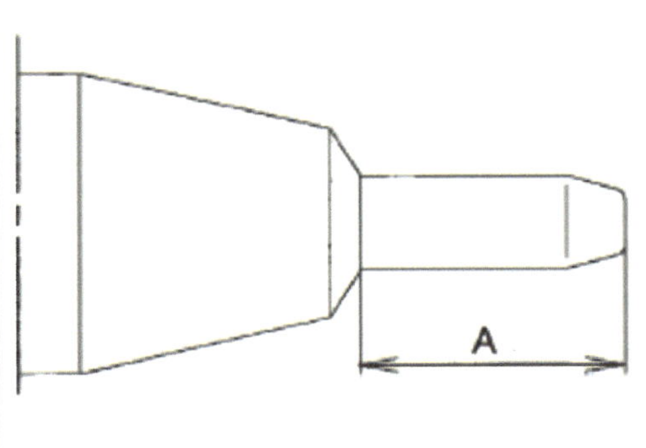

제대로 장착되었을 때의 Nozzle 길이

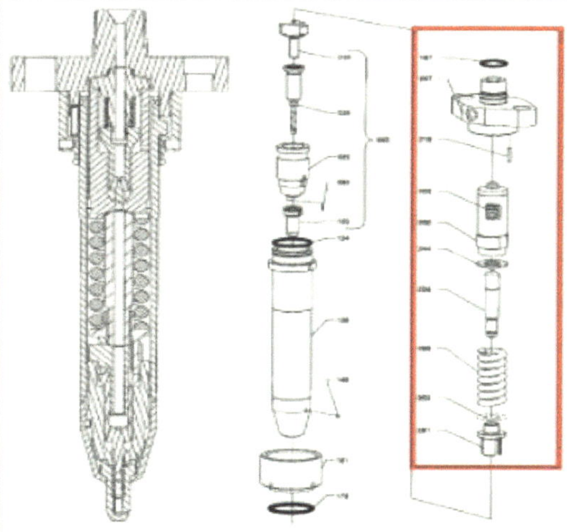

내부 부품 장착 후 Valve Head 조립

Molykote 도포

Tight 전 Valve Head Guide Pin

Tool 장착 후 Tight

Tight 후 Valve Head Guide Pin

Union Nut Tight

* 참고로 Spindle Guide(일명 Nozzle)의 경우 Complete(**90910-0143-65**) 교환한다.

보유 중인 재생품 Nozzle이 없는 경우에는 Nozzle 분해, 소제 및 분사 Test하여 사용하지만 재생품 Nozzle이 있는 경우에는 Complete로 교환한다.

정비 후 Test 시 누설하는 Nozzle의 경우 따로 모아놓고 재생 수리 신청한다.

신품 Nozzle 사용할 경우 N1(N:New), 재사용할 경우 N2로 표시하여 이력 관리한다.

신품 Nozzle의 경우 경험상 최대 N3~N4까지 분사가 되었으며, 재생품 Nozzle의 경우 R1(R:Recondition), 재사용할 경우 R2로 표시하여 이력 관리한다.

재생품 Nozzle의 경우 경험상 최대 R2까지만 분사가 되었다.

Non-return Valve 또한 동일하며 표시 및 이력 관리는 상기와 다른 방법으로 본인만의 History를 만들면 된다.

(결론 : 신품 Nozzle N3 또는 N4까지 사용, 재생품 Nozzle R1 또는 R2까지 사용 후 재생 수리)

Nozzle & Non-return Valve 분해 및 정비 방법은 다음 장을 참고한다.

(4) 스핀들 가이드 & 논-리턴 밸브 분해 및 정비

Fuel Valve Spindle Guide

Note!
This instruction is only valid for spindle guides of the slide valve design. Extreme care and accuracy should be exercised when carrying out this operation. Based on service experience it is recommended to replace the complete spindle guide after 16000 hours of operation.

1. Clean the outside of the spindle guide in pure gas oil or similar. The individual parts of the spindle guide are not interchangeable, therefore only one guide is to be disassembled at a time.

Note!
The spindle guide, thrust piece and spindle are matched parts and may not be re-placed individually.

2. Place the spindle guide in a bench vice provided with "soft" jaws, and use the brass mandrel as shown to disassemble the spindle guide.

3. Mount the pulling tool around the fuel nozzle on the spindle guide. Turn the nut to pull the fuel nozzle off the spindle guide.

Note!
The pulling tool is not standard for all engines, but may be delivered as an optional extra.

4. If no pulling tool is available, the fuel nozzle can be dismantled from the spindle guide using two screwdrivers.

   Place the screwdrivers opposite each other in the small gap between the fuel nozzle and the spindle guide and very carefully force the fuel nozzle off the spindle guide.

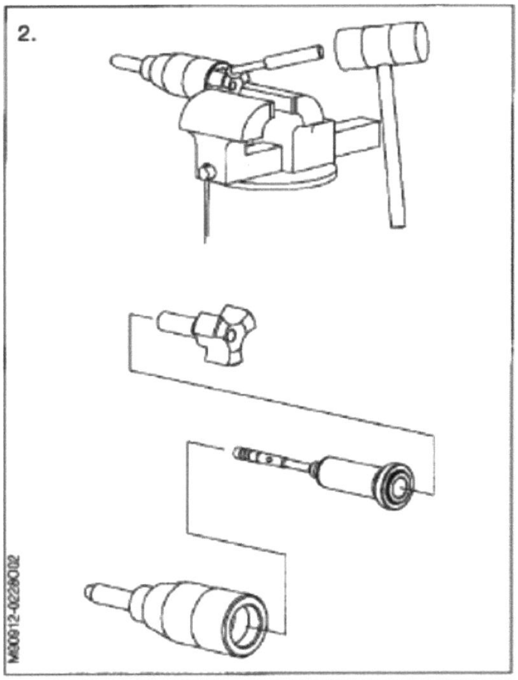

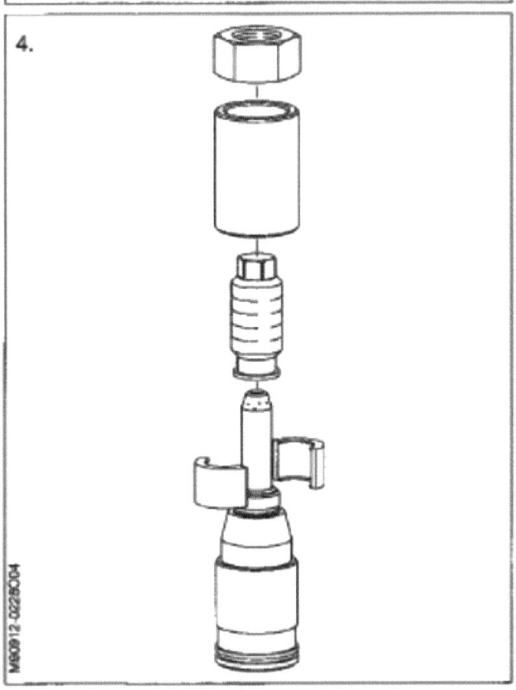

5.

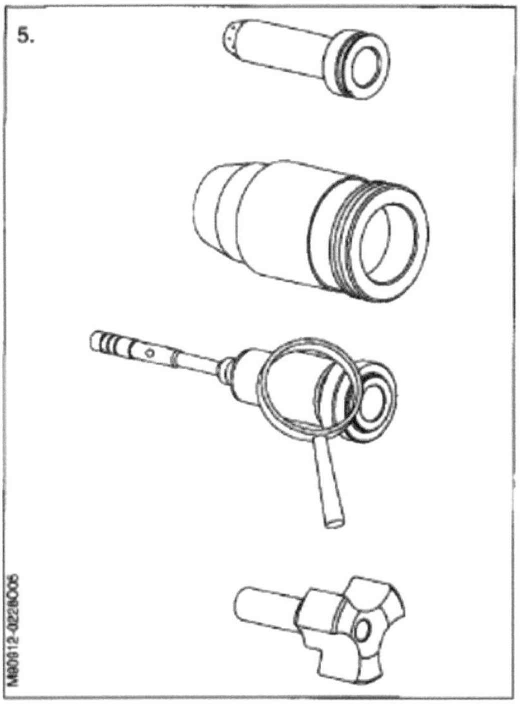

6.

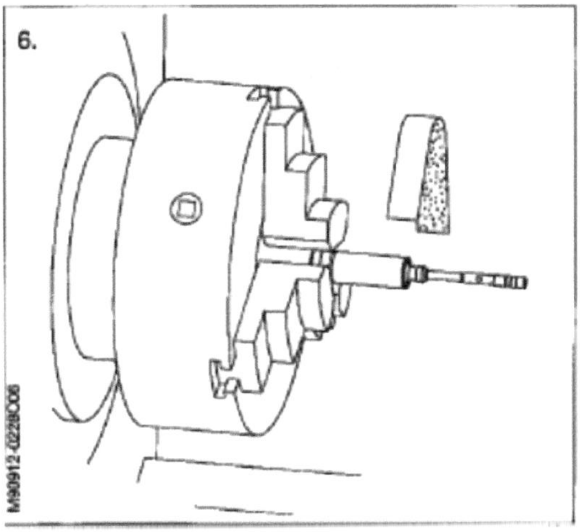

5. Clean all the parts of the spindle guide in gas oil and wipe dry with a clean piece of cloth.

   Clean all parts again in gas oil or 'Electro-cleaner' and wipe dry with a clean piece of cloth.

   Place all the parts on a clean, soft, lint-free cloth and examine them through an 8-10 times magnification magnifying glass and an inspection lamp.

   During the examination, pay special attention to the seating surfaces and sliding surfaces of the parts.

6. Remove any deposits or very fine scratches by placing the spindle, thrust piece or spindle guide respectively in a lathe, as shown, and polishing with a very fine conventional polishing linen 'grade 360'.

   Use also a little oil for the polishing (a coarser polishing linen must **absolutely not** be used).

Note!
The sliding surface of the cut-off slide may only be polished **VERY** carefully. The sliding surface must not be damaged.

After polishing, clean the parts again and re-check the seat on thrust piece/spindle, the seat on slide valve/spindle, and the seat on spindle/guide. Use an inspection lamp and an 8-10 times enlargement magnifying glass.

If the seats are not in order, i.e. if there are pressing-in marks or similar on the seats, the complete spindle guide must be discarded.

7. Clean any carbon deposits from the central bore of the fuel nozzle by means of the special brass brush. Clean the fuel nozzle with gas oil and wipe dry with a clean cloth.

8. Read the nominal hole size on the cylindrical part of the nozzle, as shown in the figure. The hole size is given in 1/100 mm. Depending on the engine model, there can be one or more different sizes of the holes. Find the correct drill size and matching test pin in the table, *D09-70, on the data sheet*, via the nominal hole size.

   Clean the spray holes, using gas oil and the special drills supplied.

Note!
During this operation be very careful not to push the drill too far to avoid scratching the snug-fit surface on the inside of the fuel nozzle.

   Then test the spray holes with the test pin. If the test pin is able to enter just one of the holes, the fuel nozzle must be discarded.

   This also applies to nozzles with oval holes (can be ascertained with a magnifying glass).

   Check the fuel nozzle before mounting on the spindle guide, the cut-off slide must be able to move freely inside the nozzle.

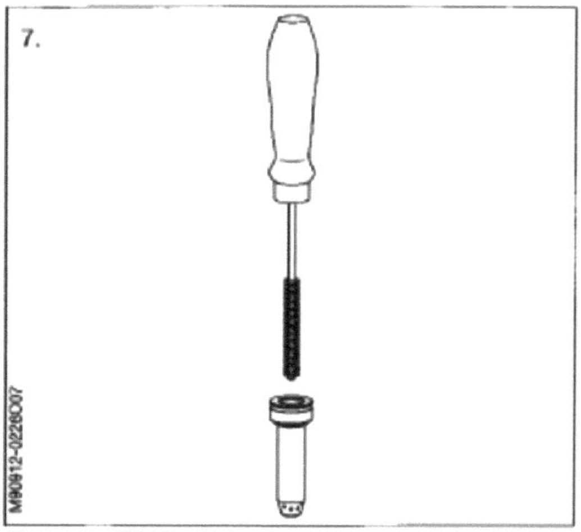

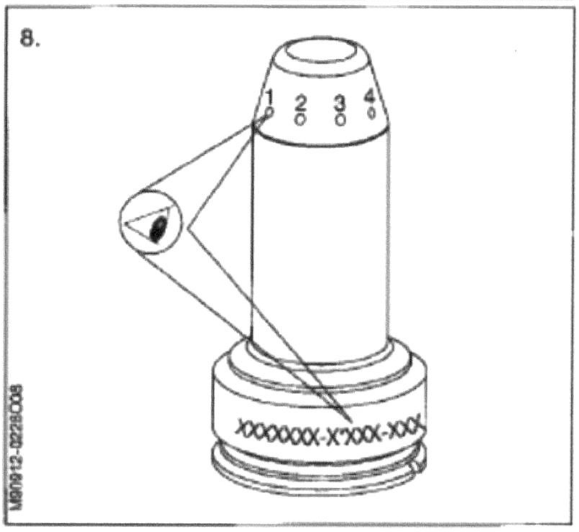

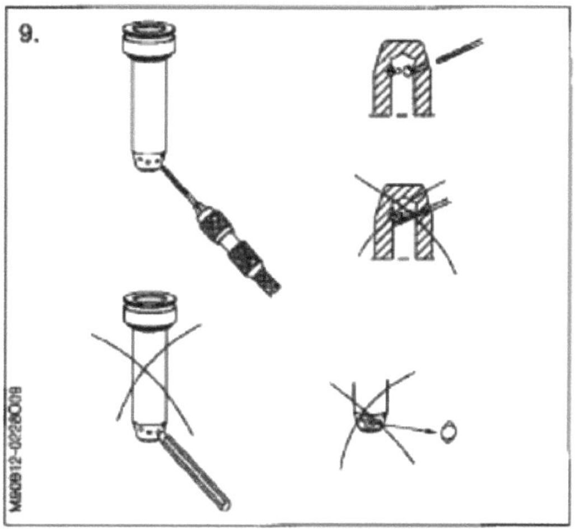

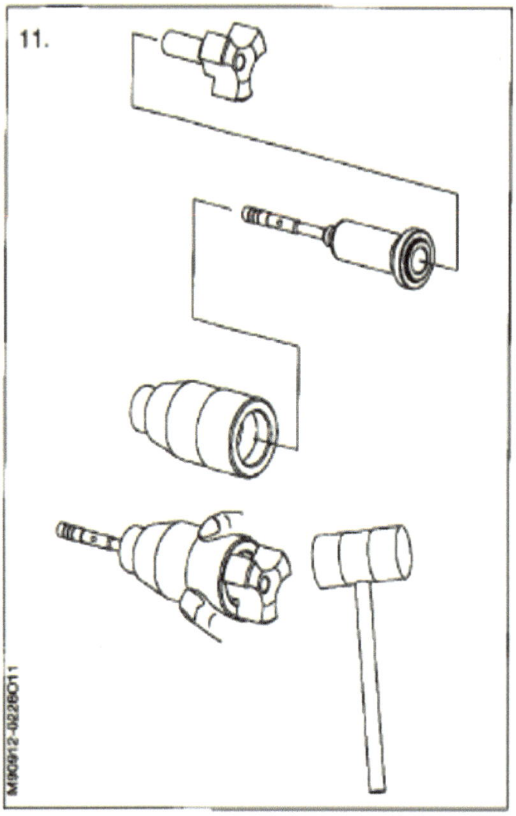

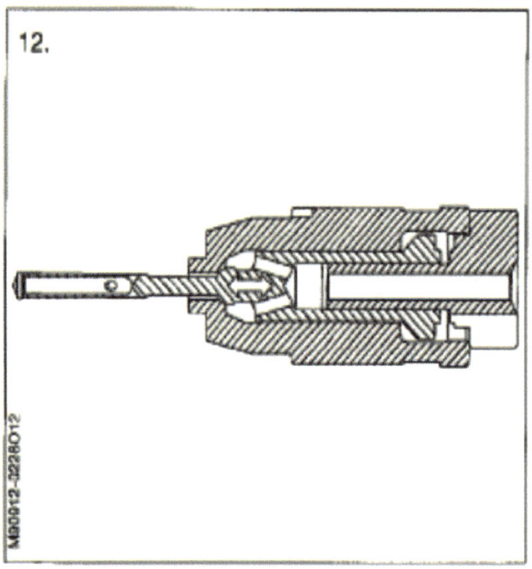

10. Lubricate the spindle and the thrust piece with the cut-off slide with a little Molybdenum Disulphide (MoS_2), *see Procedure 913-11.*

11. Assemble the thrust piece, the spindle and the spindle guide and carefully knock the parts together using a soft hammer.

12. Shake the spindle guide back and forth. The spindle with the cut-off slide must be able to slide freely back and forth inside the spindle guide, with a 'clicking' sound.

13. Lubricate the sliding surfaces of the nozzle and the spindle with a little Molybdenum Disulphide (MoS_2). *See Procedure 913-11.*

Mount the nozzle on the spindle guide.

Make sure that the half-circle shaped groove on the nozzle and the spring pin are in line.

If nozzle and spindle guide are provided with scratch marks opposite the spring pin, they must be in line.

Place the parts on the plane of a drilling machine or hydraulic press and position the mounting tool over the parts. Make sure that all the parts are perfectly aligned.

Press the nozzle on to the spindle guide.

Note!
The mounting tools are not standard for all engines but may be delivered as optional extras.

If no mounting tools are available, the nozzle can be mounted on the spindle guide using a short piece of pipe.

Place the pipe around the nozzle, so that the lower end of the pipe rests on the 'foot' of the fuel nozzle. Then press the parts together the same way as when using the mounting tools.

Check that the spindle inside the spindle guide is able to move freely as in step 10.

14. If the spindle guide is not to be mounted in a fuel valve immediately after the overhaul, cover all openings of the spindle guide with plastic to prevent dirt from entering the spindle guide during storage.

13.

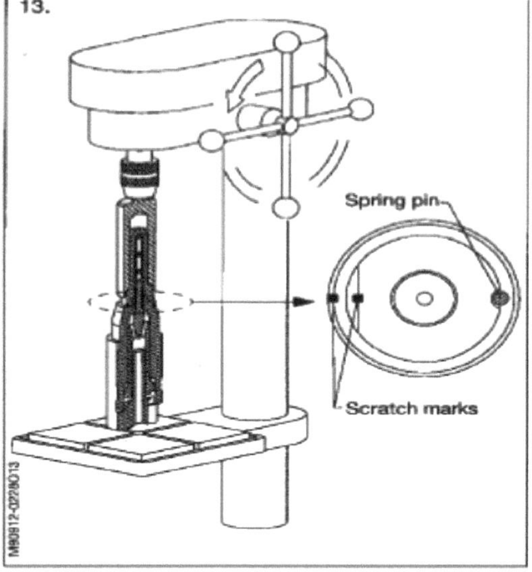

Nozzel spray hole cleaning tool table

Nozzel spray hole	Cleaning drill	Test pin	Nozzel spray hole	Cleaning drill	Test pin
0.90	0.85	0.94	1.60	1.55	1.65
0.95	0.90	0.99	1.65	1.60	1.70
1.00	0.95	1.04	1.70	1.65	1.74
1.05	1.00	1.09	1.75	1.70	1.80
1.10	1.05	1.14	1.80	1.75	1.85
1.15	1.10	1.20	1.85	1.80	1.89
1.20	1.15	1.24	1.90	1.85	1.94
1.25	1.20	1.30	1.95	1.90	1.99
1.30	1.25	1.34	2.00	1.95	2.05
1.35	1.30	1.39	2.05	2.00	2.09
1.40	1.35	1.44	2.10	2.05	2.14
1.45	1.40	1.49	2.15	2.10	2.19
1.50	1.45	1.54	2.20	2.15	2.24
1.55	1.50	1.59	2.25	2.20	2.29

Non-return valve

Note!
Extreme care and accuracy should be exercised when carrying out this operation.

1. Clean the outside of the non-return valve with pure gas oil. The individual parts are not interchangeable, therefore only one guide is to be disassembled at a time.

Note!
Except for the slide valve spring, defective parts cannot be replaced individually by new ones.

Place the non-return valve as shown in Fig. 1 in a bench vice provided with "soft" jaws, and disassemble the non-return valve, using the disassembling tool and a hammer as shown.

2. Clean all the parts for the spindle guide in gas oil and wipe them dry with a clean piece of cloth.

Finally, clean in either gas oil, kerosene or 'Electrocleaner', and wipe the parts dry with a clean piece of cloth.

3. Now place the parts on clean, lint-free rags and examine with an 8-10 times enlargement magnifying glass, and an inspection lamp with magnifying glass as shown in Fig. 3.

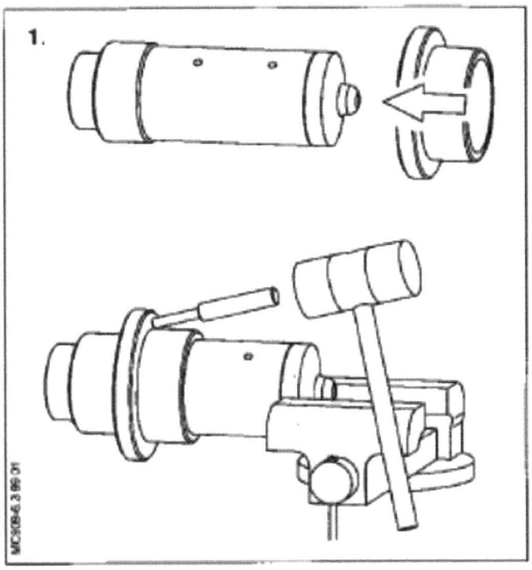

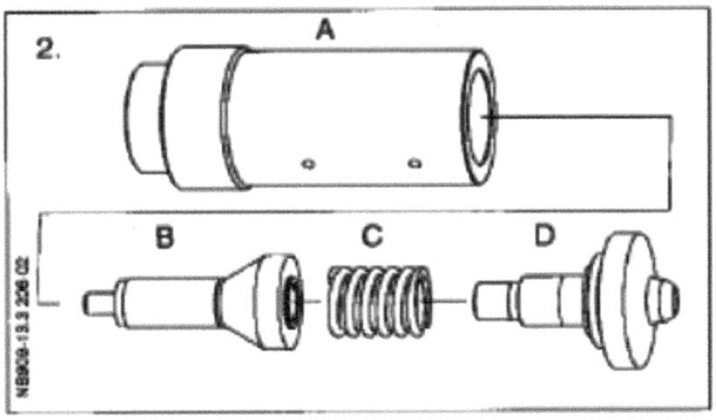

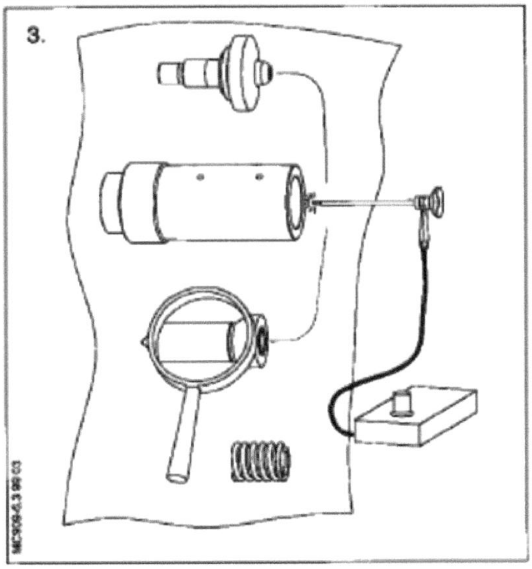

4.

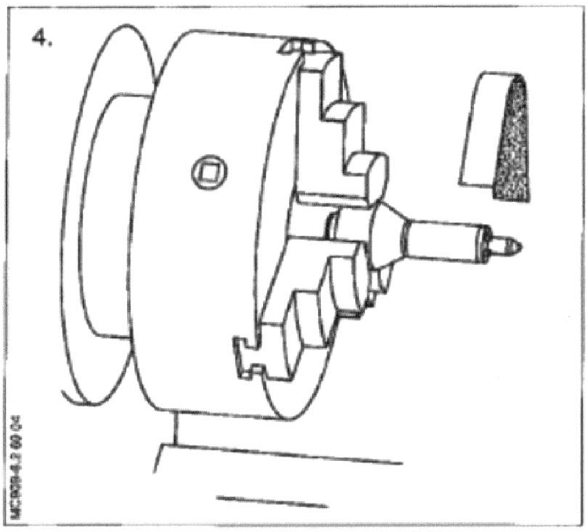

5.

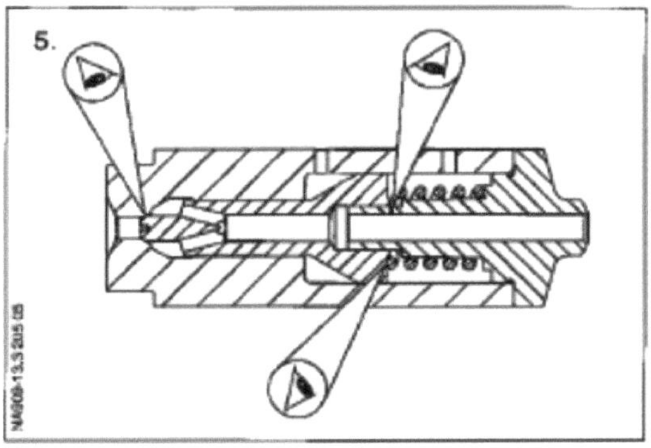

6.

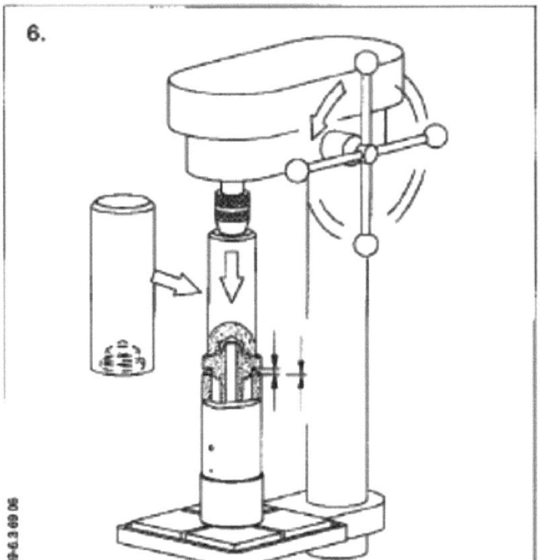

4. Examine the slide faces of movable parts for coating. Vent slide **B**/housing **A** will be too tight if there is a coating. (The letters **A-D** refer to the letters/parts in Fig. 2).

   Fix vent slide **B** and, subsequently, housing **A** in a lathe as shown in Fig. 4 and remove the coating by means of very fine conventional polishing linen 'grade 360'.

   Also a little oil should be used (a coarser polishing linen must **absolutely not** be used).

5. Check spring **C** for the thrust piece for outside wear marks. If defective, it should be exchanged.

   Check the seat on thrust piece **D**/vent slide, and the seat on vent slide/housing. Use an inspection lamp and an 8-10 times enlargement magnifying glass.

   If the seats are not in order, i.e. if there are pressing-in marks or similar on the seats, the complete spindle guide must be discarded.

6. Mount the non-return valve as follows:

   - Lubricate all movable parts with molybdenum disulphide (MoS_2).
   - Place the loosely-assembled non-return valve on the plane of a drilling machine, with the tool positioned as shown in Fig. 6.
   - Make sure that the thrust piece and the other parts are perfectly aligned and that the thrust piece is guided in the vent slide.
   - Press the handle until the housing and thrust piece meet.

7. If the non-return valve is not to be mounted in a fuel valve immediately after the overhaul, cover all openings of the non-return valve with plastic to prevent dirt from entering the valve during storage.

(5) 연료 분사 밸브 테스터

F.I.V Tester 작동하기 전, L.O Tank Level 확인하고 부족 시 보충한다.(HYD'OIL 32)

작동 전 Tester 앞에 적혀 있는 설명을 먼저 충분히 숙지한다.

Main Power On, Air Valve Open하면 Tester 작동 준비는 완료된다.

분사 압력 : 300~380Bar

Tester에 Fuel Injection Valve 장착

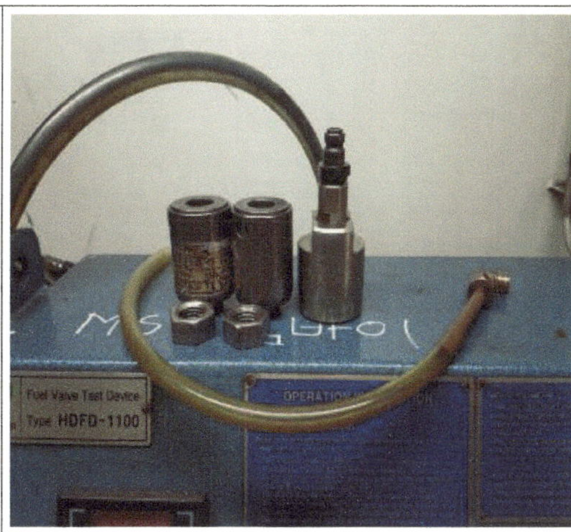

Spring Housing, HYD´ Hose 연결 잭, Drain Line

Spring Housing 평평하도록 Nut Tight

Spring Housing 평평하게

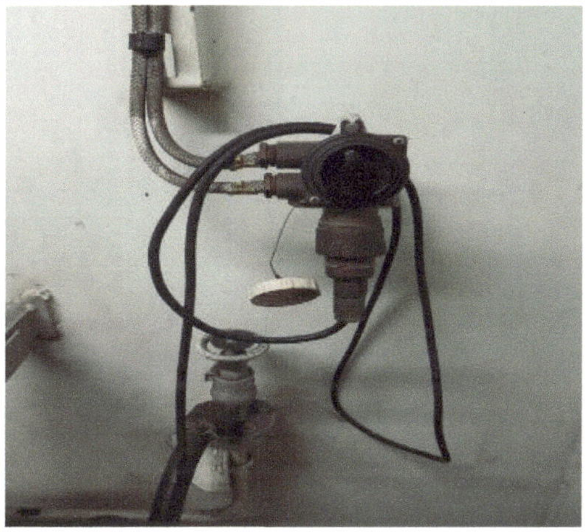

Power Switch 'On' / Air Supply Valve 'Open'

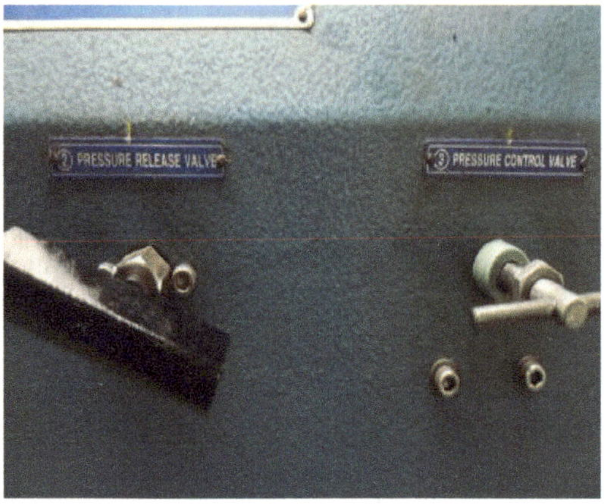

Pressure Release Valve Close
Pressure Control Valve Close하면서 분사 Test

분사 Test

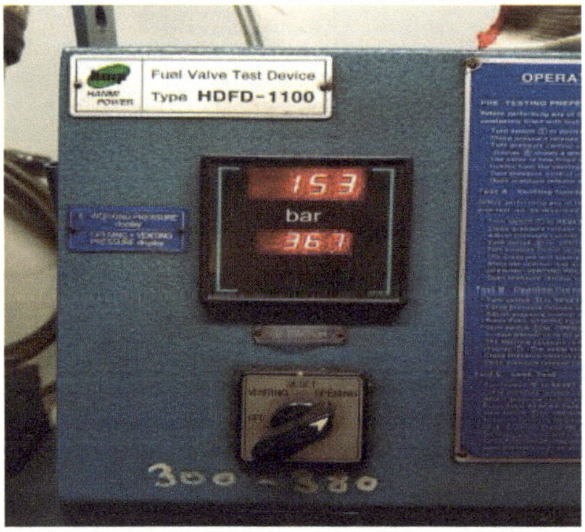

'Reset' Position 두면 계속 분사 Test되고,
'Opening' Position 두면 한번 분사 Test된다.
만약에 분사 압력이 높거나 낮으면 다시 분해하여
Additional Disc(90910-0143-950, Thim Plate)를
집어넣거나 빼서 분사 압력을 재 조절한다.
분사 압력(300~380Bar)이 범위 내 있다면
(가급적 높은 쪽이 좋으며 낮으면 누설 확률증가)
Tester에 연결된 액세서리 분해하고
Paint Makar로 분사 압력 및 정비 날짜 입력 후
Nozzle Tip Sealing하여 보관한다.

2) 연료유 고압관

정비 주기 : 8000H

High-Pressure Pipes

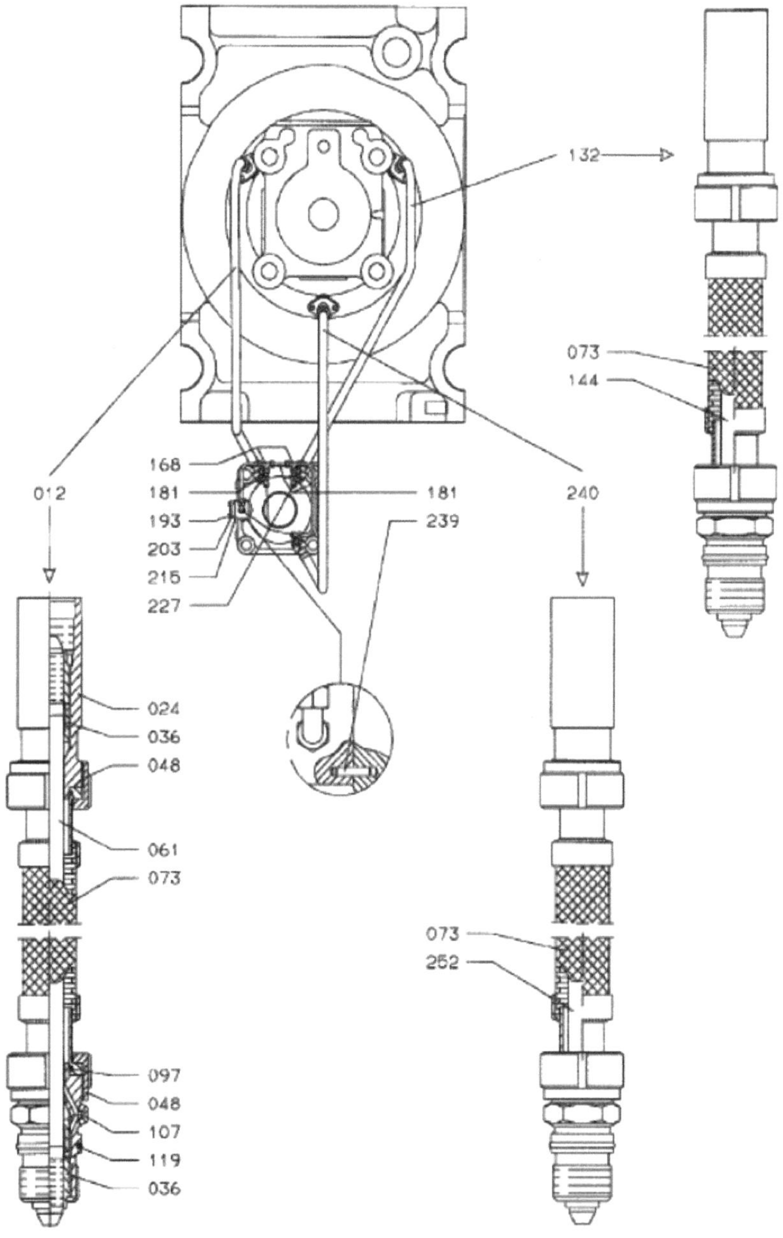

Fuel Oil High-Pressure Pipe

SAFETY PRECAUTIONS *For detailed sketch, see 900-2*

X	Stopped engine
X	Shut off starting air supply – *At starting air receiver*
X	Block the main starting valve
X	Shut off starting air distributor/distributing system supply
X	Shut off safety air supply – *Not ME engines*
X	Shut off control air supply
	Shut off air supply to exhaust valve – *Only with stopped lubricating oil pumps*
X	Engage turning gear
	Shut off cooling water
X	Shut off fuel oil
	Stop lubricating oil supply
	Lock the turbocharger rotors

Data

Ref.	Description	Value	Unit
D09-47	Fuel oil high-pressure pipe, tightening torque	300	Nm
D09-51	Distance from pipe end to thrust bushing lower edge	14	mm
D09-52	Fuel oil pipe	20	kg

교환 방법은 앞서 설명한 F.I.V에 포함되므로 생략한다.

엔진 운전 중에 F.O High Pressure Pipe 자체에서 미량 누설하는 경우가 종종 있다.

그럴 경우 가까운 항구에서 정박 중에 교환하고, Pipe 재생 수리 신청 또는 신품 청구한다.

다시 언급하지만 상기 표시한 14mm(Distance from pipe end to thrust bushing lower edge)가 맞는지 조립 전에 반드시 확인한다.

14mm 미만일 경우 Pipe와 F.I.V 연결 부 또는 Pipe와 F.O.P 연결부에서 누설한다.

Union Nipple Full Tight하고 Union Nut를 Hook Spanner 이용하여 Hand Tight한다.

Instruction Book상으로 Union Nipple 300Nm이다.

*** 분해 및 정비 방법**

분해 및 정비 방법은 아래의 Instruction Book 설명을 참고하고 중요한 부분만 추가로 설명하겠다.

아래 방법을 참조하여 분해 시 Brush 이용하여 깨끗하게 소제하면 된다.

장착된 내부 O-ring**(90913-0130-097)**을 취외하고 내부 소제 후 신품으로 교환한다.

Thrust Bushing**(90913-0130-036)**은 일반적으로 사용하는 오른 나사(Right-hand thread)와 다르게 왼 나사(Left-hand thread)이다.

따라서 Close 방향으로 Tight하면 분해되고 Open 방향으로 Tight하면 조립된다.

Digital Caliper 또는 일반 Caliper(일명 버니어 캘리퍼스) 사용하여 **14mm** 조절한다.

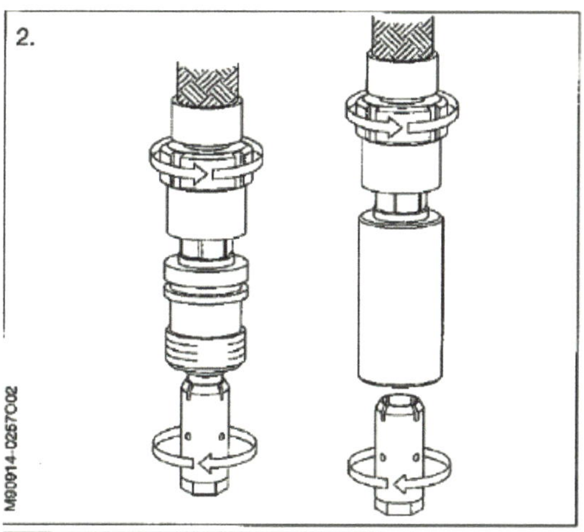

- Lift up the union nut with fl exible protective hose, together with the coupling nut, on the high-pressure pipe and screw off the thrust bushing.
- Remove the coupling nut and, when the thrust bushings at both ends of the pipe have been dismounted, pull the fl exible protective hose with union nuts off the high-pressure pipe.

3. Shape-up the threads on the pipe ends with the nut die.

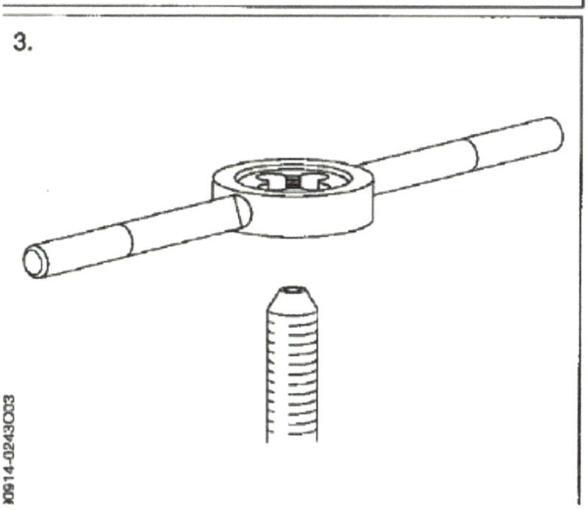

4. Mount the guide on the pipe end, place the miller in the guide and lightly screw on the union nut.

5. Turn the miller with, for instance, a tap wrench while lightly tightening the union nut to provide a suitable pressure between miller and pipe end. During the milling, add drilling oil emulsion liberally.

   After completing the milling, carefully clean the high-pressure pipe with diesel oil and wipe dry with a clean piece of cloth.

6. When assembling the high-pressure pipe, screw the thrust bushings so high up on the thread of the pressure pipe that the distance between the pipe end and the bottom edge of the thrust bushing is as stated in Data.

   Replace the O-rings.

   Before mounting a high-pressure pipe, check the centre distances between pipe ends and seats, and lubricate the threads of the union nuts with molybdenum disulphide (MoS_2).

   It is of great importance for the mounting, and a condition for obtaining tight joints, that the pipes fi t the seats accurately.

   For reconditioning of seat in fuel valve, see Procedure 909-11.3.

   For reconditioning of seat in fuel pump top cover, see Procedure 909-5.3.

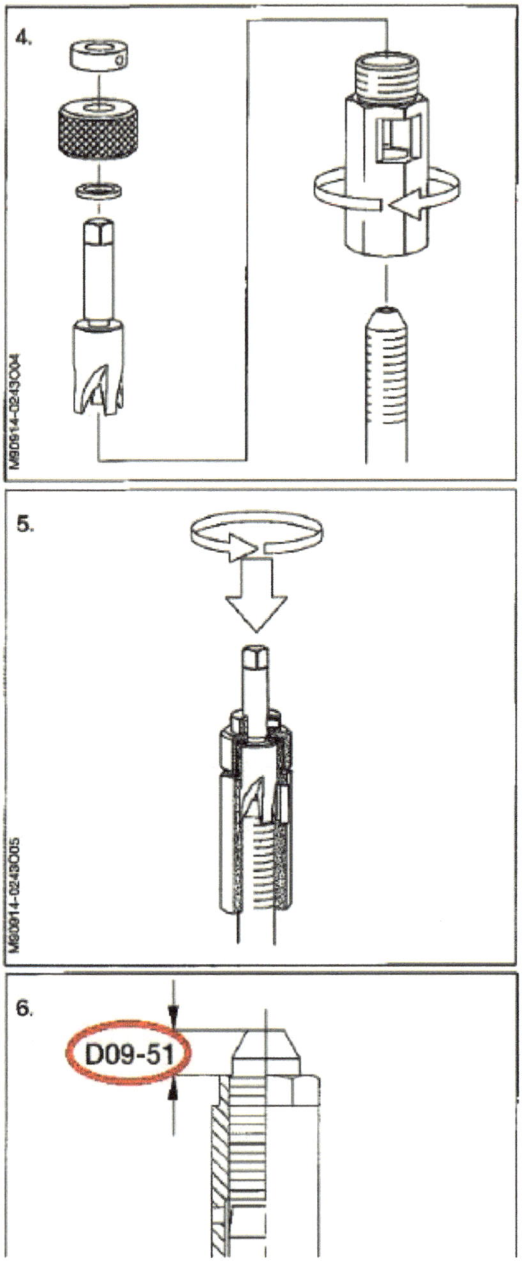

O-ring**(90913-0130-119)** 조립한다. **(교환 시에 조립해도 된다.)**

소제 및 조립이 완료되면 Pipe 양측 끝 단에 이물질 및 Dust 혼입되지 않도록 밀봉한다.

승선 중에 전 실린더 F.O Pipe 분해 및 정비하면 컨디션 관리가 용이하다.

3) 연료 펌프

정비 주기 : 16000H

Fuel Pump

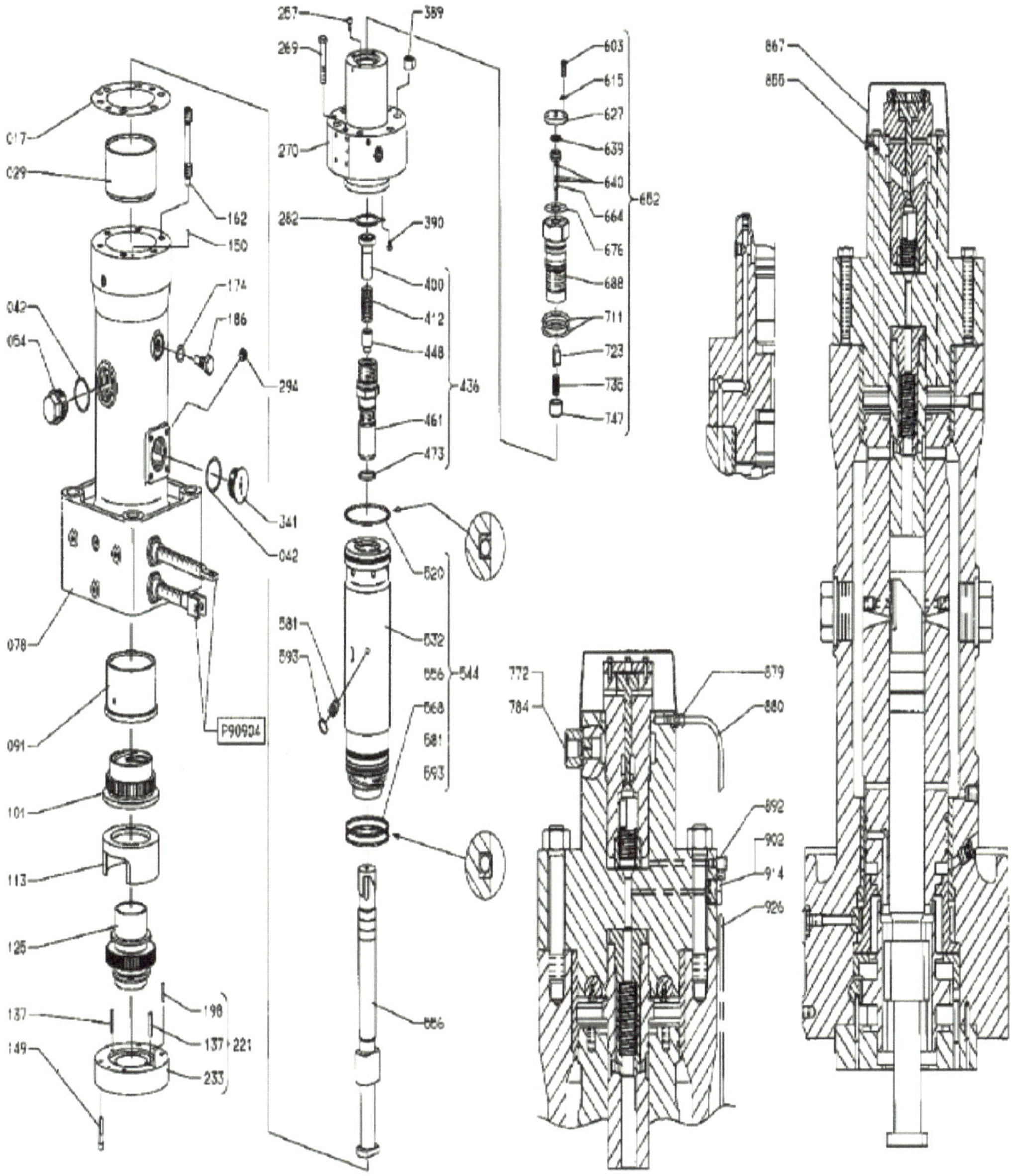

Fuel Valve and Fuel Pump Panel

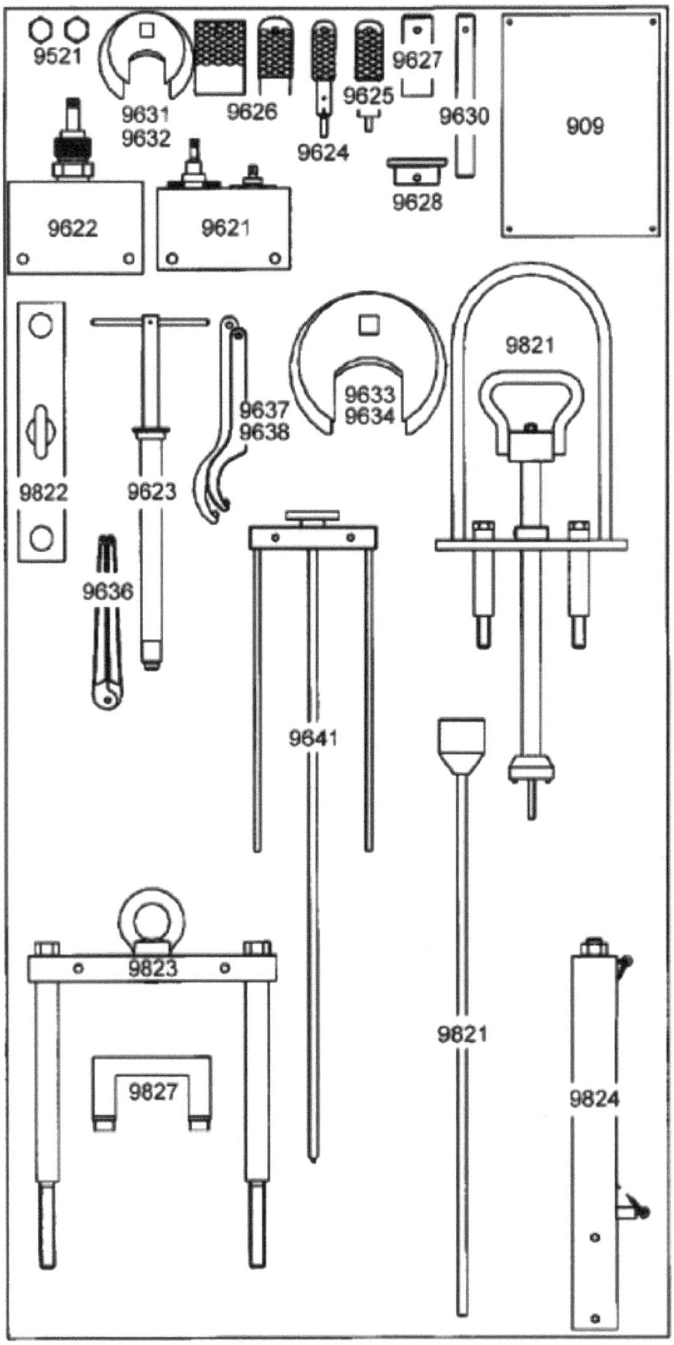

909 : Tool panel

9521 : Screw

9621 : Milling tool for fuel oil pipe seat

9622 : Milling tool for fuel oil pipe

9623 : Grinding mandrel for holder, inside

9624 : Grinding mandrel for valve head

9625 : Grinding mandrel for thrust spindle

9626 : Grinding mandrel for holder, outside

9627 : Drift for non-return valve

9628 : Flange for dismantling of non-return valve

9630 : Drift for spindle guide

9631, 9632, 9633, 9634 : Crowfoot wrench head

9636 : Pin Wrench, special

9637, 9638 : Hook wrench

9641 : Measuring tool for fuel pump lead

9821, 9823 : Lifting tool for fuel pump

9822 : Lifting tool for fuel pump housing

9824 : Tool for P.max rack

9827 : Assembling tool

Fuel Pump - Details

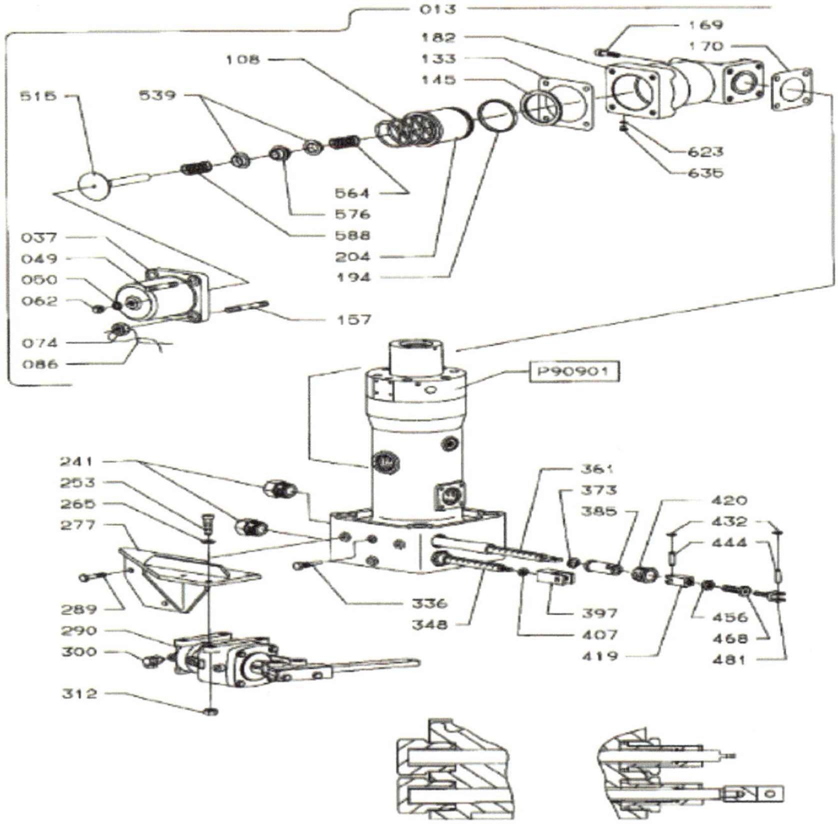

Fuel Pump

SAFETY PRECAUTIONS *For detailed sketch, see 900-2*

X	Stopped engine
X	Shut off starting air supply – *At starting air receiver*
X	Block the main starting valve
X	Shut off starting air distributor/distributing system supply
X	Shut off safety air supply – *Not ME engines*
X	Shut off control air supply
	Shut off air supply to exhaust valve – *Only with stopped lubricating oil pumps*
X	Engage turning gear
	Shut off cooling water
X	Shut off fuel oil
X	Stop lubricating oil supply
	Lock the turbocharger rotors

Data

Ref.	Description	Value	Unit
D09-18	Pump housing nuts, tightening torque	2100	Nm
D09-19	Pump housing nuts, tightening angle	100/55	Nm/°
D09-20	Pump housing	560	kg
D09-21	Bottom plate	35	kg
D09-39	Shock absorber	90	kg

(1) PLUNGER & BARREL 분해 방법

F.O.P Plunger & Barrel(**90901-0192-556 & 532**) 교환 방법이다.

Roller Guide, Reversing Air Cylinder Lifting Pin 및 Cam 점검이 아니면 Plunger & Barrel 취외 후 F.O.P Housing 내부 소제하고 신품 또는 정비품 Plunger & Barrel 교환하는 것이 작업하는데 있어 수월하며 작업시간도 단축된다.

F.O.P 교환 시 가장 중요한 부분은 Plunger & Barrel이며 그 중에서도 Barrel Sealing 교환이다.

① F.W.E(Finished With Engine)후 Control Air & Safety Air Valve Close 한다.

F.O Pump Stop 후 F.O In Out Valve Close 한다.
(Alarm Monitoring System에서 M/E F.O Low Pressure Alarm 발생한다.)
그 다음 F.O.P Inlet Valve Close, Drain Valve Open하여 Pump 내부 F.O Drain 한다.
(단, M/E & G/E F.O Line Common Line이라면 Pump Stop 하지 않고 F.O Line Inlet & Outlet Valve Close 후 F.O.P 앞의 Inlet Valve Close 한다.)

*** 만약, 상기와 같은 방법으로 F.O Pressure Drop 되지 않는다면 F.O Valve 누설하는 것이므로 Pipe Line 확인한 다음 누설 부 Valve 교환한다.**

② F.O High Pressure Pipe(3EA), Top Cover Bolt 및 연결된 Pipe 및 액세서리 취외한다.

③ Top Cover Lifting Tool 설치하고 Overhead Crane, Chain Block, Lifting Tool 연결한다.

Top Cover(**90901-0192-270**)를 분해한다.

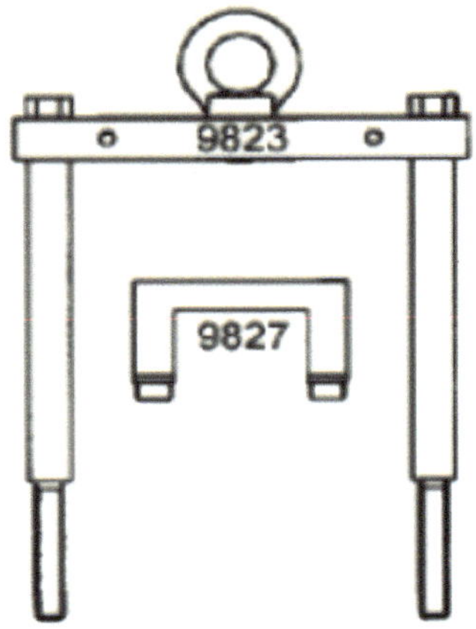

< Top Cover Lifting Tool : 90904-0052-9823 >

④ VIT Rack & Regulation Rack Position을 기록하고 Rack 고정 Nut와 연결 부분리한 다음 Rack Position을 전부 빼낸다.(**= Rack Position '0'로 맞춘다.**)

Regulation Rack(**90904-0052-348**)는 Plunger와 연결, VIT Rack(**90904-0052-361**)는 Barrel과 연결되며, VIT Rack의 경우 Barrel 무게 때문에 쉽게 빠지지 않으므로 Tool을 사용한다.

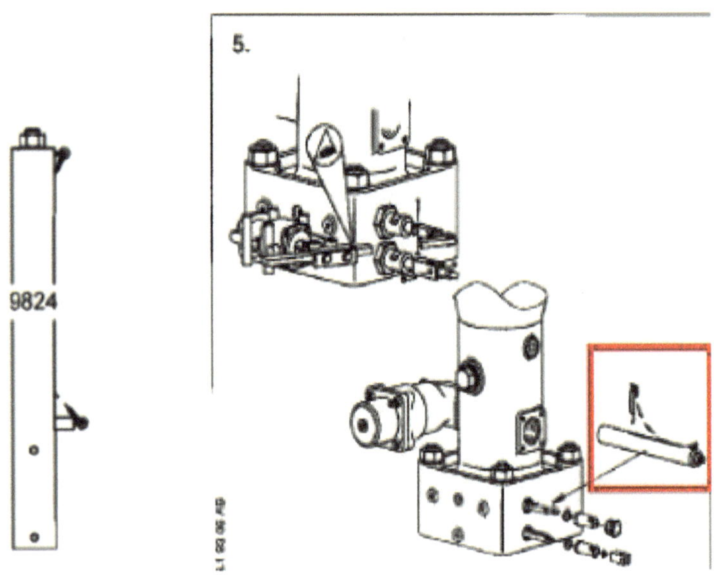

< VIT Rack Tool : 90951-0422-9824 >

⑤ Plunger & Barrel에 Lifting Tool 설치하고 Overhead Crane Hook, Chain Block & Lifting Tool을 연결하여 Chain Block에 Tension을 만든다.

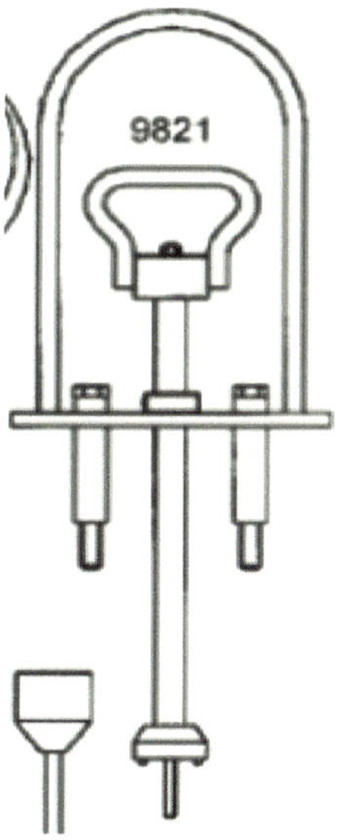

< Plunger & Barrel Lifting Tool: 90951-0422-9821 >

⑥ 이 상태에서 Guide Screw(**90901-0192-186**), Packing(**90901-0192-174**)을 취외한다.

이런 순서로 하지 않고 Guide Screw 먼저 취외하여 Lifting Tool을 설치할 때 Plunger & Barrel이 움직여버리면 Housing 내부에서 걸려 취외하지 못한다.

또한 VIT Rack & Regulation Rack 전부 뺀 상태(= **Rack Position '0'**)에서 Plunger & Barrel을 취외 할 수 있으며 취외 시 정확히 **'T'**모양이 된 것을 확인할 수 있다.

(정확하게 'T'모양을 맞춰야 빼낼 수 있다.)

만약에 이런 방법으로 빠지지 않는다면 Guide Screw 빼낸 부분에 Barrel이 틀어지지 않았는지 확인 후 Barrel을 원래 Position**(내부 홈이 보임)**으로 조정한 다음에 취외한다.

이렇게도 안 된다면 Plunger & Barrel Lifting Tool에 Barrel만 연결하여 먼저 취외 후 다시 Tool 사용하여 Plunger에 연결하여 취외하면 된다.

즉, Plunger & Barrel 개별적으로 취외하는 것이다.

또는 Barrel 하단부가 Screw 형식이므로 돌리면서 빼내도 되지만 이것은 시간이 오래 소요되어 추천하는 방법은 아니다.

결론적으로 Complete로 취외하지 않아도 다양한 방법들이 있으니 당황할 필요 없다.

⑦ F.O.P 내부를 깨끗이 소제하고 VIT Rack 및 Regulation Rack를 충분히 Free-up 한다.

(2) PLUNGER & BARREL 조립 방법

조립은 분해의 역순이다.

① VIT Rack 및 Regulation Rack Position 전부 빼놓는다.(= **Rack Position '0', 내부 'T'** 모양)

조립 시 Plunger & Barrel 및 F.O.P Housing 내부 Diesel Oil을 충분히 도포한다.

② 정비품 또는 신품 Plunger & Barrel에 Lifting Tool을 설치하고 Overhead Crane과 Chain Block을 이용하여 천천히 조립한다.
(조립할 때 Sealing Ring 부분 확인하면서 조립한다.)

③ Guide Screw 부분을 확인하여 Barrel이 제대로 안착 되는지 확인한다.

Guide Screw 확인하면서 Zinc로 Barrel을 아래 방향으로 쳐서 조립하되, 더 이상 내려가지 않을 경우 VIT Rack 부분을 Zinc를 이용하여 조금씩 밀어 넣으면 Barrel이 내려간다.

정확히 Center가 맞아야 Guide Screw 손으로 Tight할 때 부드럽게 조립되고, 조금이라도 걸리는 느낌이 든다면 Guide Screw를 취외하고 Lifting Tool을 움직여 Center를 다시 맞춰야 한다.

만약에 Plunger 조립하고 Regulation Rack가 움직이지 않는다면 조립이 잘못된 것이며, 다시 취외하여 Plunger & Barrel을 천천히 내리면서 Lifting Tool Plunger를 잡아주는 부분과 Regulation Rack를 움직여 제대로 조립되도록 한다.

④ Guide Screw가 조립되면 VIT Rack 및 Regulation Rack 앞, 뒤로 조절하였을 때 각각 Barrel 및 Plunger가 움직이는 것을 확인한다.

VIT Rack는 Barrel 무게 때문에 Zinc를 이용하여 확인하고, Plunger는 손으로 Regulation Rack 또는 Lifting Tool Plunger를 잡아주는 부분을 돌리면 Regulation Rack가 움직인다.

(VIT Rack : Barrel 움직여 분사 타이밍 조절한다.
Regulation Rack : Plunger 움직여 연료 토출량 조절한다.)

⑤ Plunger & Barrel Lifting Tool 제거한 뒤 VIT Rack 및 Regulation Rack 취외할 때의 Position에 맞게 조립한다.

⑥ Top Cover Gasket(**90901-0192-017**)에 Molykote 도포하고 정비된 Top Cover와 함께 조립한다.

Top Cover Suction Valve Sealing Ring(**90901-0192-473**)부분에 Diesel Oil을 충분히 도포하여 조립 시 Sealing Ring 손상되지 않도록 한다.

Top Cover Nut(**90901-0192-389**) : **500Nm**

⑦ 연결된 Pipe, 액세서리 및 F.O High Pressure Pipe 조립한다.

Control Air & Safety Air Line 복구하고 Air Pipe 누설 여부 확인한다.

F.O Line 복구하고 Line Venting을 위해 F.O Pump Drain Valve Open하여 Venting 한다.

Venting 후 Drain Valve Close 한다.

⑧ Engine 시동 시 현장에서 누설 부 또는 이상 소음 확인하고 RPM 변동에 따른 Exhaust Gas의 상태를 확인한다.

(3) 분해 및 조립 사진

F.O Line In & Out Valve

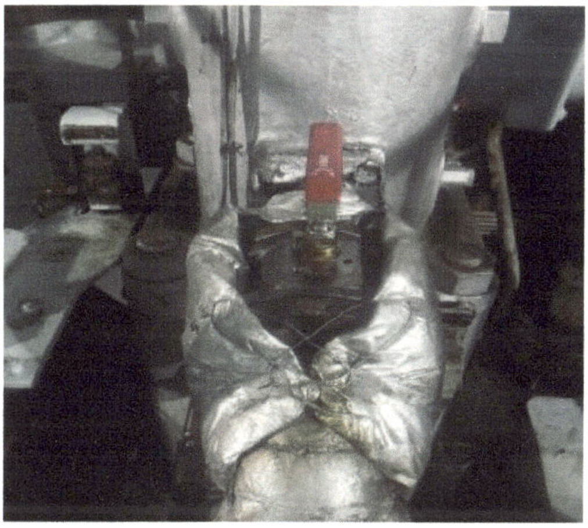

F.O.P Inlet Valve

F.O.P Drain Valve 및 Drain Position

Control Air & Safety Air Drain

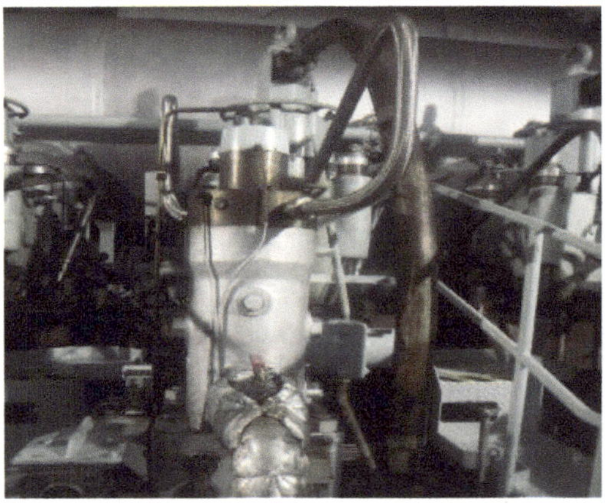

F.O.P High Pressure Pipe 및 액세서리 분해

Top Cover 분해 및 예비품 준비
Top Cover Lifting Tool : 90904-0052-9823

F.O.P Plunger & Barrel Lifting Tool
(90951-0422-9821)

F.O.P Lifting Tool 설치

VIT & Regulation Rack 연결부 및 Nut 분해

VIT Rack Tool : 90951-0422-9824

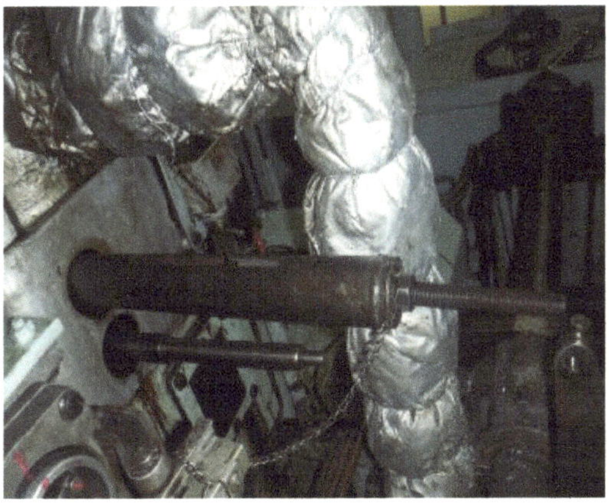

상기와 같이 설치하여 Rack 취외

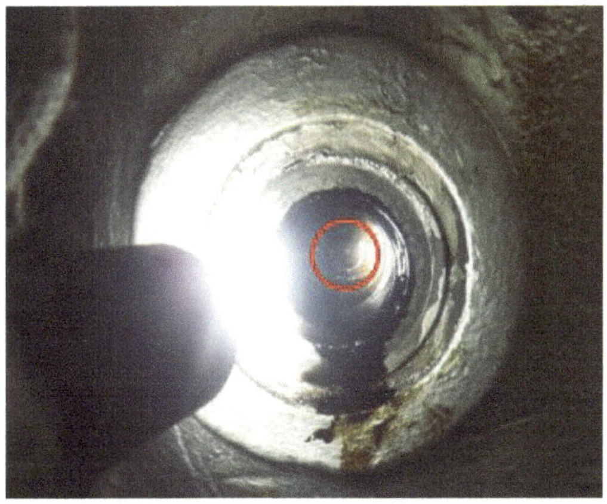

Rack 전부 뺀 상태에서 Guide Screw 취외하여
Barrel Position 확인 후 Plunger & Barrel 취외

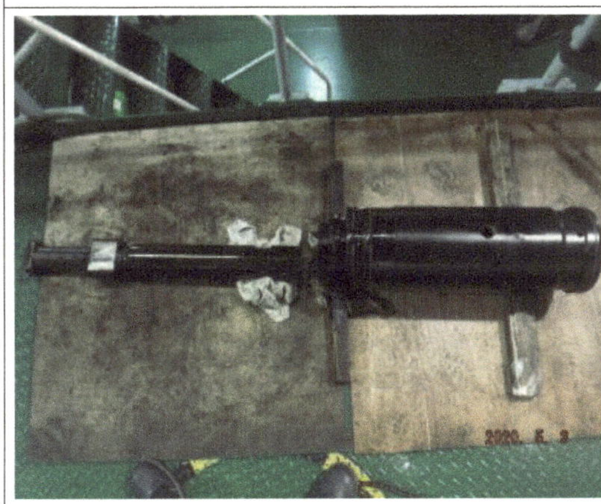

Plunger & Barrel 취외 후 분해하여 계측
(신품 기준 Plunger 외경, Barrel 내경: 84mm)

Rack 전부 취외하면 상기와 같이 일직선으로 됨.
Plunger & Barrel 'T'모양이 되도록 조립

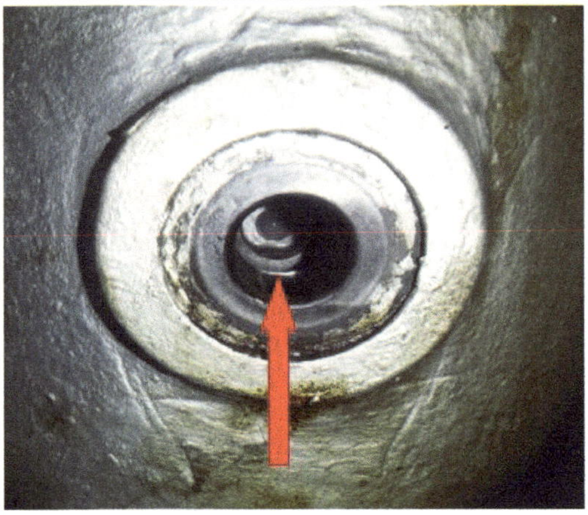

Plunger & Barrel 조립 시 정확히 Center 맞춘 다음 Guide Screw를 손으로 조립 시 원활하게 조립되어야 제대로 장착된 것

Top Cover 조립 및 Nut 500Nm

앞서 설명한 바와 같이 Plunger & Barrel 조립 할 때 VIT & Regulation Rack 전부 뺀 상태 (=Rack Position '0')에서 조립한다.
Plunger를 넣고 Regulation Rack를 움직여서 제대로 조립되는 것을 확인한다.
Zinc로 Barrel 조금씩 치면서 조립하되 더 이상 들어가지 않을 경우 VIT Rack를 Zinc로 넣으면 Barrel이 아래로 들어간다.
상기 그림에서 Center가 정확히 맞을 경우에는 Guide Screw가 손으로 부드럽게 조립된다. 그렇지 않으면 다시 분해, 조립한다.

(4) 연료 펌프 BARREL SEALING RING 조립

Mounting Tools - Pump Barrel Seals

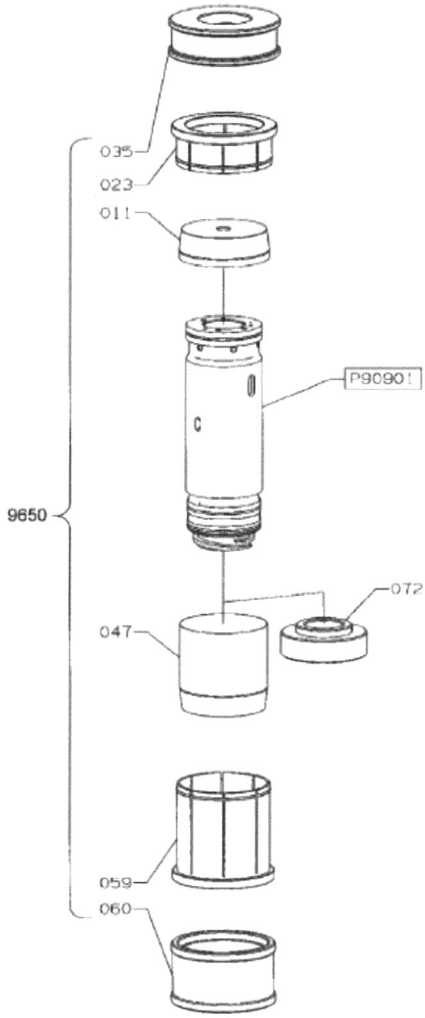

011, 047 : Cone

023, 059 : Pushing tool

035, 060 : Compression tool

072 : Spacer ring

9650 : Mounting tools - pump barrel seals

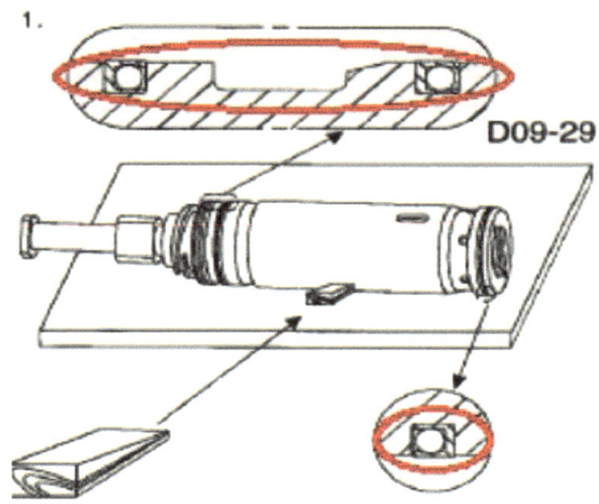

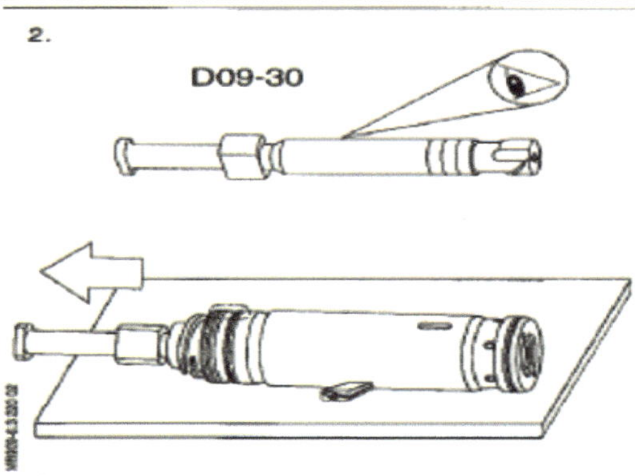

Plunger 분해하고 마모도 계측하고 Plunger 상부 Cavitation Mark 확인 및 정상일 경우 부드러운 천과 Air 이용하여 깨끗이 소제한다.

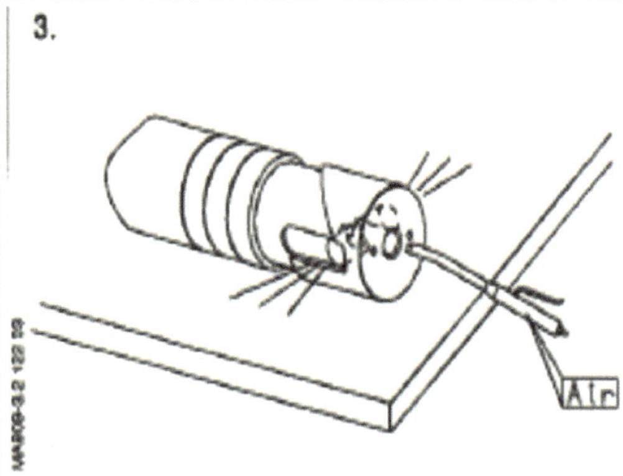

Plunger와 마찬가지로 Barrel 깨끗이 소제하고 상부의 Seizure Mark 확인한다.

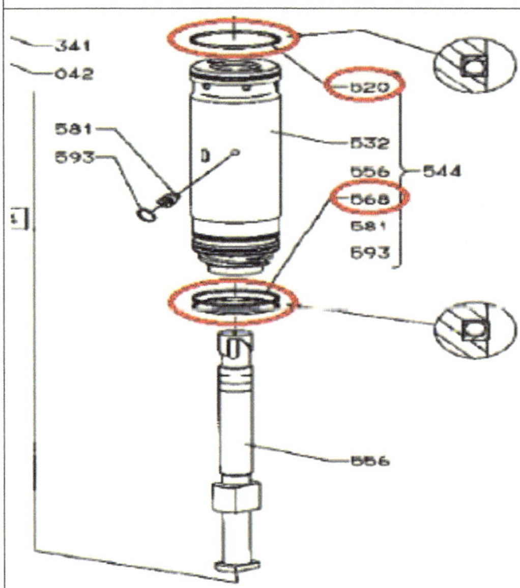

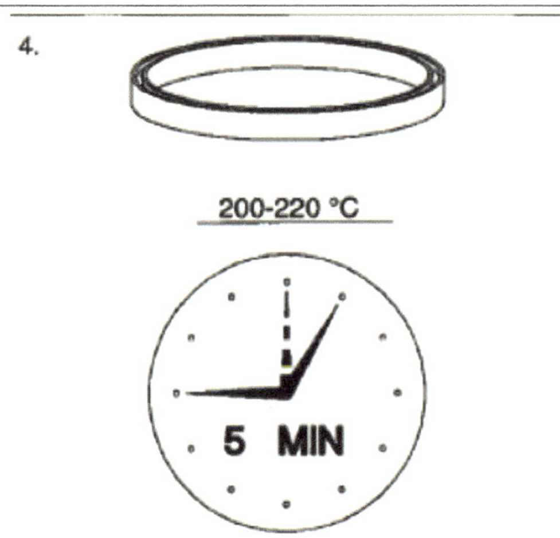

Sealing Ring을 200~220℃에서 적어도 5분 Heating 한다.

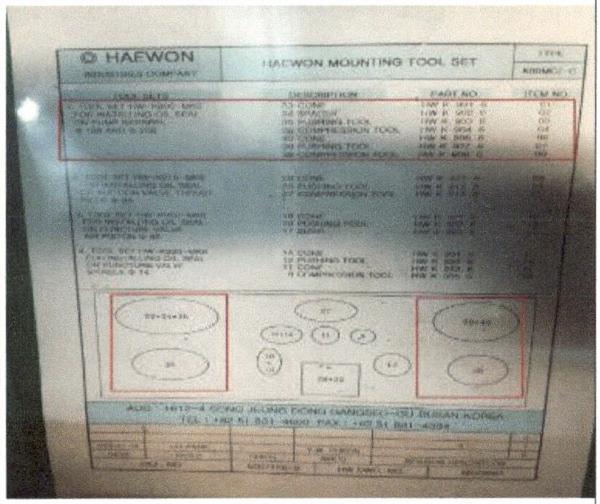

5.

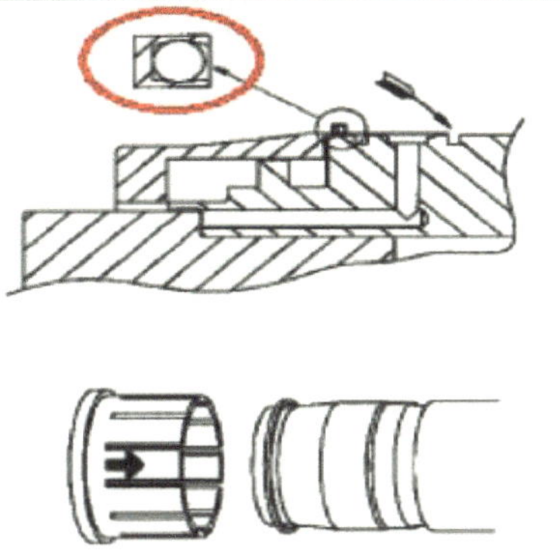

Barrel의 아래쪽은 상부부터 Sealing 조립한다. Barrel안 Guide Stick 설치 후 Cone 설치한다. 그림과 같이 스프링이 위쪽 향하도록 Sealing을 설치하고 Pushing Tool로 조립한다.

6.

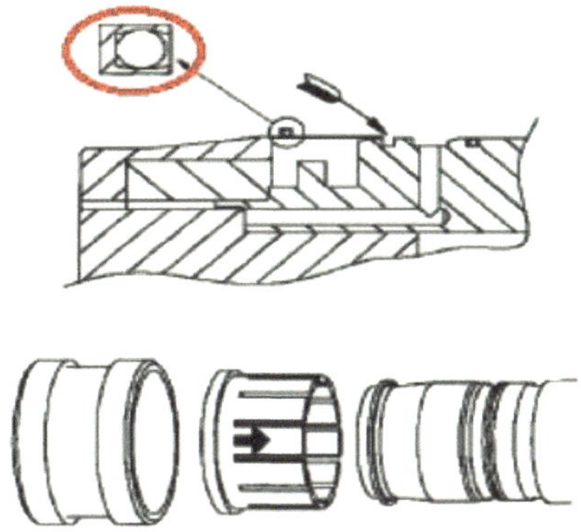

아래의 바깥쪽 홈에 Sealing Ring을 끼울 때는 홈과의 거리를 정확히 맞추기 위하여 Cone 안에 Spacer 설치한 다음, 위 순서를 반복한다. 그 후 Seizer 이용하여 압축한다.

7.

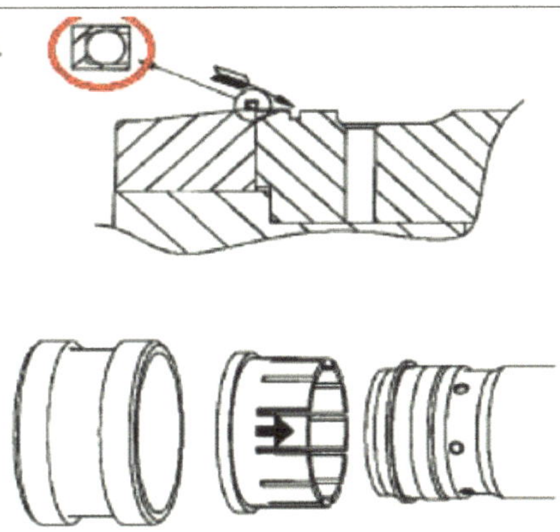

Barrel의 상부에 Guide Stick과 Cone 설치하여 위와 같은 방법으로 Sealing Ring을 설치하되 그림과 같이 스프링이 아래로 향하도록 한다.

8.

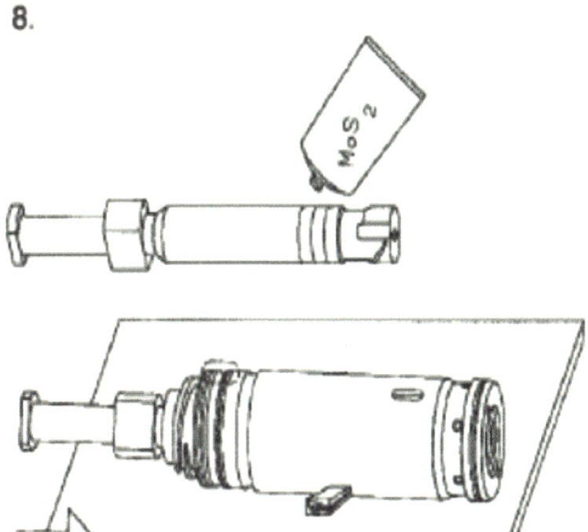

(5) 연료 펌프 COMPLETE 교환

1.

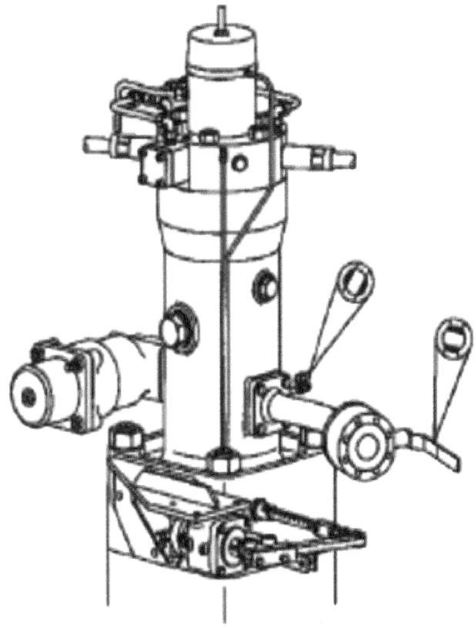

F.W.E 후 Control & Safety Air Close,
Main L.O Pump Stop,
F.O Valve Close 및 F.O Drain 한다.

4.

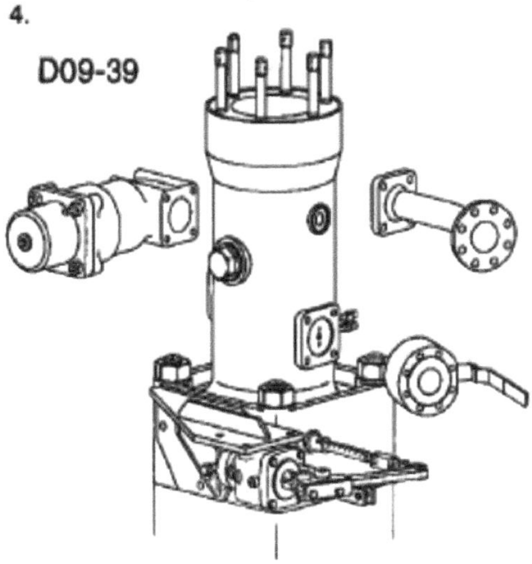

Top Cover, Plunger & Barrel 분해한다.
F.O In, & Out Pipe Shock Absorber 분해한다.

6.

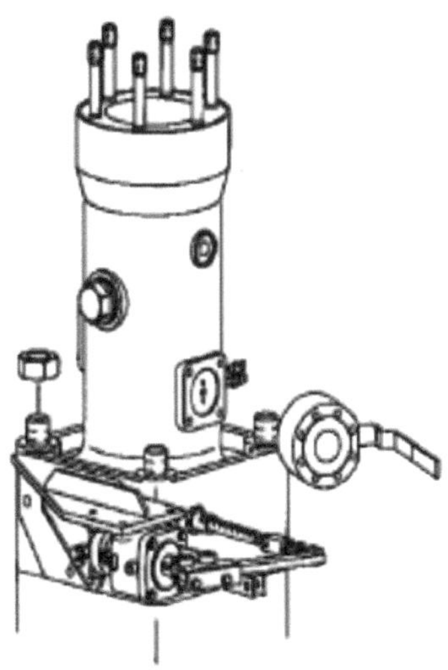

Drain Pipe & Actuator(For VIT index arm)의 Control Air Pipe들 분해한다.
F.O.P Housing의 Nut 4개소 분해한다.

7.

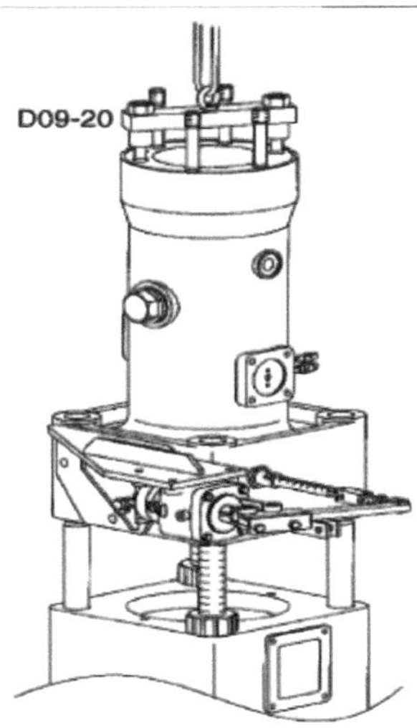

Top Cover Studs & Nuts에 Lifting Tool 설치하여 Thread 손상되지 않게 F.O.P Housing 분해하고 나무판자를 준비하여 그 위에 올린다.

2.

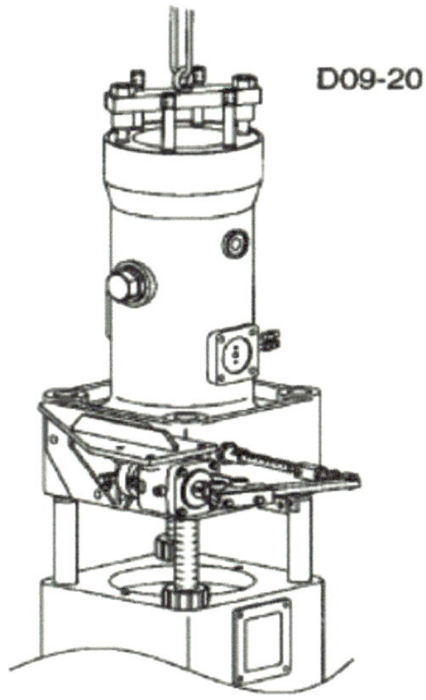

Housing 터치면 및 Thread Brush 소제 후
Housing 분해할 때와 같이 조립한다.

3.

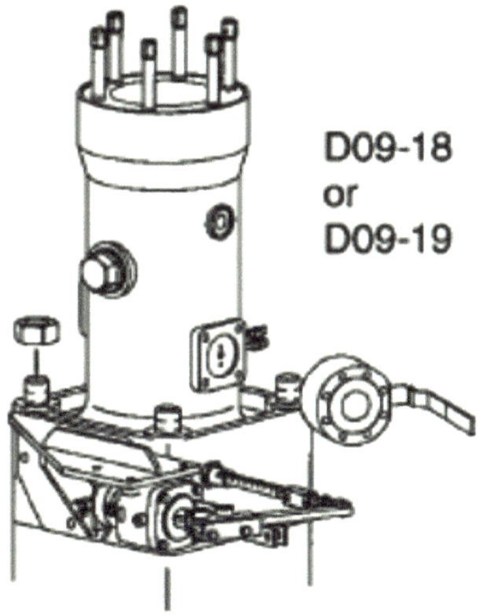

Housing Nut 4개소 조립한다.
(Tight Torque : 2100Nm)

5.

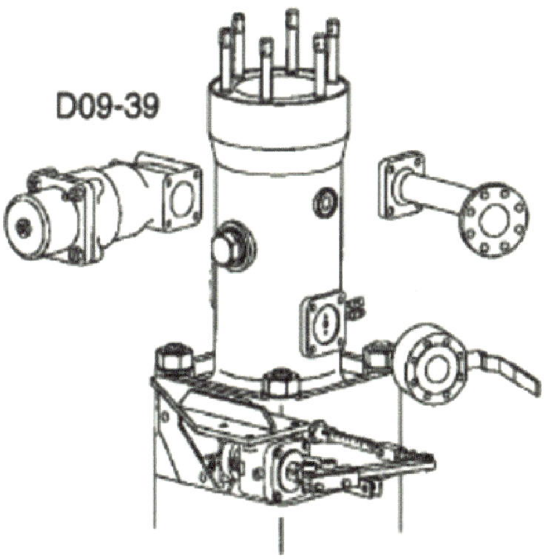

분해한 Pipe들(F.O Pipes, Control Air Pipes)
& Shock Absorber 조립한다.
연결 부에 장착되는 Gasket 누락되지 않도록
주의한다.

8.

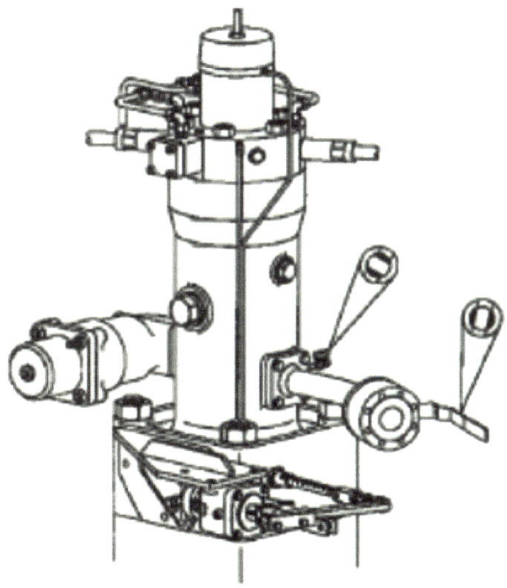

앞서 설명한 Plunger & Barrel과 Top Cover
그리고 주변 액세서리 및 Pipe 조립한다.
F.O Line 복구하고 Drain Valve Open하여
Line Venting 후 Drain Valve Close 한다.
Control Air & Safety Air Line복구하고
Main L.O Pump Start한다.
주변 누설 부 확인한다.

(6) 연료 펌프 분해 및 정비

F.O.P OVERHAUL의 경우 아래 Instruction book 자료를 참고하면 쉽게 이해할 수 있다.

1. If the fuel index arm or the VIT index arm of the fuel pump is moving too tightly, the regulating and timing systems of the fuel pump must be overhauled.

2. In order to carry out this overhaul, the top cover and plunger/barrel assembly of the fuel pump must have been removed, and the fuel pump must have been dismantled from the pump base and landed on wooden planks.
   See Procedure 909-4.2.
   See Procedure 909-5.2.
   See Procedure 909-6.2.

3. Dismantle the actuator and actuator bracket.

4. Tilt the fuel pump housing into a horizontal position on the wooden planks.

5. Dismantle the screws and the bottom plate from the fuel pump housing.

   If the bottom plate is stuck, loosen it by screwing two M10x80 mm screws into the holes of the bottom plate.

3.

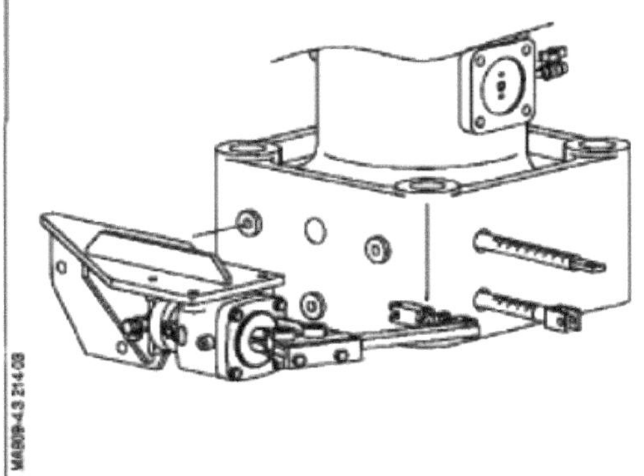

4.

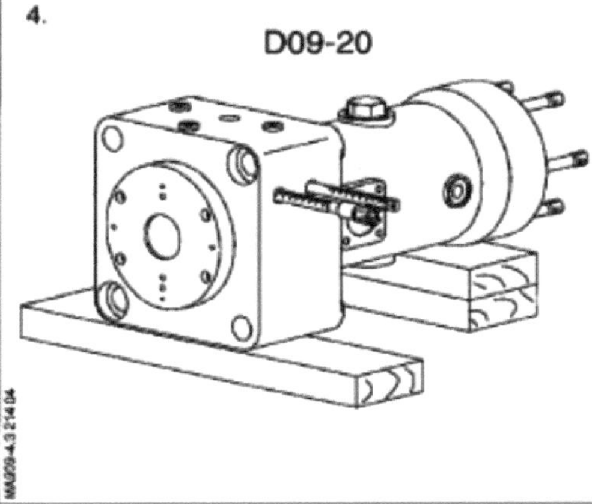

5.

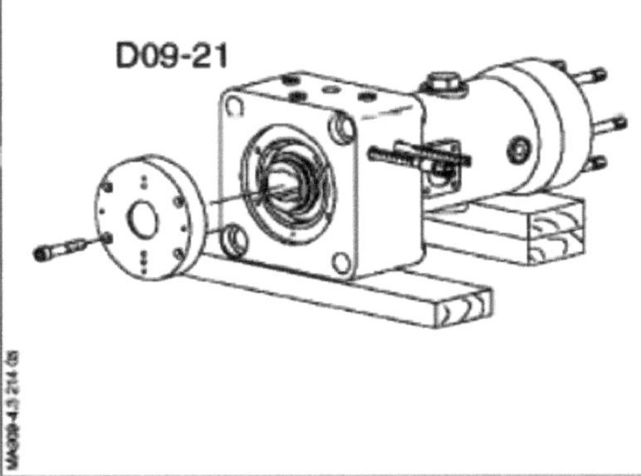

6.

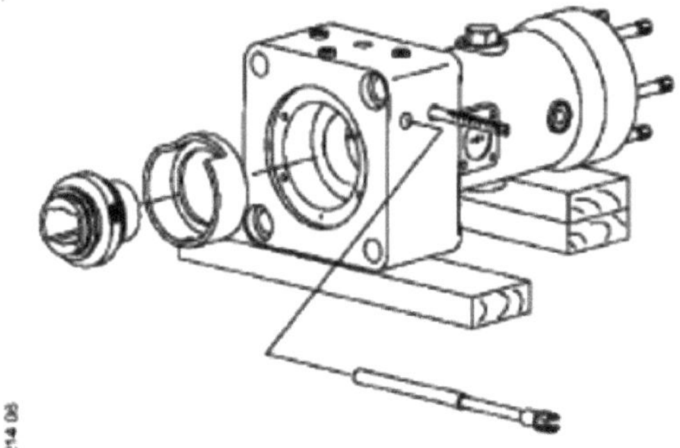

7.

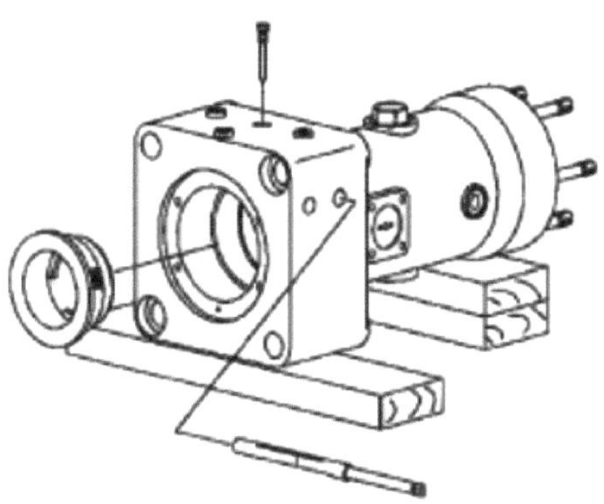

8.

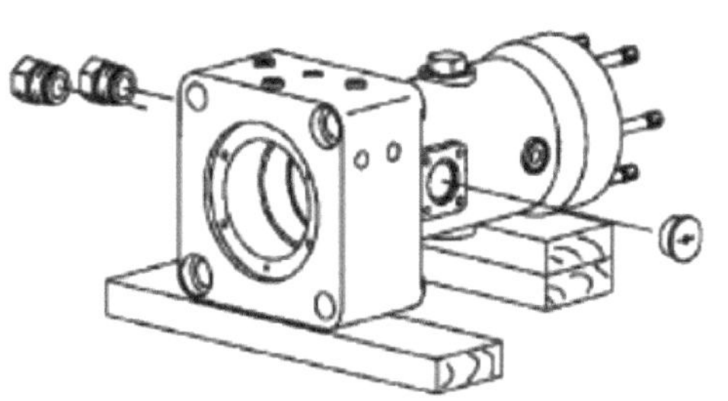

6. Dismantle the:
   - Fuel index arm
   - Regulating guide
   - Sleeve
7. Dismantle the:
   - Guide screw
   - VIT index arm
   - Timing guide
8. Remove the caps from the back of the fuel pump housing.

   Remove the orifice plug from the front side of the fuel pump housing.

   Discard all sealing rings and gaskets.

9. Dismantle and inspect the plug screws in both sides of the fuel pump housing.

   If eroded, replace the plug screws.

9.

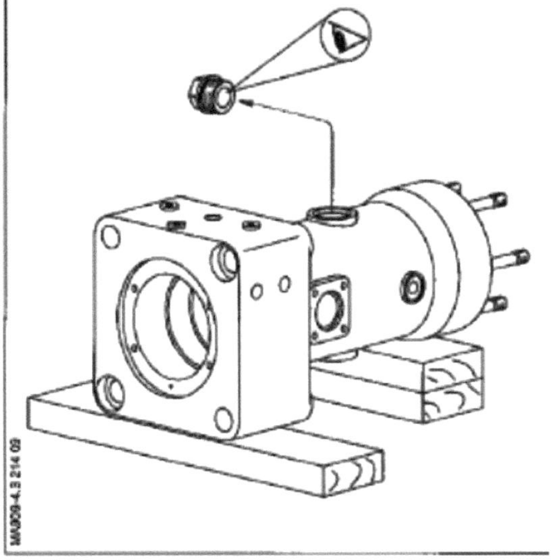

10. Clean all the dismantled parts of the fuel pump thoroughly in diesel oil or kerosene. Dry all parts carefully.

    Clean all inside surfaces of the fuel pump housing.

    Check that all the bores of the fuel pump housing are clean.

11. Mount new sealing rings and gaskets.

    Mount the caps and the orifice plug.

    Mount new sealing rings on the plug screws and mount the plug screws.

Note!
Lubricate the threads of all the parts with Molybdenum Disulphide (MoS_2) before mounting.

11.

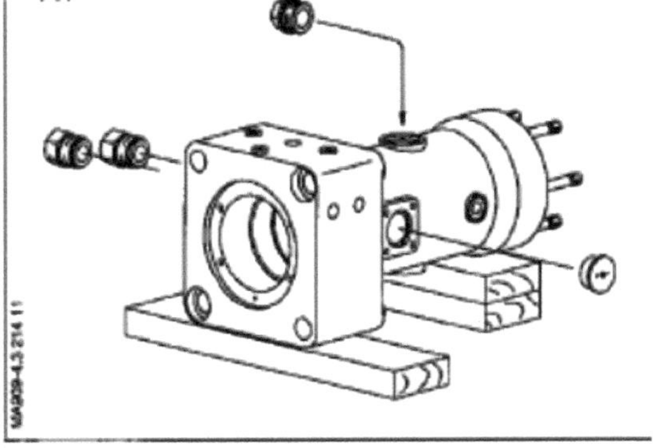

12. Mount new sealing rings in the sleeves that are provided with a pointer scratch.

Note!
Soak the sealing rings in lubricating oil before mounting.

12.

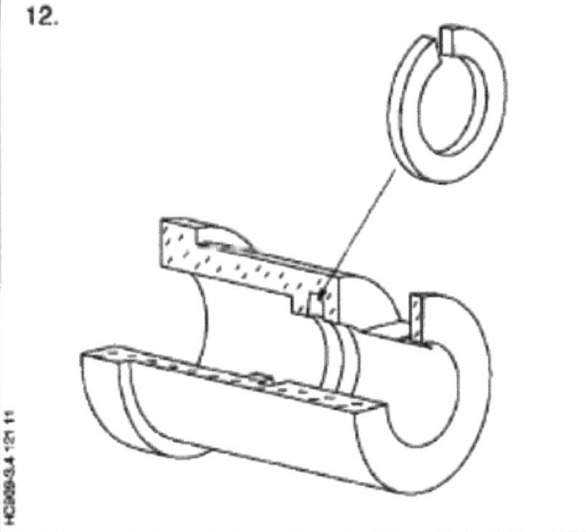

14.

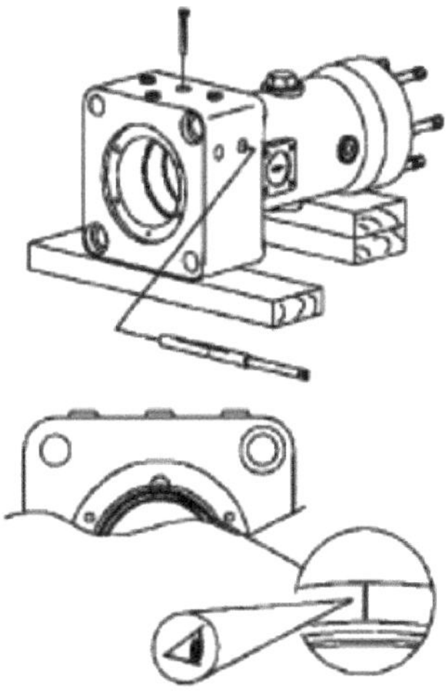

15.

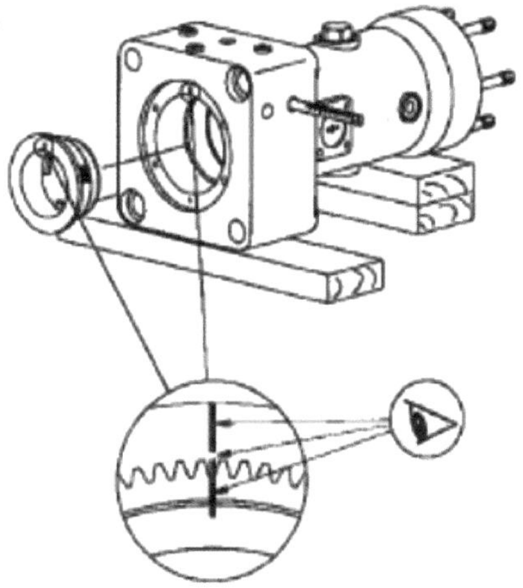

16.

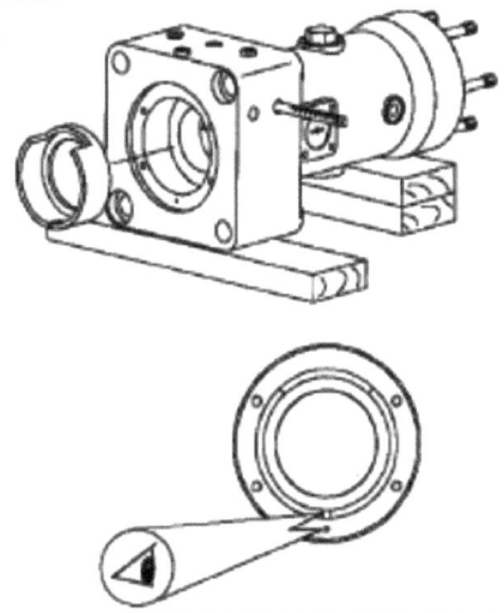

Mounting

13. Lubricate the following with molybdenum disulphide (MoS_2):

    - all internal surfaces of the fuel pump housing
    - the sliding surfaces of all parts of the regulating and timing systems.
    - the teeth of the:

      regulating guide

      timing guide

      VIT index arm

      fuel index arm.

14. Mount the VIT index arm and the guide screw.

    Adjust the position of the VIT index arm so that the scratch mark is aligned with the centre of the fuel pump housing.

15. Mount the timing guide in the fuel pump housing.

    The timing guide must be mounted in such a way that the scratch marks on the timing guide, on the VIT index arm and on the fuel pump housing are aligned.

16. Mount the sleeve in the fuel pump housing, ensuring that the notch in the side of the sleeve is aligned with the pin hole in the fuel pump housing.

17. Mount the fuel index arm.

Adjust the position of the fuel index arm so that the scratch mark on the fuel index arm is aligned with the centre of the fuel pump housing.

18. Mount the regulating guide in the fuel pump housing.

The regulating guide must be mounted so that the scratch mark on the regulating guide is aligned with the scratch mark on the fuel index arm.

19. Mount the bottom plate on the fuel pump housing.

20. Return the fuel pump to a vertical position. Mount the actuator and actuator bracket on the side of the fuel pump.

17.

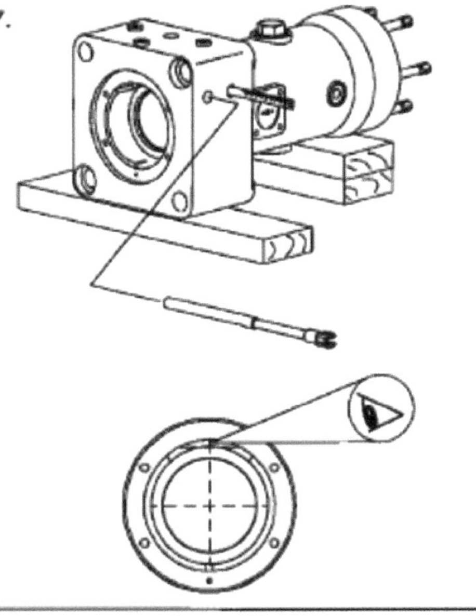

18.

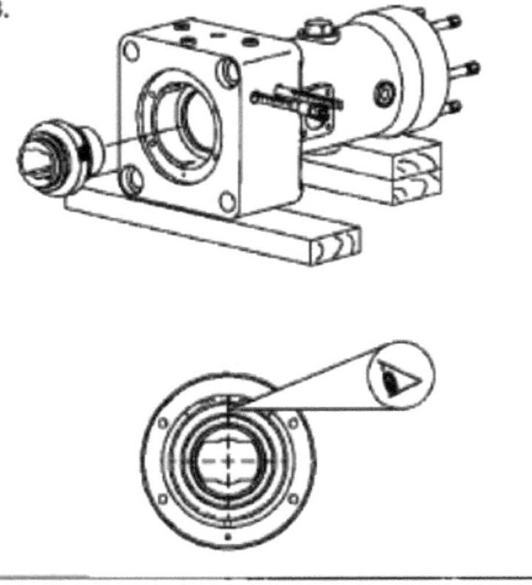

19.

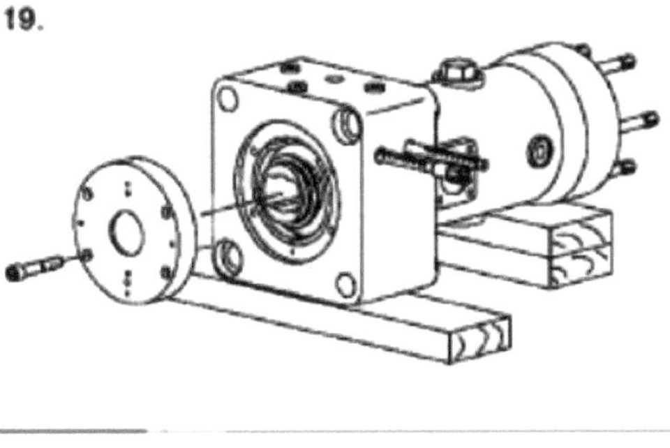

(7) 연료 펌프 롤러 가이드 분해 및 정비

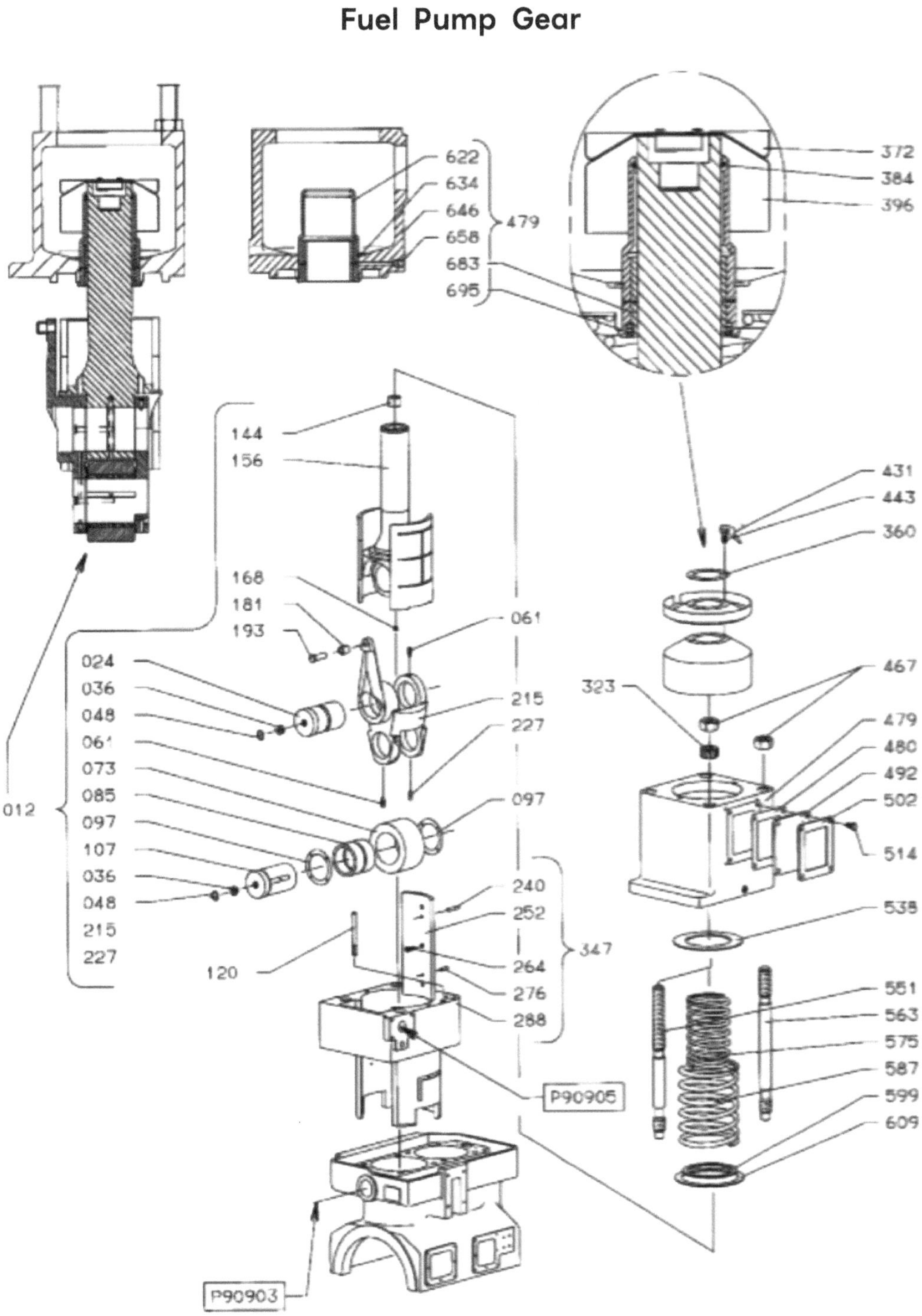

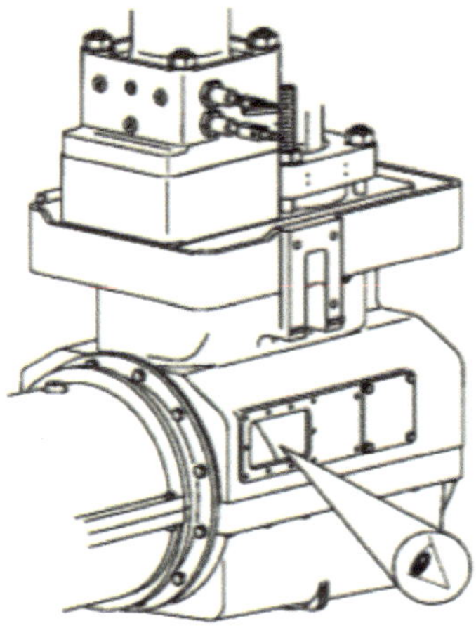

F.O.P 취외한 다음
Camshaft Cover Open 한다.

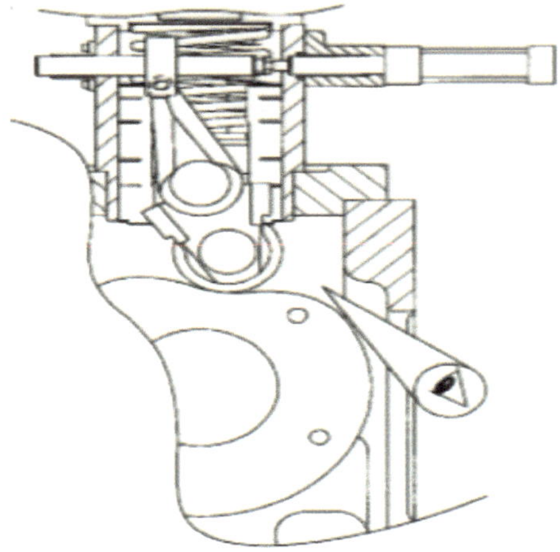

Turning Gear로 Engine Turning하여
Roller Guide를 아래로 내려놓는다.

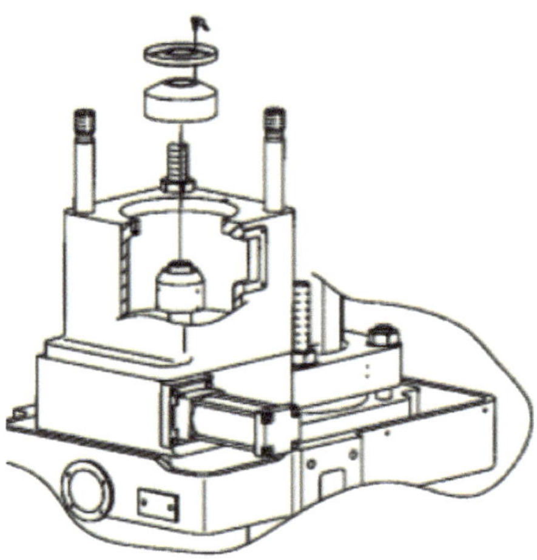

F.O Pump 분해한 다음 Pump Base 내부
Roller Guide 상부에 장착된 Locking Wire,
Screw, Disc & Sealing Cap을 분해한다.
Pump Base에 연결된
L.O Pipe & Drain Pipe 분해한다.

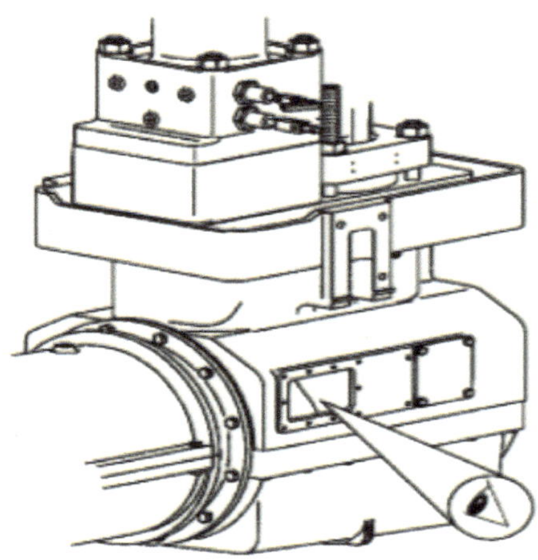

Camshaft Cover를 확인하여 Roller Guide의
Position 확인한다.
Engine Turning하여 Roller Guide를
BDC로 내려놓는다.

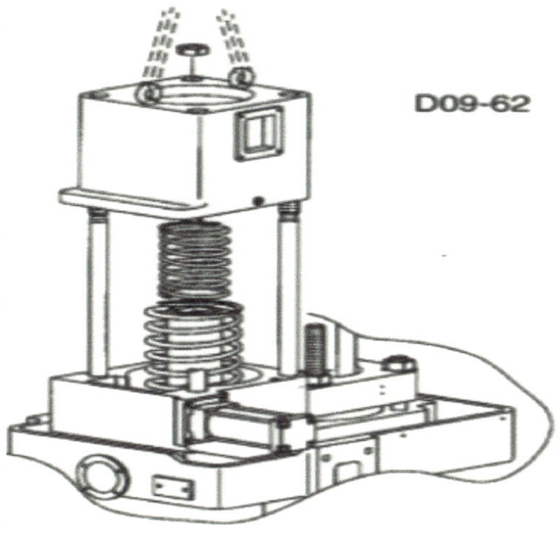

Pump Base Nut를 분해한다.
ump Base에 Eye Bolt를 연결하여 취외하고
남은 Spring 취외한다.

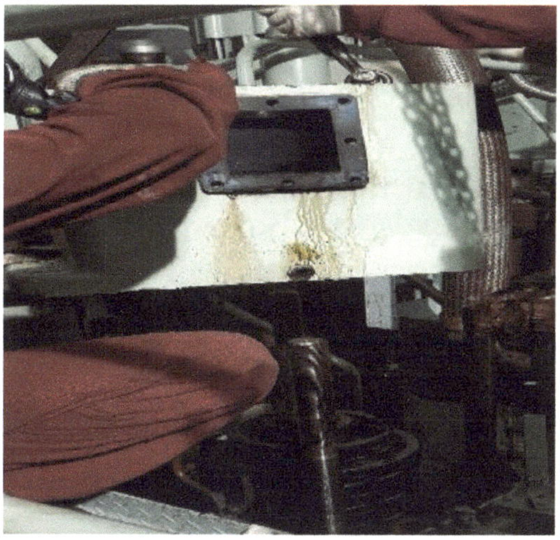

Pump Base 분해하면 아래에 놓인 Spring 보인다.

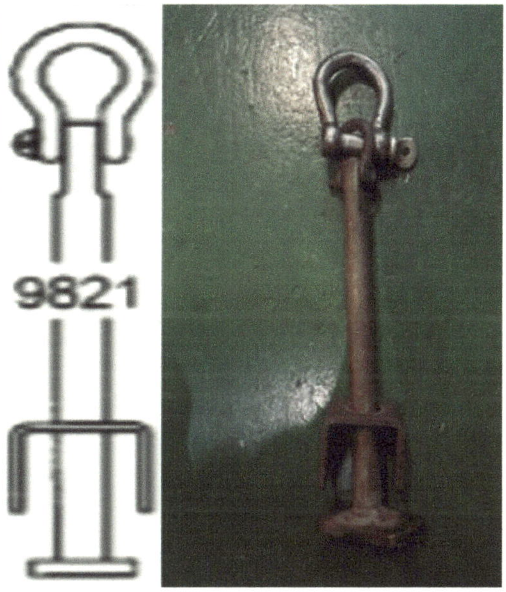

Roller Guide Lifting Tool (90851-0196-9821)

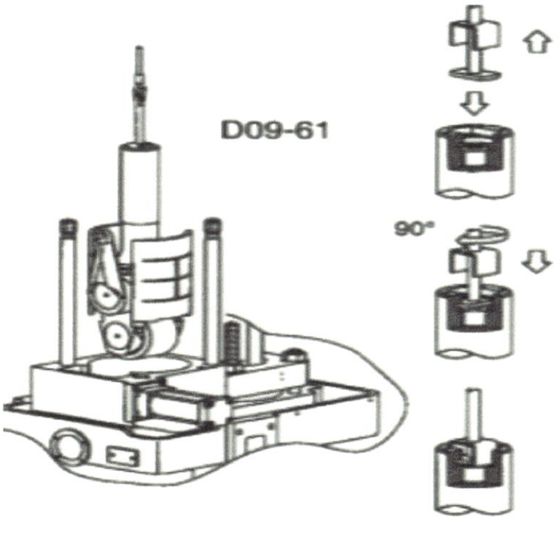

Roller Guide Lifting Tool 사용 방법

 Roller Guide 분해	 Roller Guide Roller, Damage, Seizure Mark등 전반적인 상태 점검한다.
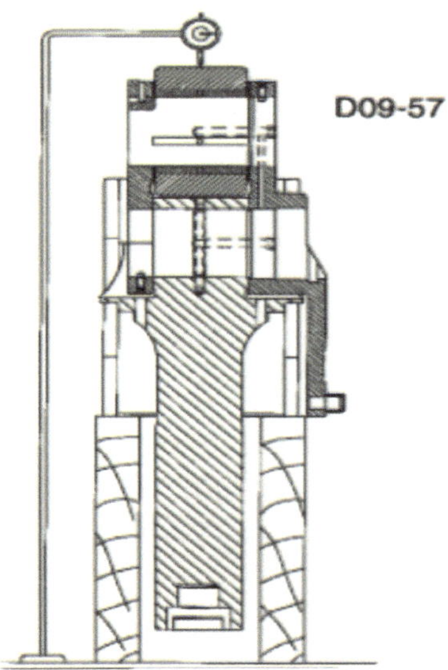 Roller/Bushing/Shaft Pin, max clearance → 0.5 mm	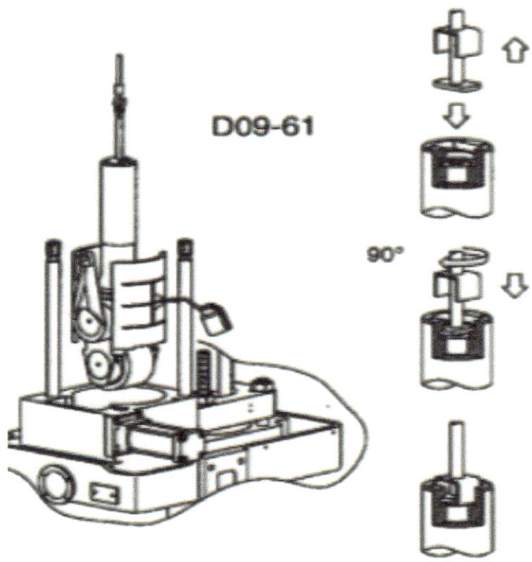 Roller Guide 조립 시 L.O를 충분히 도포하고 Roller Guide Pin이 Guide Way Pin구멍에 정확하게 맞게 조립한다. 그렇지 않으면 역전 시 시동이 되지 않는다.

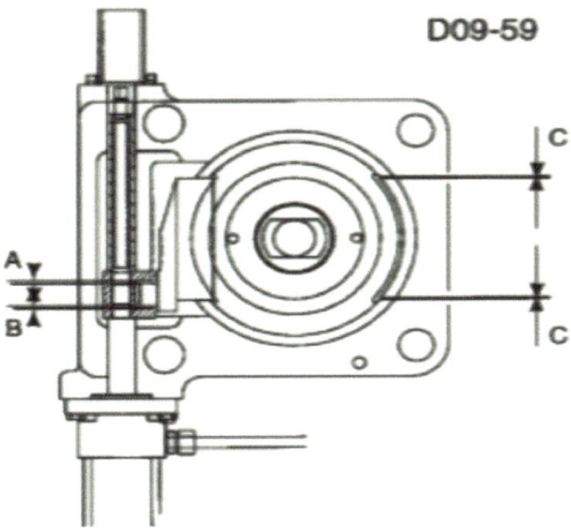

Max. deviation of clearances
→ ± 0.1 mm

After mounting the roller guide with the reversing link, check the clearance between the roller guide and the guide plate mounted in the roller guide bushing.

Turn the camshaft so that the roller guide is lifted approx. 20 mm.

The clearance **C** between the roller guide and the guide plate must be the same at both ends. *See Data.*

The clearances **A** and **B** between the guideway and the pin of the reversing link (arm) must be checked in the AHEAD and ASTERN positions.

Connect working air to the air cylinder, and check that the reversing mechanism is working smoothly.

The clearances **A** and **B** must be approx. the same in both positions.

Note that clearances **A** and **B** change position when changing from AHEAD to ASTERN.

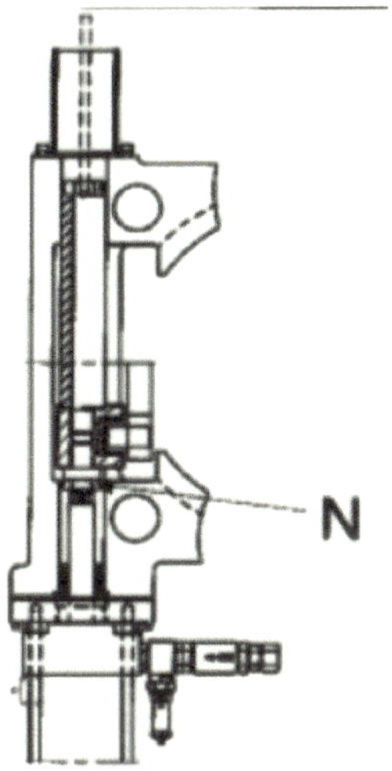

조정이 필요하다면 반드시 ASTERN Position에 놓고 Reversing Shaft의 Nut(N)을 풀고 조정한다. 조정 후에는 AHEAD Position으로 변경 후 상기 Clearance 'A' and 'B'를 다시 확인한다. 마지막으로 ASTERN으로 다시 옮긴 다음 Reversing Shaft의 Nut(N) Tight 한다.

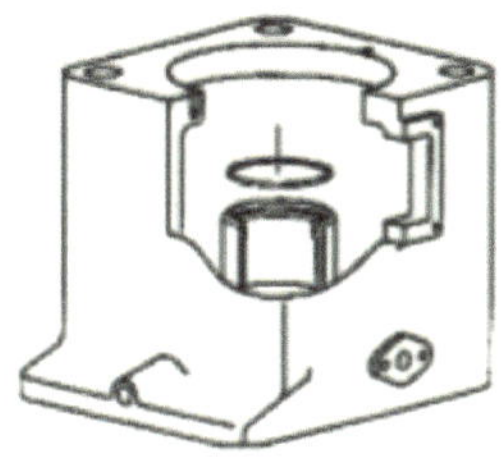

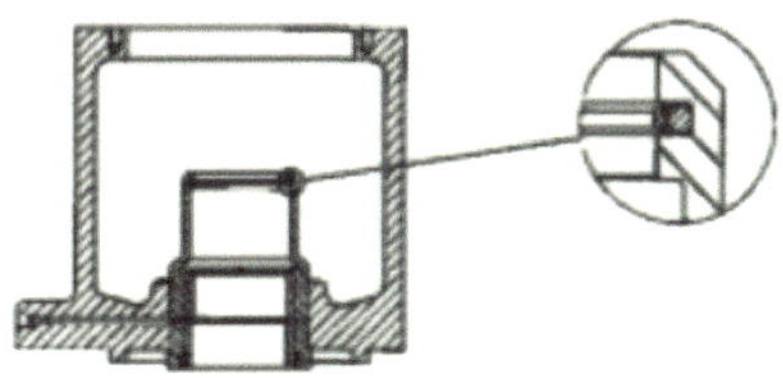

Turning Gear로 Engine Turning하여 Roller Guide를 아래로 낮춘다. Pump Base Bushing 아래에 조립되는 Sealing Ring을 방향에 맞게 조립한다.

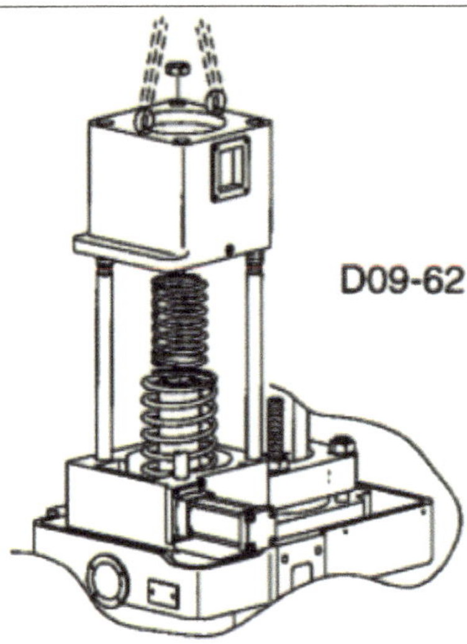

Spring 올리고 Pump Base 조립 시 Cam Position B.D.C 또는 Cam의 볼록한 부분이 아래로 되어있는 상태에서 수월하게 조립된다. (그렇지 않으면 조립되는 Spring 힘을 이기면서 Nut를 조립해야 한다.) Pump Base가 완전히 밀착될 때까지 Nut를 Tight 한다.

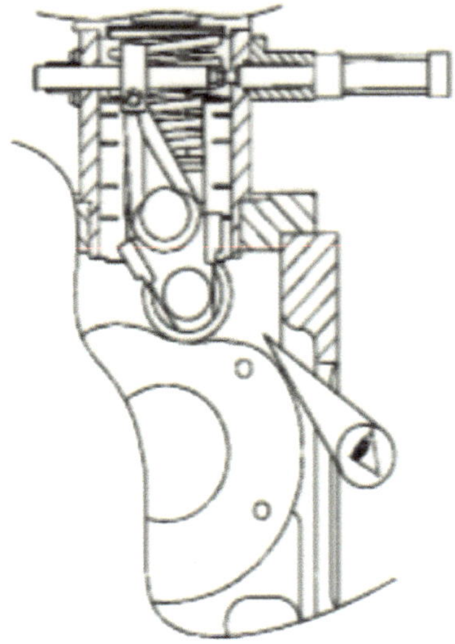

Camshaft Cover를 통해서 Reversing Link의 움직임을 확인한다.

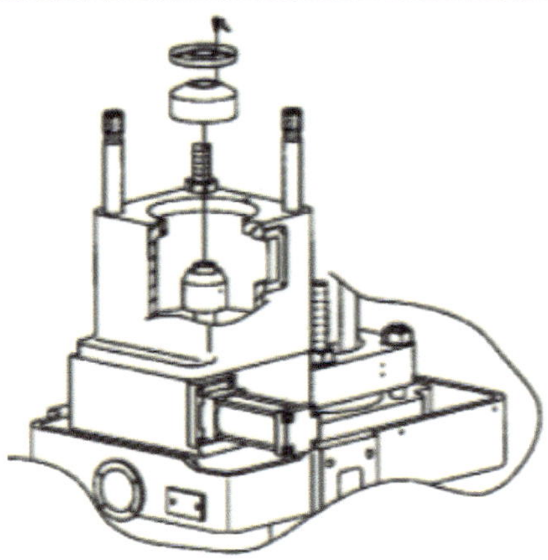

Pump base에 L.O Pipe & Drain Pipe 조립하고 Sealing Cap, Disc, Screw 조립한다. Screw Tight 후 Locking Wire로 Locking 한다.

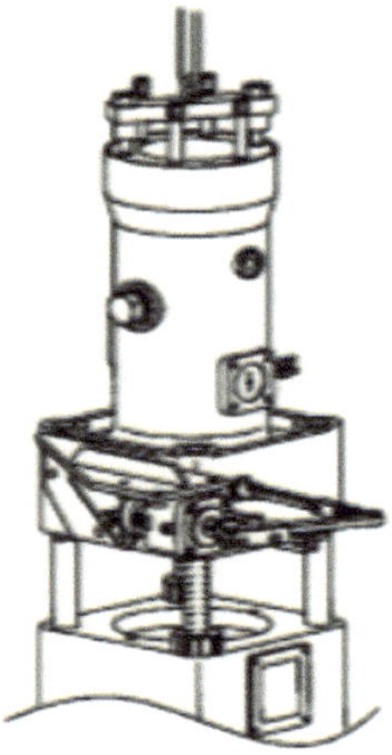

Pump Housing, Plunger & Barrel 조립하고 Fuel Cam Top Lift 값 계측한다.

*** 연료 펌프 롤러 가이드 작업 사진**

Taken out the top cover

Took out plunger with barrel

Took out F.O pump housing

Took out roller guide complete

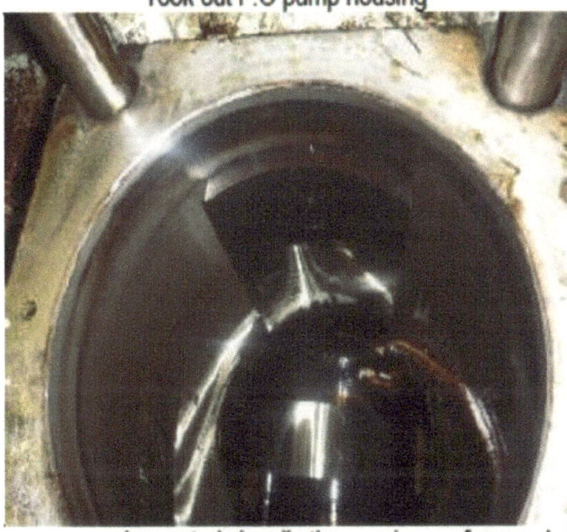

Inspected visually the running surface and checked loosening of fixed bolts for guide plate

Replaced roller guide complete with new one

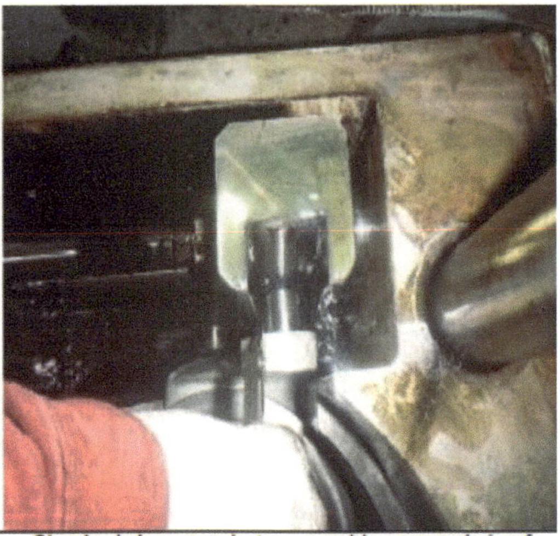

Checked clearance between guide way and pin of reversing link

Re-assembled lower housing

Replaced F.O pump housing with ship's spare

Installed new plunger with barrel

Measured top lift

Malfunction of control valve for cut-out device

Activated cut-out device during normal condition

Dismantled and cleaned F.O pump housing which took out from unit #6

Before and after cleaning of F.O pump housing

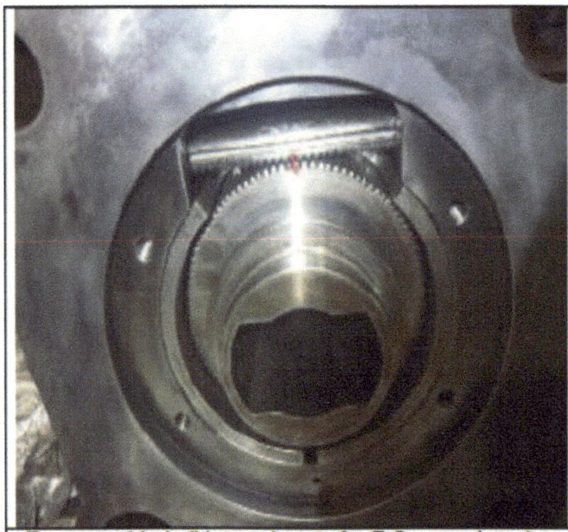

Re-assembled all internal parts for F.O pump housing

Maintained F.O pump housing for spare

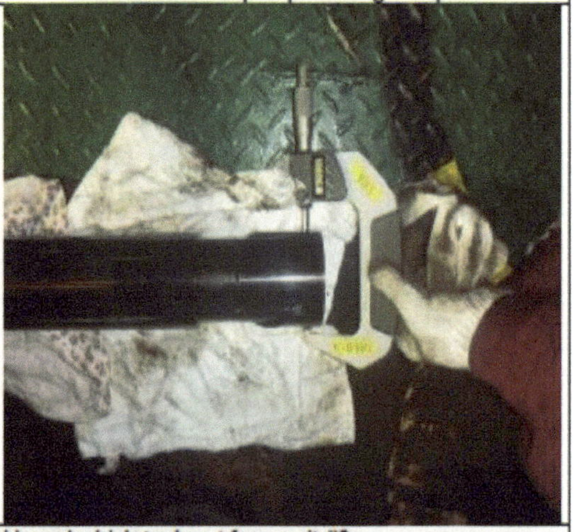

Measured wear down for plunger and barrel which took out from unit #6

Replaced plunger and barrel with new one for spare F.O pump as requested

* 외주 업체 연료 펌프 롤러 가이드 정비 사진

Landed roller complete to TES's workshop

Came out pin from reversing link

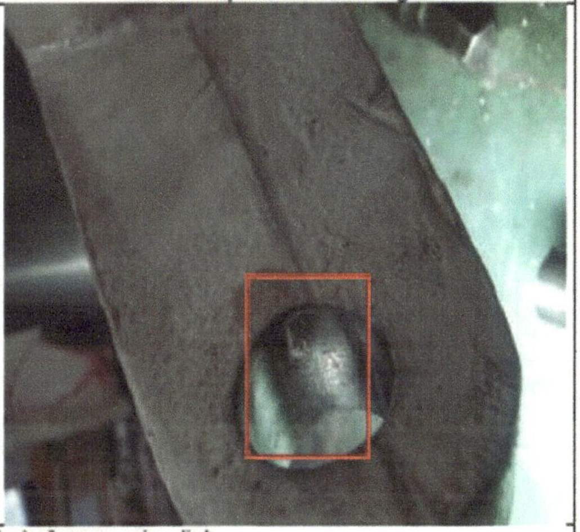

Deep scratch mark of hole for reversing link

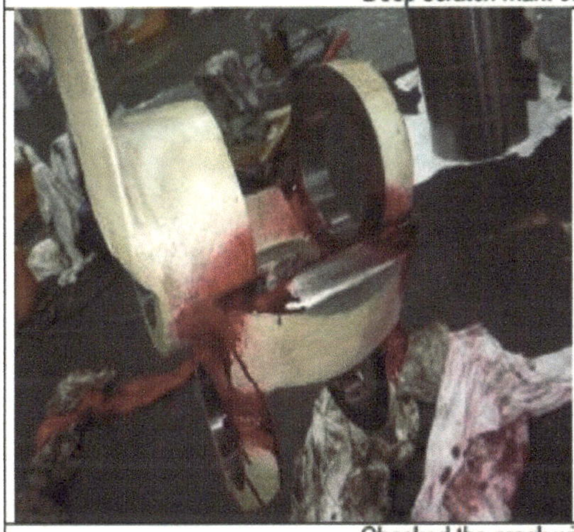

Checked the crack with PT for reversing link

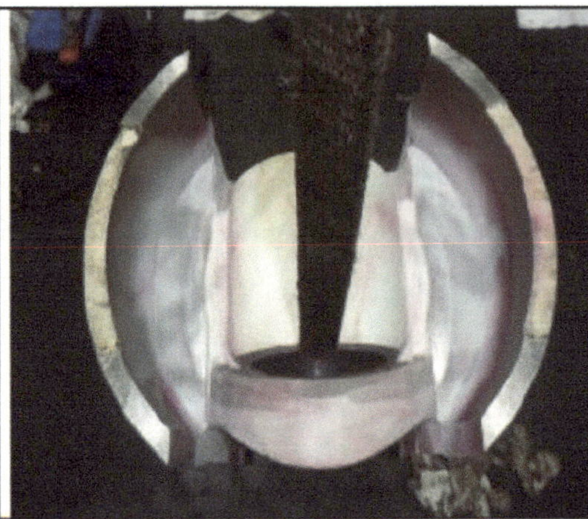

Checked the crack with PT for roller guide

Cleaned and inspected roller, bush, pins, washers

Maintained roller guide complete with new reversing link

* 연료 캠 TOP LIFT값 계측

Cam 제일 아래 B.D.C(하사점)에 있을 때 Measuring Tool, Dial Gauge 설치하고 Zero Setting 한다.

그 다음 T.D.C(상사점)로 올렸을 때 값이 Shop Test의 Fuel Cam Top Lift 값과 비교한다.

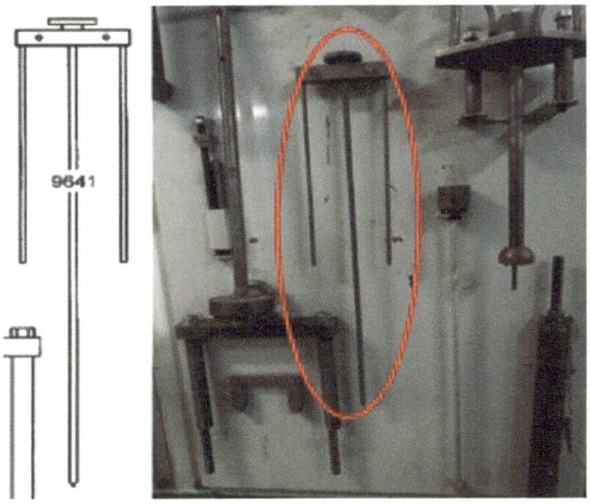

F.O.P Lead Measuring Tool(90951-0422-9641)

4. Fuel Cam (Comp. Shim thickness : 10 ㎜) (Top Lift : 22.5 ㎜)

CYLINDER No.	1	2	3	4	5	6	7	8	AVG.			
Top Lift	23.8	22.5	23.5	21.3	22.5	22.5	22.5	22.5	22.6			

4) 연료 펌프 상부 커버

정비 주기 : 8000H

Fuel Pump

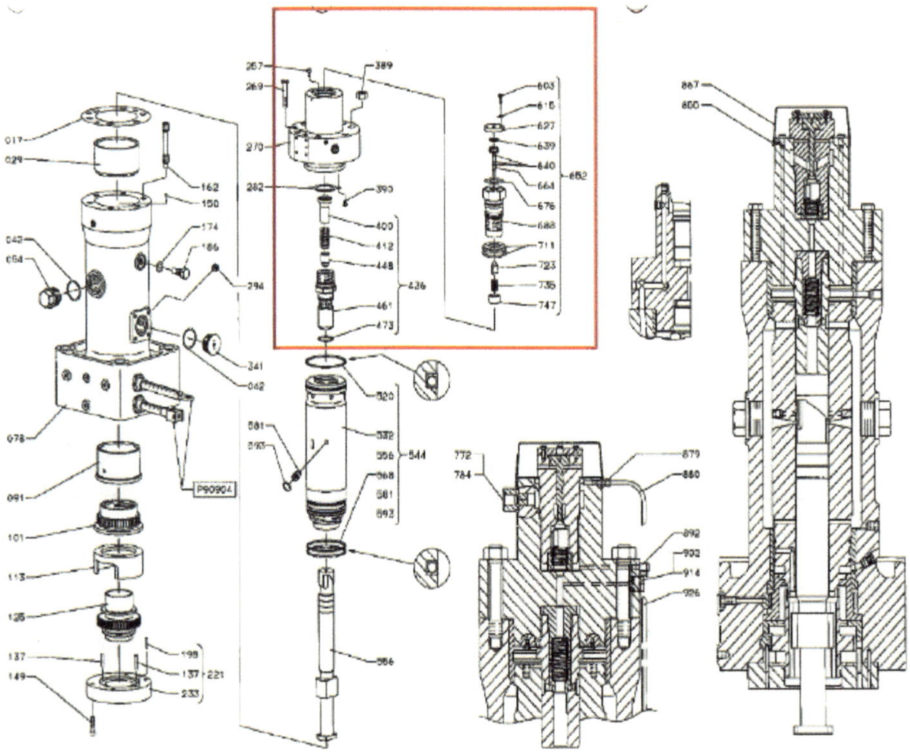

Fuel Pump Top Cover

SAFETY PRECAUTIONS *For detailed sketch, see 900-2*

X	Stopped engine
X	Shut off starting air supply – *At starting air receiver*
X	Block the main starting valve
X	Shut off starting air distributor/distributing system supply
X	Shut off safety air supply – *Not ME engines*
X	Shut off control air supply
	Shut off air supply to exhaust valve – *Only with stopped lubricating oil pumps*
X	Engage turning gear
	Shut off cooling water
X	Shut off fuel oil
X	Stop lubricating oil supply
	Lock the turbocharger rotors

Data

Ref.	Description	Value	Unit
D09-23	Top cover studs, screwing-in torque	400	Nm
D09-24	Top cover nuts, tightening torque	500	Nm
D09-26	Outlet seat, max. grinding diameter	26	mm
D09-27	Top cover	120	kg

(1) 분해 및 소제 방법

① Top Cover 교환 방법은 이전 F.O.P Plunger & Barrel 부분에서 이미 설명하였다. 취외한 Top Cover를 고정 Tool에 장착한다.

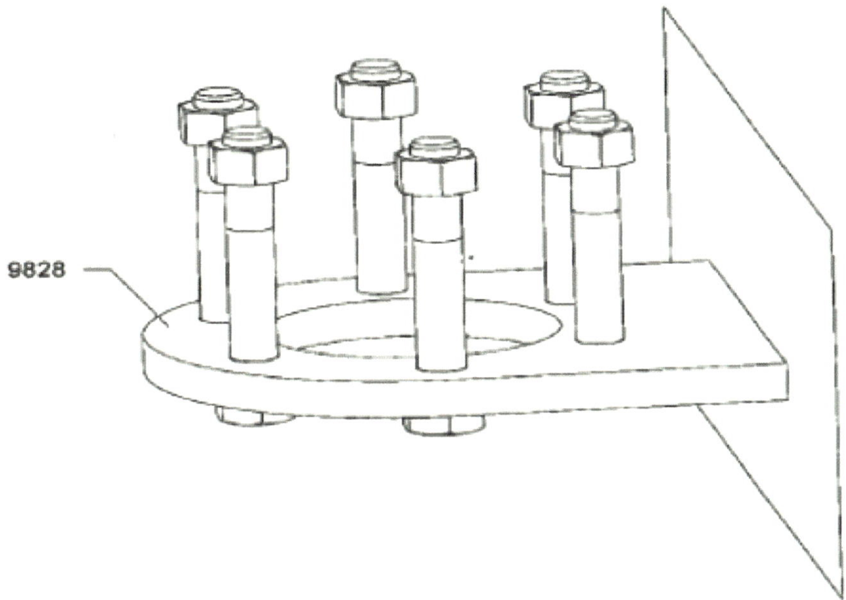

< Top Cover고정 Tool : 90967-0001-9828 >

② **Spanner와 Hammer 이용하여Top Cover의 Puncture Valve(90901-0192-652) 풀어 낸다.**

③ Suction Valve Lock Washer(**90901-0192-282**), Screw(**90901-0192-390**) 분해하고 Suction Valve(**90901-0192-436**)를 분해한다.

④ Screw(**90901-0192-25**7) 2EA 및 Guide Screw(**90901-0192-294**), 3EA 분해한다.

⑤ Plug Screw(**90901-0192-902**) 및 Disc(**90901-00192-914**)를 분해한다.

⑥ Plug Screw는 왼 나사이며 Disc 상태를 확인 후 불량 시 교환한다.

⑦ Top Cover 내부(**특히 FO High Pressure Pipe Thread 부분**)를 Diesel Oil & Gas Oil 을 이용하여 깨끗하게 소제하고 Compressor Air로 이물질을 불어낸다.

* Puncture Valve & Suction Valve 각각 **1400Nm, 1850Nm**으로 조립되므로 Size 에 맞는 Spanner를 이용하여 Hammering하여 분해한다.
조립할 때는 해당 Torque Wrench로 Tight 한다.

Top Cover 고정 Tool

Puncture Valve 분해

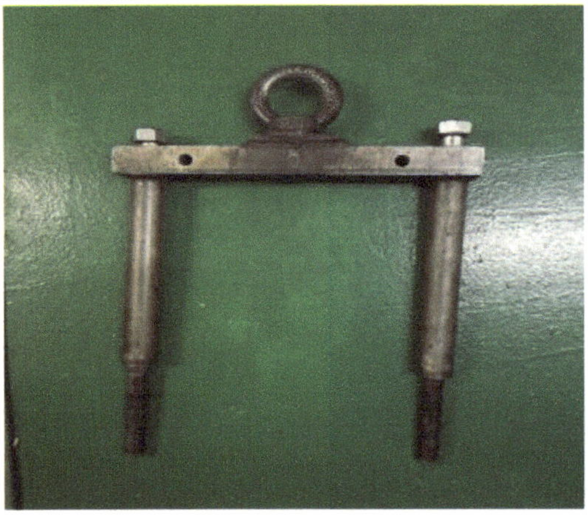
Top Cover Lifting Tool

Suction Valve Lock Washer & Screw 분해

Suction Valve Lock Washer & Screw 분해

Suction Valve 분해

(2) 조립 방법

① 소제한 Top Cover 안착시키고 정비된 Suction Valve Thread에 Molykote 도포 한다.

② Lock Washer(가급적 신품 교환)와 Screw를 정비된 Suction Valve와 함께 조립한다.

③ Suction Valve Tight Tool, Torque Wrench로 **1850NM** Tight 하고, Lock Washer

를 고정한다.

④ Top Cover Lifting Tool을 사용하여 반대로 뒤집어서 안착한다.

⑤ Puncture Valve O-ring(**90901-0192-711**) 2EA 조립, Molykote 도포하여 Tight 한다.

⑥ Puncture Valve Tight Tool, Torque Wrench로 **1400NM** Tight한다.

⑦ Screw(**90901-0192-257**) 2EA 및 Guide Screw(**90901-0192-294**), 3EA 조립한다.

⑧ Plug Screw(**90901-0192-902**) 및 Disc (**90901-00192-914**)가 불량하면 교환한다. **(Plug Screw는 왼 나사이다.)**

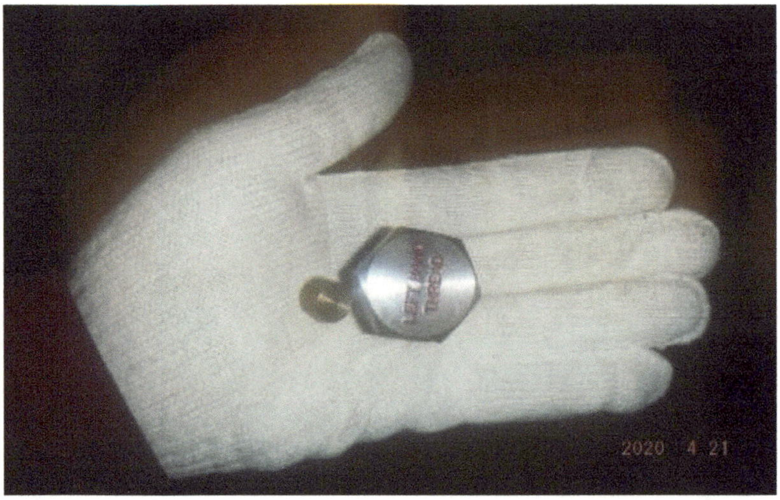

⑨ Suction Valve, Puncture Valve 및 Top Cover에 이물질이 들어가지 않도록 밀봉한다.

Tight Tool (90951-0422-9633, 9634)

Suction Valve, Lock Washer, Screw 조립

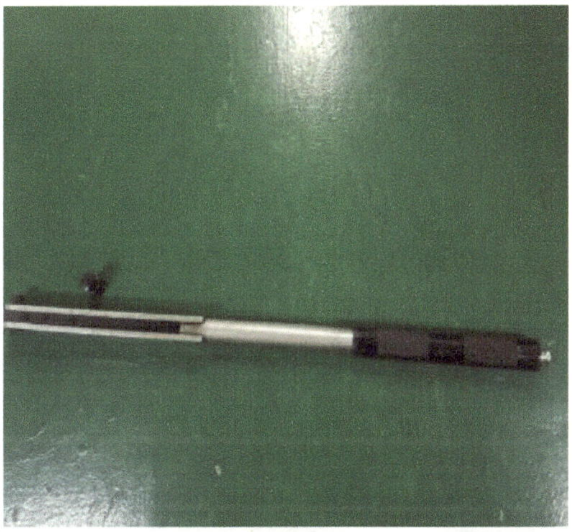

Suction Valve Torque(1850Nm)

Lock Washer Tight

Puncture Valve O-ring 조립 및 Molykote 도포

Puncture Valve Torque(1400Nm)

Screw (2EA) & Plug Screw (3EA)

5) 연료 펌프 상부 커버 압력 방출 밸브

정비 주기 : 8000H

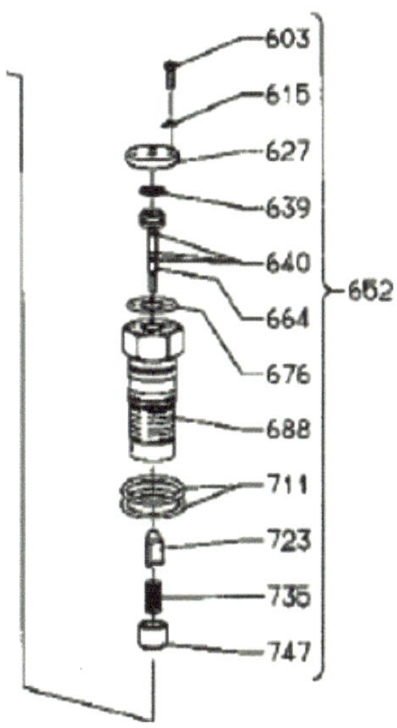

Fuel Pump Puncture Valve

SAFETY PRECAUTIONS *For detailed sketch, see 900-2*

X	Stopped engine
X	Shut off starting air supply – *At starting air receiver*
X	Block the main starting valve
X	Shut off starting air distributor/distributing system supply
X	Shut off safety air supply – *Not ME engines*
X	Shut off control air supply
	Shut off air supply to exhaust valve – *Only with stopped lubricating oil pumps*
X	Engage turning gear
	Shut off cooling water
X	Shut off fuel oil
X	Stop lubricating oil supply
	Lock the turbocharger rotors

Data

Ref.	Description	Value	Unit
D09-35	Puncture valve, tightening torque	1400	Nm
D09-36	Puncture valve	11	kg

F.O.P Top Cover 위에 Puncture Valve가 조립되어 있으며 Top Cover 상부 Drain Line으로 F.O가 누설한다면 Puncture Valve가 누설하는 것이므로 교환한다.

Puncture Valve는 공기압으로 작동되는 Air Piston(**90901-0192-664**)이 설치되어 있다.

Shut-down작동하거나 Engine 'Stop' 명령이 내려지면 Control Air가 Puncture Valve Air Piston 상부로 전달되고 Piston이 아래쪽으로 눌려지면서 F.I.V로 연료유가 공급되지 않는다.

Puncture Valve가 작동하면 F.O가 분사되지 않고 F.O CIRC′ Pump Inlet측으로 되돌아간다.

엔진 시동 시에는 Control Air가 공급되지 않고, F.O.P Cam의 동작으로 압력 형성되어 Puncture Valve의 Air Piston이 올라간다.

만약, Air Piston 고착되면 F.O가 공급되지 않아 배기 온도가 올라가지 않는다.

이럴 땐 엔진 시동 중엔 Control Air가 공급되지 않으므로 Control Air Pipe를 취외하고, Air Piston Tool을 Air Piston에 조립하고 강제로 들어 올려준다.

Puncture Valve 조립 시(**1400Nm**) O-ring(**90901-0192-711**)을 확인한다.

(1) 분해 및 소제 방법

① Screw(**90901-0192-603**), Lock Washer(**90901-0192-615**) 및 Cover(**90901-0192-627**) 분해한다.

② Air Piston(**90901-0192-664**) 분해 Tool 이용하여 분해한다.
(분해 Tool 없다면 Size 맞는 Screw 이용한다.)

Air Piston Sealing(**90901-0192-639 & 640**) 분해한다.

③ Gasket(**90901-0192-676**) 분해하고 Puncture Valve를 뒤집어서 Vice에 고정시킨다.

④ Puncture Valve 분해 Tool 과 해머를 이용하여 Slide(**90901-0192-723**), Spring (**90901-0192-735**), Spring Guide(**90901-0192-747**)를 분해한다.

< Puncture Valve 분해 Tool >

⑤ 분해한 Part 깨끗하게 소제한다.

⑥ Valve Slide(**90901-0192-723**)와 Housing(**90901-0192-688**) Seat 마모나 스크래치 확인한다.
만약, 마모나 스크래치가 심하면 폐기한다.

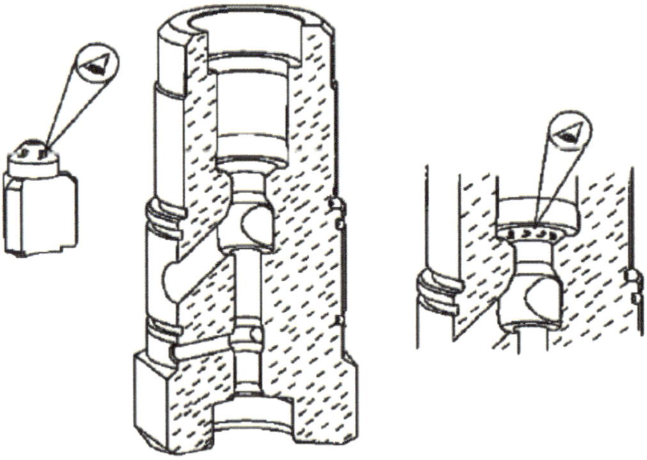

Lock Washer & Screw 분해

Cover & Gasket

Air Piston 분해 Tool

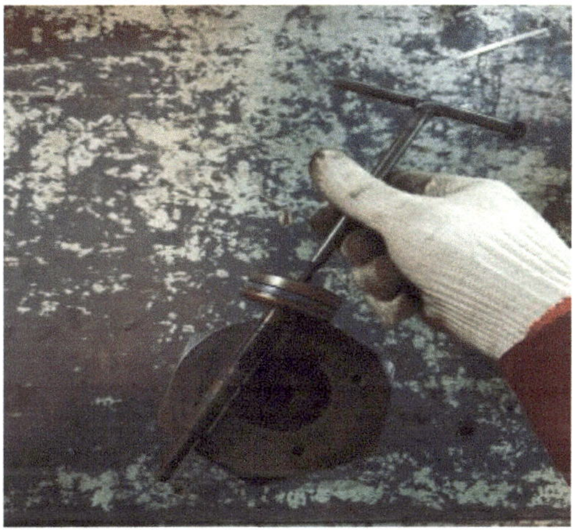

Air Piston 분해

내부 부품 분해

(2) 조립 방법

① Puncture Valve를 뒤집은 상태에서 Slide(**90901-0192-723**), Spring(**90901-0192-735**), Spring Guide(**90901-0192-747**)에 L.O를 도포하여 Zinc나 해머를 이용하여 조립한다.

② Puncture Valve를 다시 뒤집어 정비된 Air Piston(**90901-0192-664**)를 조립한다.

Air Piston 과 Housing(**90901-0192-688**)에 L.O를 충분히 도포하여 조립 시 Sealing 손상되지 않도록 조립한다.

(Air Piston 정비는 아래 사진자료를 참조)

③ Gasket(**90901-0192-676**)에 Molykote 도포하여 조립한다.

④ Cover, Lock Washer, Screw(**90901-0192-627 & 615 & 603**)를 조립한다.

⑤ O-ring(**90901-0192-711**) 2EA를 조립한다.

⑥ 정비된 Puncture Valve는 Wrap을 씌워서 보관한다.

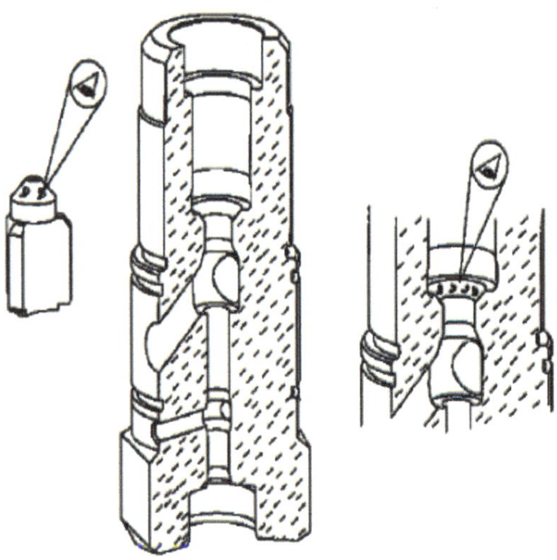	
Slide, housing Seat 상태 육안으로 확인한다.	L.O를 Slide, Spring, Spring Guide와 Housing 안쪽에 도포하여 조립이 원활하도록 한다.

Air Piston Sealing에 L.O를 도포하여 조립한다.

Air Piston sealing Ring조립 확인한다.

Gasket에 Molykote 도포하여 조립한다.

Cover를 조립하고 Lock washer & Screw Tight한다.	Wrap을 씌워 보관한다.

* 에어 피스톤 정비 방법

Mounting Tools - Air Piston Seals

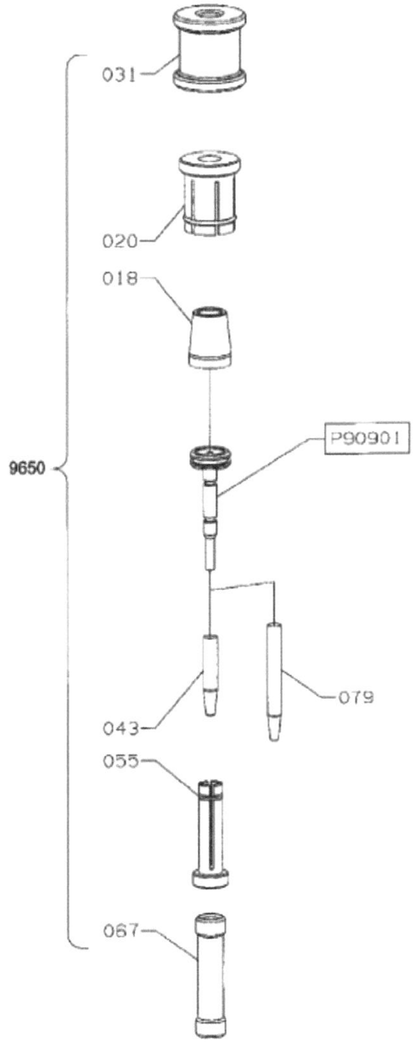

018, 043, 079 : Cone

020, 055 : Pushing tool

031, 067 : Compression tool

9650 : Mounting tools - air piston seals

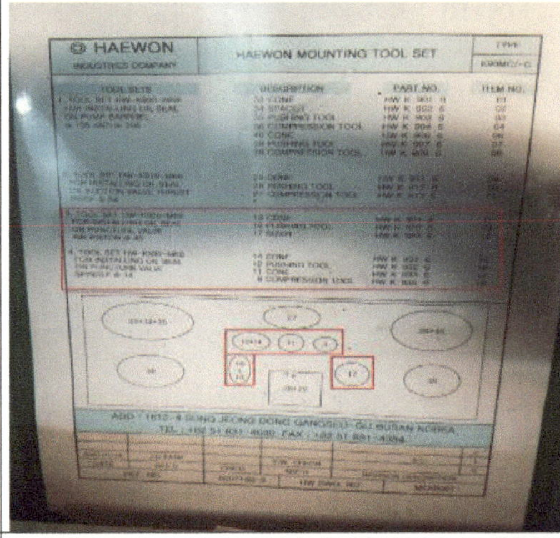

Cone, Pushing Tool, Compression Tool

Air Piston Sealing Ring

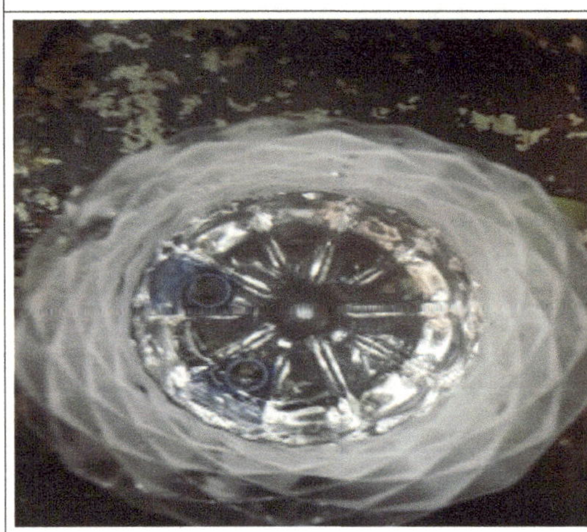

뜨거운 물에 Sealing Ring을 담근다.

Cone을 설치하고 Sealing Ring을 올린다.

Pushing Tool 이용하여 Sealing Ring 설치한다.

WD-40을 이용하여 Sealing Ring 냉각시킨다.

Compression Tool을 이용하여 압축시킨다.

Cone, Pushing Tool, Compression Tool

Air Piston Sealing Ring

뜨거운 물에 Sealing Ring 담근다.

Cone을 설치하고 Sealing Ring을 올린다.

Pushing Tool 이용하여 Sealing Ring설치한다.

WD-40을 이용하여 Sealing Ring 냉각시킨다.

Cone을 설치하고 Sealing Ring을 올린다.

Compression Tool을 이용하여 압축시킨다.

6) 연료 펌프 상부 커버 흡입 밸브

정비 주기 : 8000H

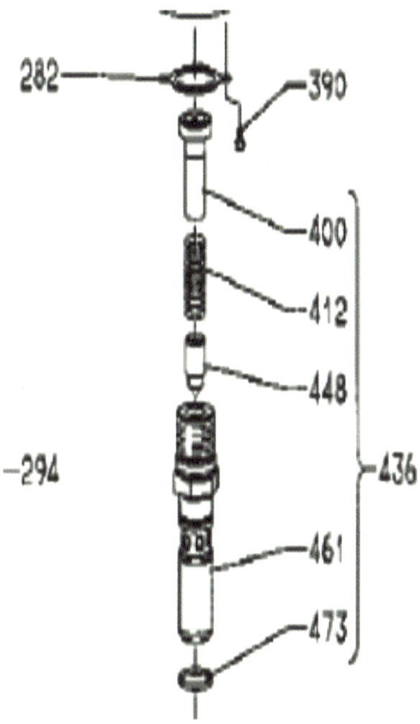

Fuel Pump Suction Valve

SAFETY PRECAUTIONS *For detailed sketch, see 900-2*

X	Stopped engine
X	Shut off starting air supply - *At starting air receiver*
X	Block the main starting valve
X	Shut off starting air distributor/distributing system supply
X	Shut off safety air supply - *Not ME engines*
X	Shut off control air supply
	Shut off air supply to exhaust valve - *Only with stopped lubricating oil pumps*
X	Engage turning gear
	Shut off cooling water
X	Shut off fuel oil
X	Stop lubricating oil supply
	Lock the turbocharger rotors

Data

Ref.	Description	Value	Unit
D09-33	Suction valve, tightening torque	1850	Nm
D09-34	Suction valve	10	kg

F.O.P Top Cover 아래에 Suction Valve가 조립되어 있으며 Top Cover 하부 Drain Line으로 F.O가 나온다면 Suction Valve가 누설하는 것이므로 교환한다.

Suction Valve 바깥에 Sealing Ring이 있으며 Top Cover 내부로 **조립 시 Sealing Ring이 손상되지 않도록 주의하며 조립한다.**

Suction Valve : 1850Nm

(1) 분해 및 소제 방법

① Suction Valve를 아래와 같이 뒤집어 Vice에 고정시킨다.

② Suction Valve 분해 Tool 및 해머를 이용한다.

Spring Guide(90901-0192-400), Spring(90901-0192-412), Slide(90901-0192-448) 분해한다.

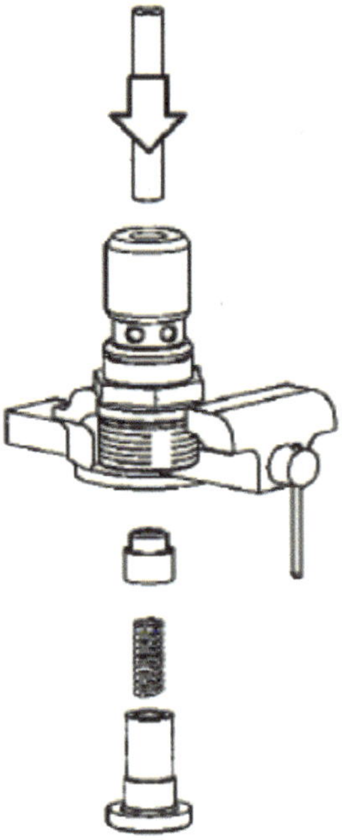

< Suction Valve 분해 Tool >

③ Valve Slide(**90901-0192-448**)와 Housing Seat에 마모나 스크래치 확인한다.

만약 마모나 스크래치가 심하면 폐기한다.

2.

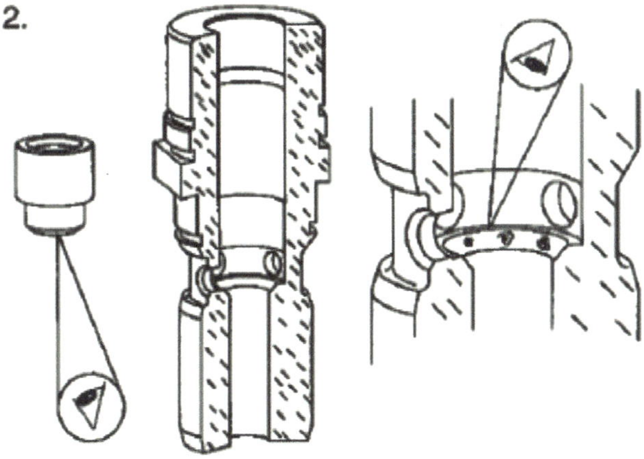

④ 분해된 Part를 소제한다.

Suction Valve 분해 Tool

내부 부품 분해

(2) 조립 방법

Mounting Tools – Suction Valve Seals

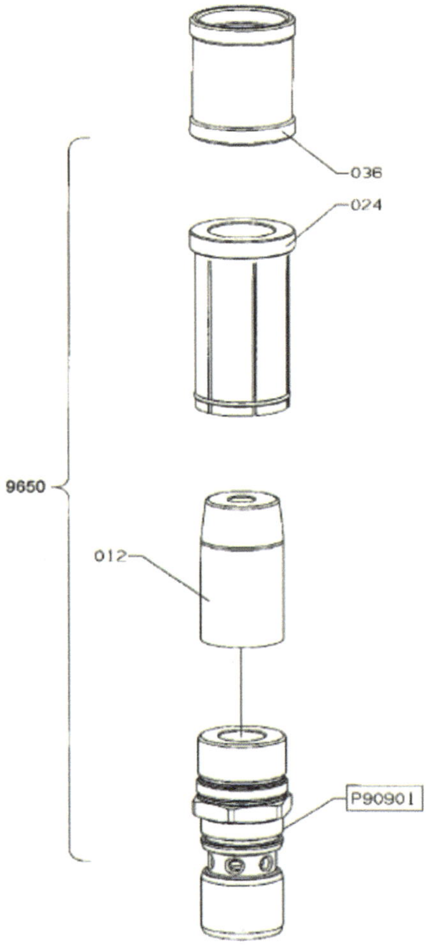

012 : Cone

024 : Pushing tool

036 : Compression tool

9650 : Mounting tools – suction valve seals

① Suction Valve를 뒤집은 다음 Slide(**90901-0192-448**), Spring(**90901-0192-412**), Spring Guide(**90901-0192-400**)에 L.O를 도포하여 Zinc나 Hammer 이용하여 조립한다.

② Puncture Valve를 다시 뒤집어 아래 그림과 같이 Diesel Oil 이용하여 **누설 유무를 확인한다.**

누설하면 조립을 잘못하였거나 Slide 및 Housing을 재확인한다.

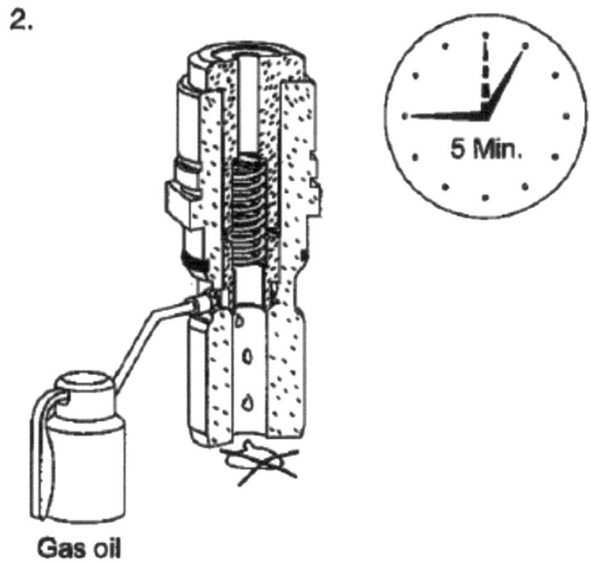

③ 아래와 같이 Sealing(90901-0192-473)을 조립한다.

Puncture Valve Air Piston Sealing과 같이 뜨거운 물을 받아서 Sealing을 팽창시키면 보다 쉽게 조립을 할 수 있다.

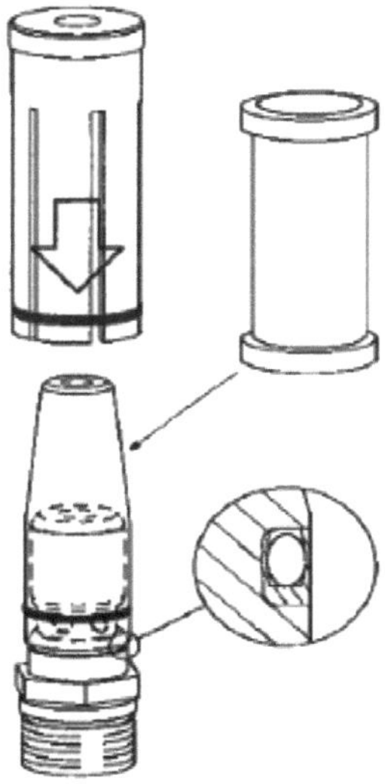

④ 정비된 Suction Valve는 Wrap을 씌워서 보관한다.

2.

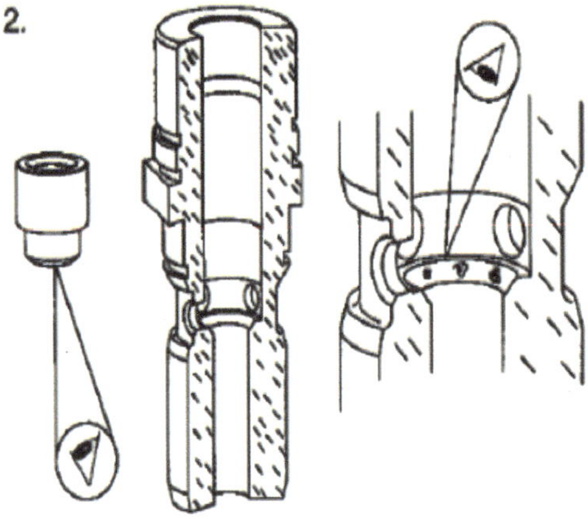

Slide, housing Seat 상태 육안으로 확인한다.

Slide, Spring, Spring Guide
L.O를 도포하여 조립한다.

2.

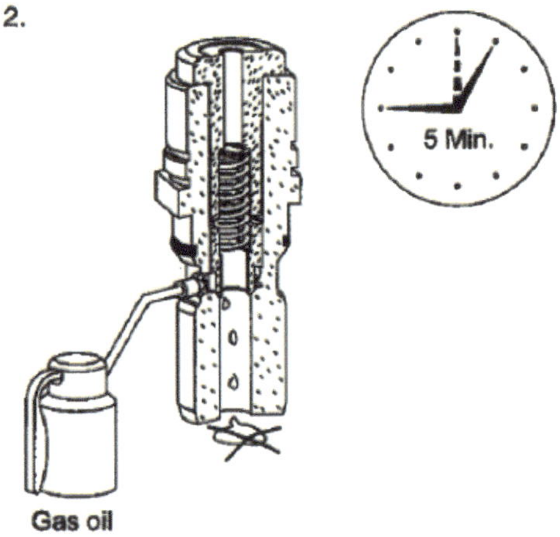

Slide, housing Seat사이로 Oil 누설 확인한다.

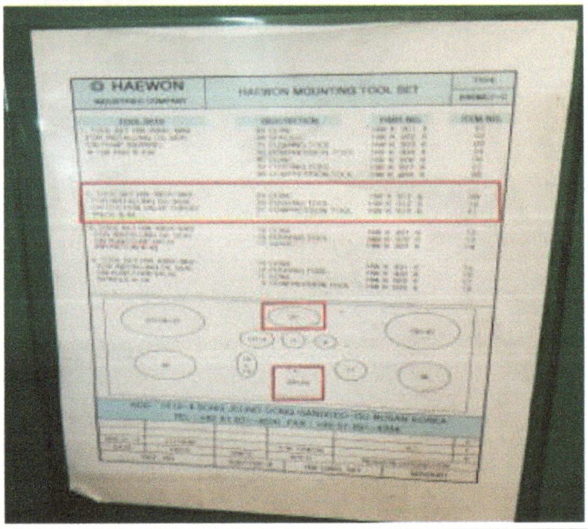

뜨거운 물에 Sealing Ring을 담근다.

Cone을 설치하고 Sealing Ring을 올린다.

Pushing Tool 이용하여 Sealing Ring설치한다.

WD-40 골고루 분사하여 수축시킨다.

Pushing Tool, Cone, Compression Tool

Compression Tool을 이용하여 압축시킨다.

Wrap을 씌워 보관한다.

7) 연료 펌프 충격 흡수기

정비 주기 : 8000H

Fuel Pump - Details

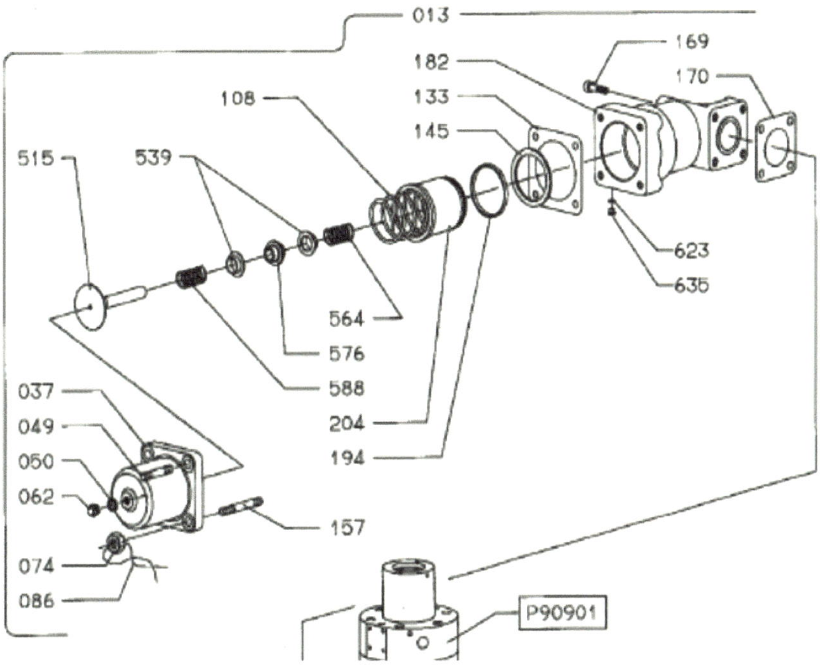

Shock Absorber는 토출 행정 끝날 때마다 Plunger가 차단 구멍을 Open할 때 발생하는 충격을 흡수하는 것이다.

스프링 제어식 Piston과 Cylinder로 구성되며, Piston은 토출 쪽에서 나온 여분의 오일이 펌프의 Barrel 입구 공간으로 빠져나가도록 밀어준다.

주요 문제는 **Spring(90904-0052-108 & 588 & 564)들의 손상이다.**

누설하는 Shock Absorber 경우는 아래와 같다.

Piston(90904-0052-204) 손상, Housing(90904-0052-182) 내부 손상

Piston Wearing(90904-0052-194), Housing Sealing Ring(90904-0052-145) 손상

따라서 매일 순찰 때마다 **Drain Pipe에서 누설 유무를 확인**하는 것이 매우 중요하다.

(1) 교환 방법

① 엔진 사용 종료(Finished With Engine) 후, F.O Pump Stop 하고 F.O In & Out Valve Close 한다.

(Alarm Monitoring System에서 M/E F.O Low Pressure Alarm 발생한다.)

그 다음 F.O.P Inlet Valve Close, Drain Valve Open하여 Pump 내부 F.O Drain한다.

(단, M/E & G/E F.O Line Common Line이라면 Pump Stop 하지 않고 F.O Line Inlet & Outlet Valve Close 후 F.O.P 앞의 Inlet Valve Close 한다.)

*** 만약, 상기와 같은 방법으로 F.O Pressure Drop 되지 않는다면 F.O Valve가 누설하는 것이므로, Pipe Line 확인한 다음 누설 부 Valve 교환한다.**

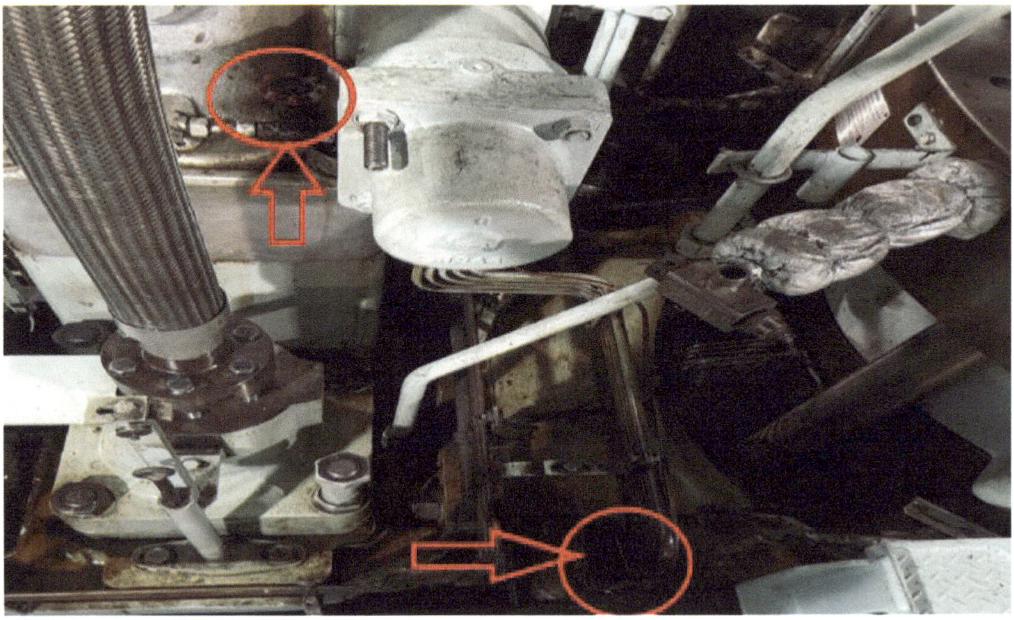

< Drain Valve와 Drain Place >

② Shock Absorber Drain Pipe 및 Nipple 분해한다.

③ Overhead Crane, Sling Rope & Shock Absorber 연결하고 Screw(**90904-0052-169**) 분해한다.

④ Shock Absorber를 분해 및 Gasket(**90904-0052-170**)을 분해한다.

⑤ F.O.P와 Shock Absorber 연결 부 Brush 소제한다.

⑥ 취외품의 Drain Nipple을 정비품 Shock Absorber에 Packing과 같이 조립한다.

E/R Overhead Crane 이용하여 장착하되 Gasket(**90904-0052-170**) 조립 누락되지 않도록 한다.

취외품 Drain Nipple 분해한다.	정비품에 Drain Nipple 조립한다.
	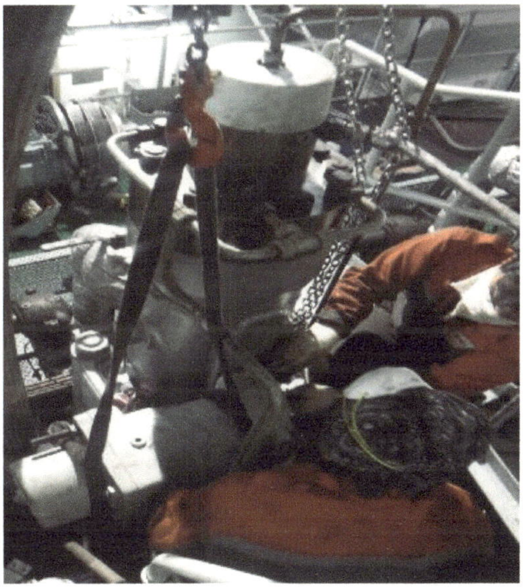
Overhead Crane과 Chain Block 이용하여 Shock Absorber & Gasket 조립한다.	Shock Absorber Screw Tight 한다.

* Shock Absorber 교환 시 Gasket(90904-0052-170) 조립 누락하지 않도록 주의한다.

(2) 분해 및 소제 방법

① End-cover Nut(**90904-0052-074**) 분해한다.

Plug Screw(**90904-0052-062**)를 분해하고 Gasket(**90904-0052-050**)을 취외한다.

② 취외한 곳에 아래 사진과 같이 알맞은 Size의 Long Thread(Piston과 End-Cover사이)를 넣고 Nut를 Tight 한다.

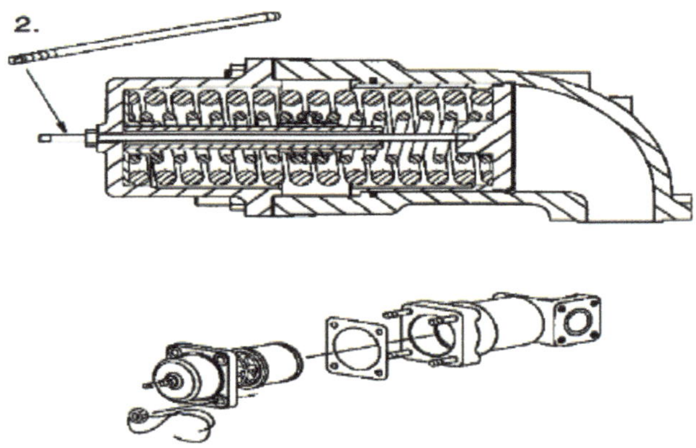

③ End-cover Nut(**90904-0052-074**)를 분해하고 아래 그림과 같이 전부 취외한다.

만약 Piston 쉽게 분해되지 않는다면 Piston 제외 전부 취외한다.

다시 End-Cover닫고 Thread Tight하여 Piston를 취외한다.

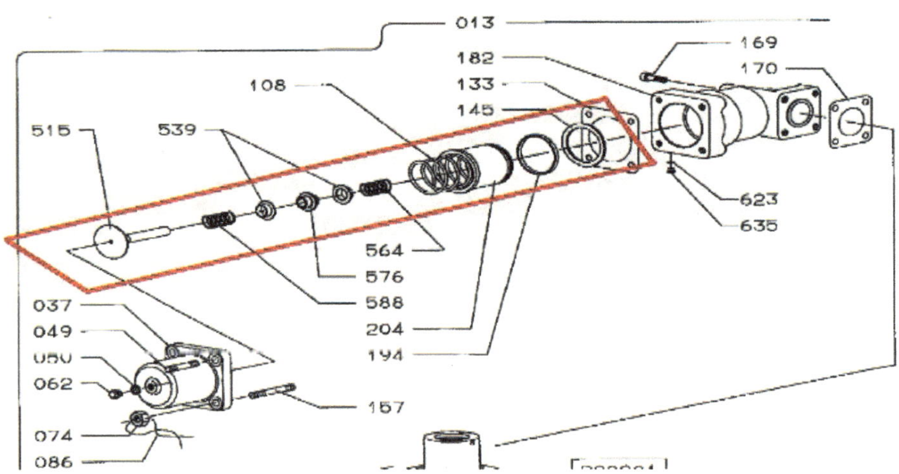

< Spring Guide, Spring, Disc, Slide, Piston, Wear Ring, Sealing Ring, Gasket >

④ 내부를 깨끗이 소제한다.

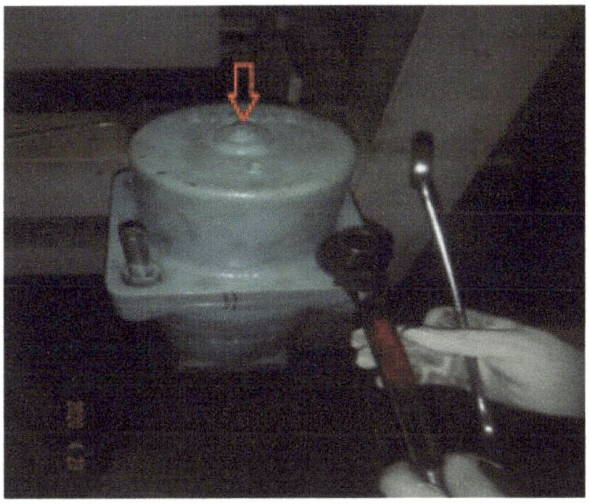

End-cover Nut, Plug Screw & Gasket
분해한다.

Plug Screw취외한 곳에 알맞은 Size의
Long Screw를 집어넣고 Nut Tight 한다.

End-cover 분해한다.

Spring Guide

바깥쪽 & 안쪽 Spring

Slide & Disc

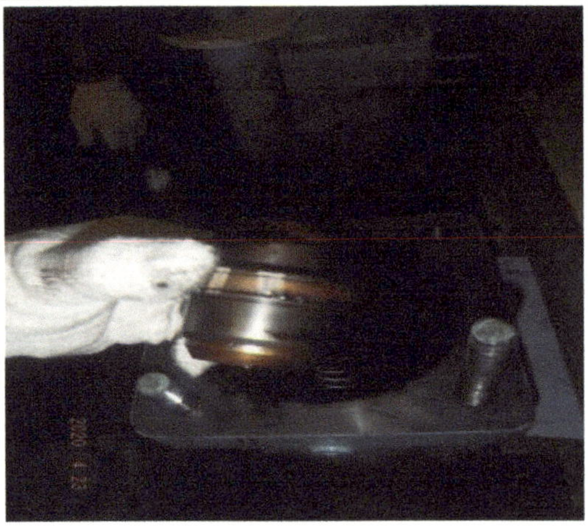

Slide & Disc

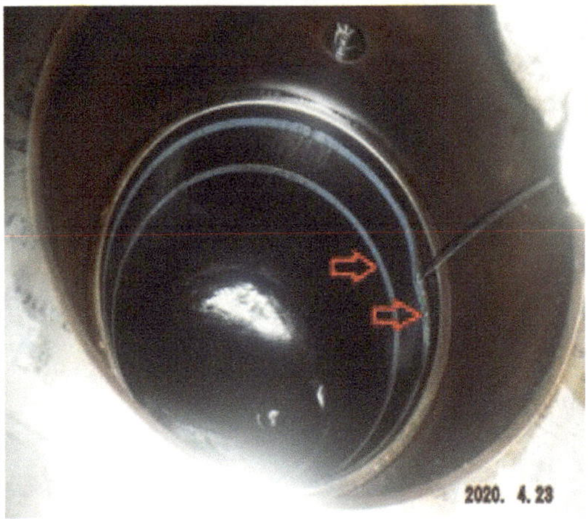

Housing Sealing Ring

Piston 상처 부위

Housing 상처 부위

(3) 조립 방법

① Housing Sealing Ring(**90904-0052-145**) 2EA, Piston Wear Ring(**90904-0052-194**) 조립한다.

Housing 내부, Piston 외부 L.O를 충분히 도포하고, Piston 내부 Thread에 알맞은 Size의 Bolt 조립한 다음 Zinc로 Bolt를 쳐서 조립한다.

이때 하부 측에서 Piston 조립될 때 상기 Sealing Ring, Wear Ring이 손상되지 않는지 반드시 확인한다.

(Piston을 Zinc로 조립할 때 Shock Absorber 아래 부분에서 Wear Ring, Sealing Ring 빠져나오는지 확인한다. 만약, 빠져나오면 취외 후 다시 조립한다.)

② Spring, Disc, Slide, Spring Guide를 도면과 같이 순서대로 조립한다.

③ Gasket(**90904-0052-133**)을 조립한다.

④ End-cover(**90904-0052-037**)와 Nut(**90904-0052-074**)를 조립한다.

*** End-cover Nut : 200Nm**

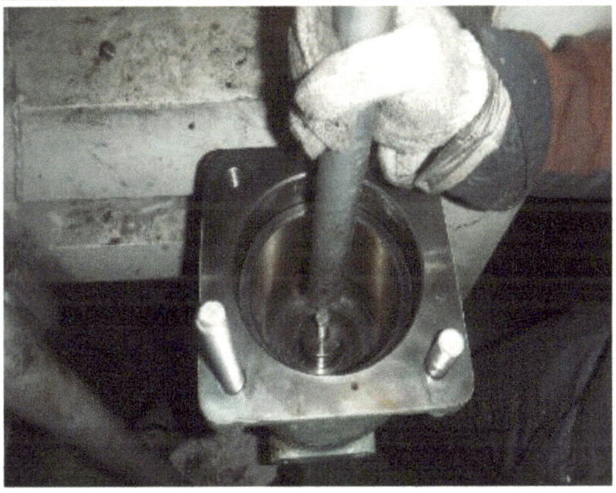

Piston 내부 알맞은 Size의 Bolt를 조립하고
Zinc로 쳐서 Piston을 조립한다.
아래에서 Sealing Ring 조립 상태 확인한다.

Gasket에 Molykote 도포하고 조립한다.

200NM Torque로 End-cover Nut Tight 한다.

8) 리버싱 메커니즘

앞서 설명했듯이 Cam Position이 Ahead / Astern방향으로 정확히 가지 않을 경우, **"WRONG WAY/FUEL CAM POSITION ABNR"**와 같은 Alarm이 AMS(Alarm Monitoring System)에 발생하며, Indicator Panel의 **Ahead / Astern** Lamp에 불이 들어오지 않고, C/R Control Panel 내부 Safety System의 **Fuel Cam Position**에 **빨간불**이 들어온다.

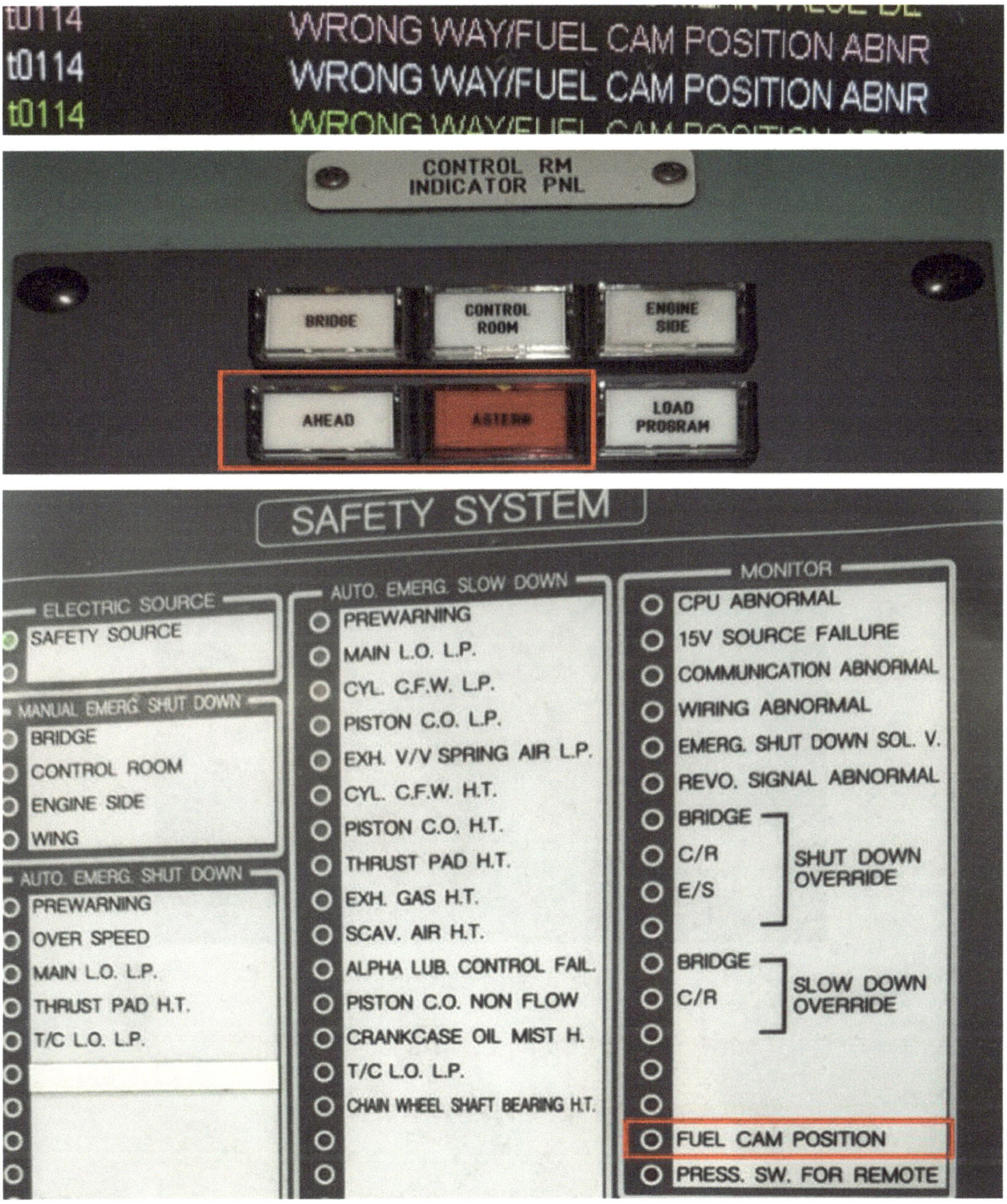

상기와 같이 Alarm이 발생할 경우 실제 Local에서 Cam Position이 변경되지 않은 Cylinder를 확인한 다음 Engine Stop후 다시 재시동을 한다.

빈번하게 발생되는 Cylinder의 경우 Reversing Air Cylinder 분해 점검 또는 Cam 점검한다.

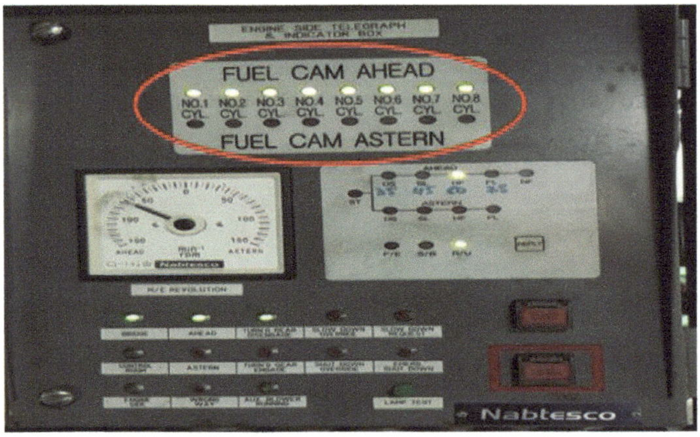

< Engine Side Telegraph & Indicator Box >

→ **현재 Cam Position 전부 Ahead 상태이다.**

< F.O.P Cam Position >

상기 그림과 같이 **Cam Position이 빠져나오면 'Ahead Position'**

Cam Position이 들어가면 'Astern Position'

<위쪽 Ahead / 아래쪽 Astern>

만약에 Indicator Panel의 Ahead / Astern Lamp가 깜빡이거나 점등되지 않을 경우, Reversing Air Cylinder의 Clamping Piece 분해하여 소제한 다음 조립한다.

상태가 동일하면 Connector Re-tight, 내부 Cable 확인 Junction Box Cable Re-tight 한다.

본선 예비품이 있다면 Clamping Piece, Position Switch, Connector까지 전부 교환하고 Cable 상태 불량 시 Cable을 교환한다.

구성은 아래와 같다.

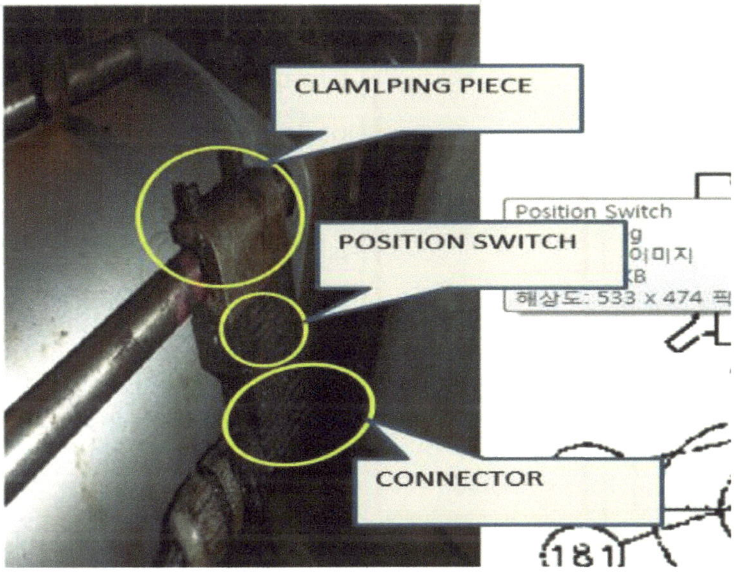

5 . **CABLE TYPE**

AT THE WIRING DIAGRAM, THE CABLE TYPE OF THE WIRING AT ENGINE SIDE ARE TO BE SPECIFIED BY ABRIDGED LETTERS AT INDICATED AT THE BELOW TABLE. THIS CABLE IS FRAME RETARDMENT CABLE.
BUT, IN CASE THE SPECIAL WIRE IS EMPLOYED, IT IS NOTED AT EACH WIRING DIAGRAM.

ABRIDGED	KIND OF CABLE	NO. OF CONDUCTOR	CABLE SIZE (mm2)	GLAND SIZE(PG)	TYPE OF CABLE
LMS(102)	FFCS	2	0.75	9	FFCS
LMS(103)		3	0.75	9	1.CONDUCTOR: TINNED, FINE WIRE
LMS(105)		5	0.75	11	2.TWISTING: CORES OPTIMAL STRANDED IN LAYER
LMS(107)		7	0.75	11	3.INNER SHEATH: THERMOPLASTIC MATERIAL FILLER SHEATH
LMS(112)		12	0.75	13.5	4.SCREEN: BRAID MADE OF TINNED COPPER ROUND WIRE
LMS(116)		16	0.75	16	5.FLAME RETARDARNT(IEC-332-1/3)
LMS(202)		2	1.5	9	
LMS(203)		3	1.5	9	
LMS(205)		5	1.5	11	
LMS(207)		7	1.5	13.5	
LMS(302)		2	2.0	13.5	
LMS(303)		3	2.0	13.5	
LMS(403)		3	2.5	13.5	

6 . **APPARATUSES AND DEVICE USED FOR WIRING OF ENGINE SIDE**

THE CABLES ARE TO BE EMPLOYED AS ENGINE MAKER'S STANDARD.
PLEASE SEE EATH WIRING DIAGRAM.

7 . **WIRING AND CONNECTIONS**

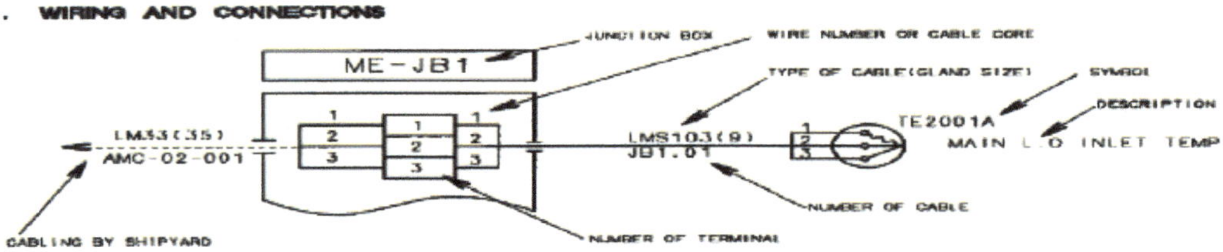

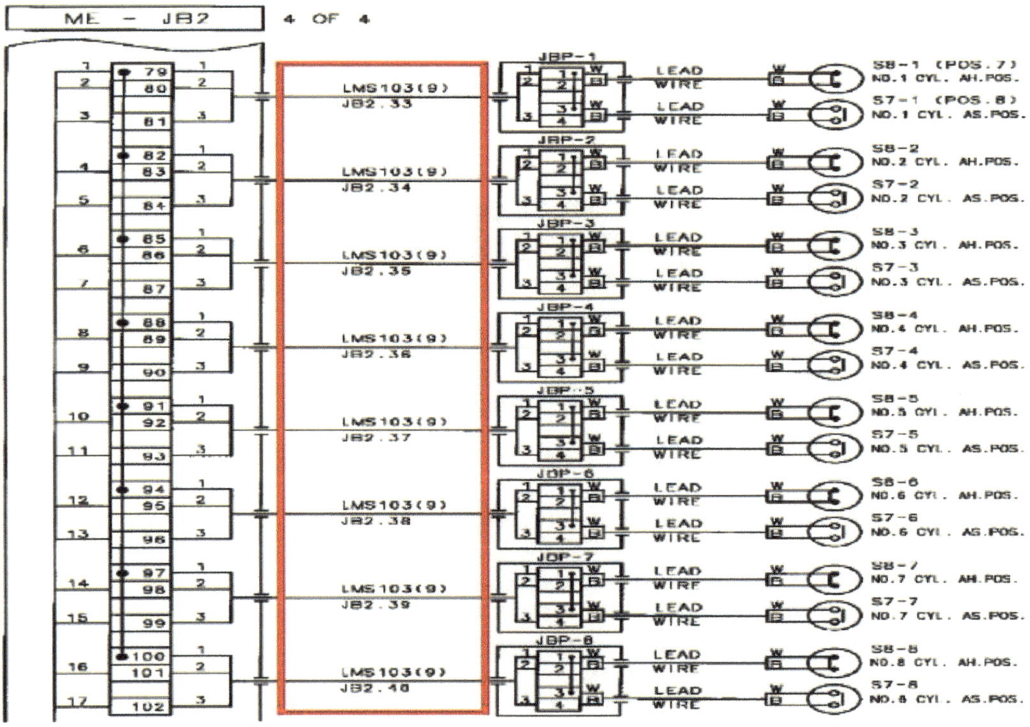

상기 도면을 참조하여 Cable 청구 및 점검하도록 한다.

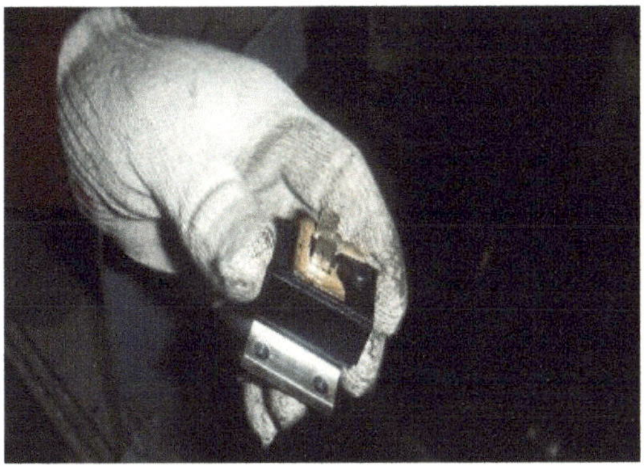

취외품이며 오랜 사용으로 Packing 경화

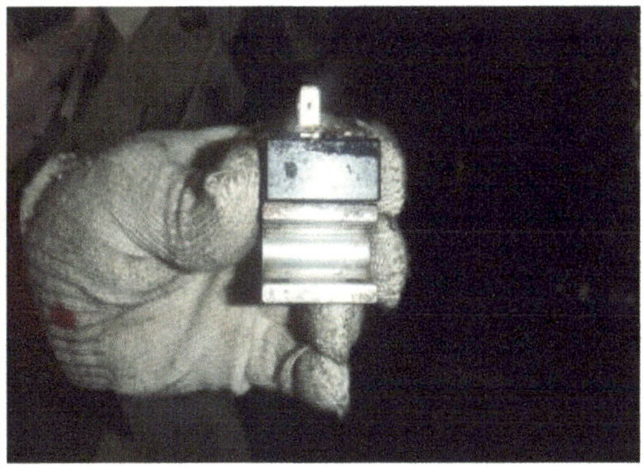

Position Switch, Clamping Piece

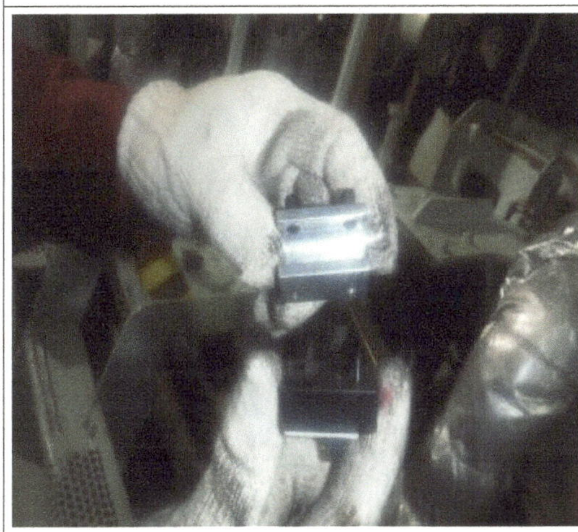

Position Switch와 Clamping piece 조립

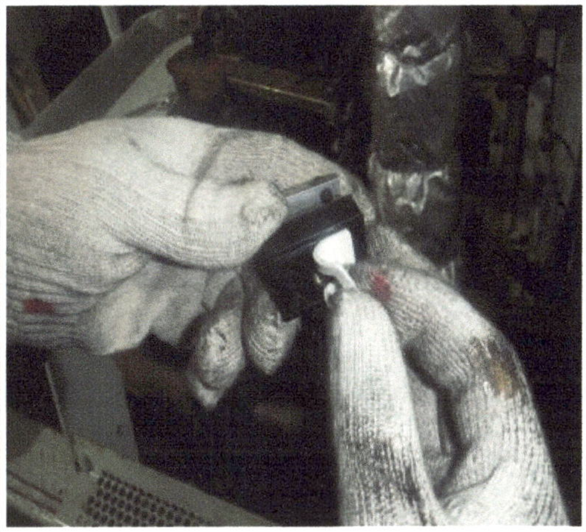

Position Switch와 Clamping piece 연결하는 Hex. Screw Tight

교환 방법은 아래와 같다.

Air Cylinder에 조립된 Clamping Piece와 Position Switch를 Paint Makar로 표시한다.

Connector 분해하고 Clamping Piece Hex. Screw 분해하여 Clamping Piece와 Position Switch를 함께 취외한다.

취외한 부분 깨끗한 천으로 소제하고 신품으로 조립한다.

조립 완료 후 Engine Side Telegraph & Indicator Box에서 Lamp 점등된 것을 확인한다.

실제로 Cam의 역전(Astern)이 되지 않는다면 Reversing Air Cylinder에 연결된 Starting Air Line의 Non-return Valve 고착, Reversing Air Cylinder 분해 점검 후 Line Venting 하지 않았을 경우이다.

Line Venting은 Engine Stop 후 Ahead & Astern Starting Air Line 연결 부 분해한 다음 Engine을 짧게 Ahead Astern 시동하면 Line Venting된다.

Non-return Valve가 고착된 경우라면 Engine Start시에만 Starting Air가 공급되므로, 운항 중이면 Starting Air 공급되지 않으니 Non-return Valve 취외하여 정비 및 신품 교환하거나 Engine Stop후 정비 및 신품 교환한다.

수동으로 Cam을 작동하는 방법이 있다.

Engine Stop 후 Main L.O Pump Stop Camshaft Cover Open 한다.

Reversing Air Cylinder Ahead & Astern Air Line으로 Compressor Air 공급하면서 Engine Turning하면 Camshaft 회전하면서 Air Distributor의 순서에 맞게 Cam Ahead & Astern으로 변경된다.

Cam의 오목하게 패인 부분이 F.O Pump의 Roller와 마주칠 때 Cam의 Position이 변경된다.

(F.O Pump Roller와 Cam이 분리될 때 = Air Distributor의 순서에 맞을 때)
(Camshaft Cover를 통해서 Cam 변경되는 Position을 확인할 수 있다.)

Local(Emergency) Control Stand에서도 수동으로 Cam을 작동시킬 수 있다.

Engine Stop 후 Main L.O Pump Stop Camshaft Cover Open 한다.

앞서 언급한 Local Engine Start하는 방법에서 Blocking Arm을 Emergency 위치에 놓고, Impact wheel 'P'를 반 시계 방향으로 돌리면 Governor와 분리되고 Engine Side Lamp 점등된다.

Ahead & Astern을 설정하고 Remote에서 Emergency로 변경하면 Starting Air 공급된다.

Engine Turning하면 Camshaft 회전하면서 Air Distributor의 순서에 맞게 Cam Ahead & Astern로 변경되며 Cam의 오목하게 패인 부분이 F.O Pump Roller와 마주칠 때 Cam Position변경된다.

*** F.O Pump Roller와 Cam이 분리될 때 = Air Distributor의 순서에 맞을 때**

(Camshaft Cover를 통해서 Cam 변경되는 Position을 확인할 수 있다.)

* 리버싱 에어 실린더 파이프 라인 & 리페어 키트

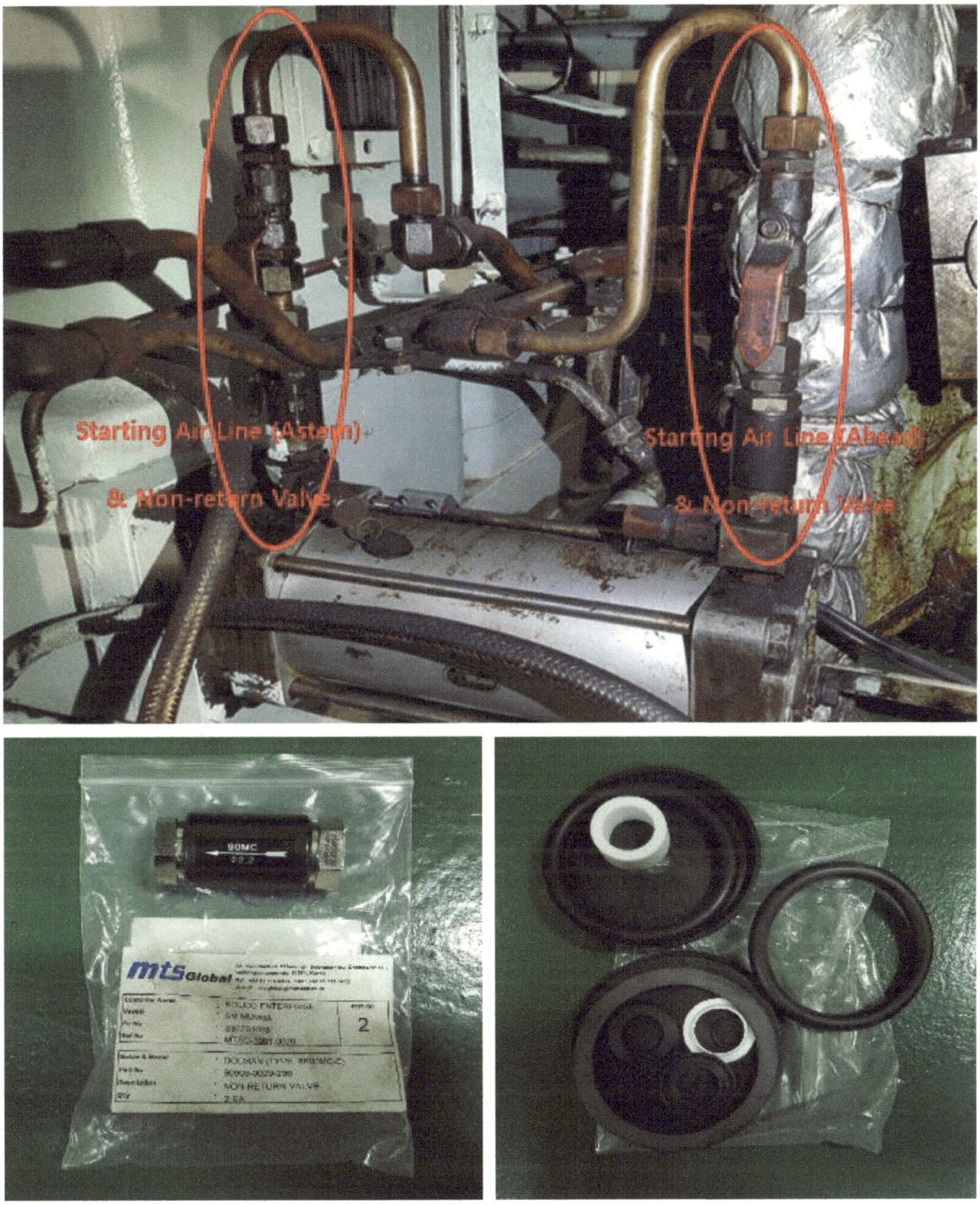

< 리버싱 에어 실린더 파이프 라인 & 리페어 키트 >

* 리버싱 에어 실린더

* 리버싱 에어 실린더 스타팅 에어 라인으로 캠 위치 변경 방법

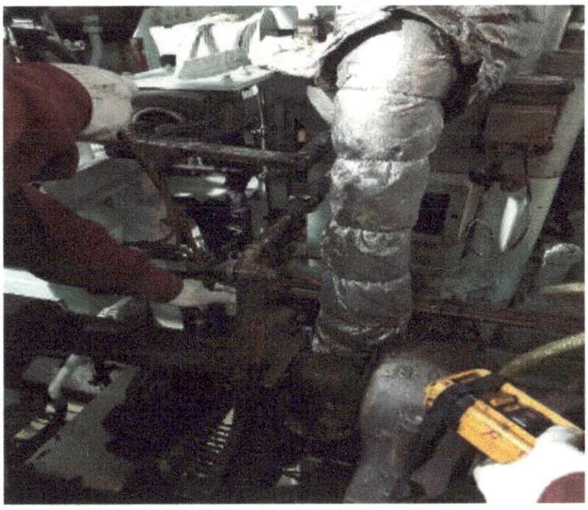

Compressor Air 공급하면서 Engine Turning

Compressor Air 공급하면서 Engine Turning

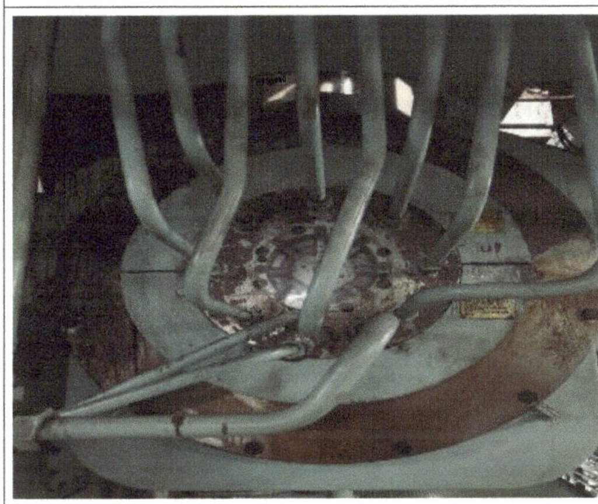

Air Distributor

Camshaft Cover

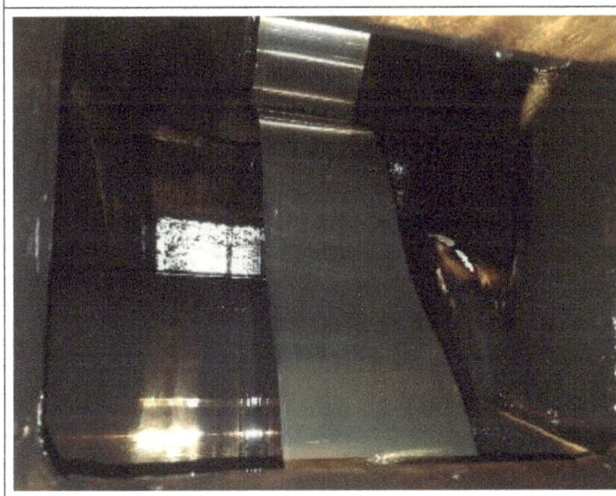

Fuel Cam

Control Room Position

* 로컬(비상) 컨트롤 스탠드에서 캠위치(CAM POSITION) 변경 방법

Camshaft Cover

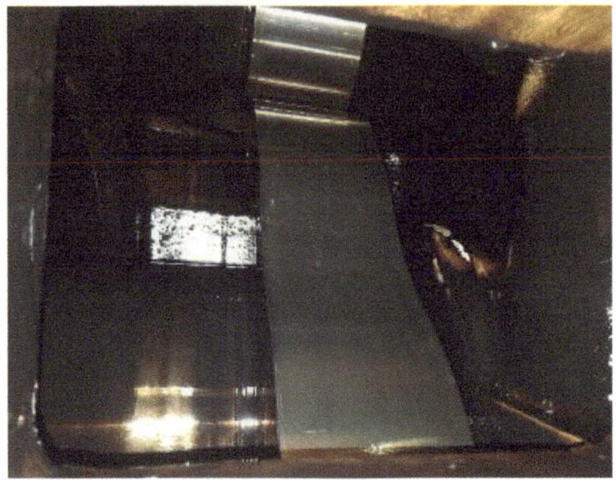

Fuel Cam

Governor 분리

Ahead & Astern 설정 후 Emergency로 변경

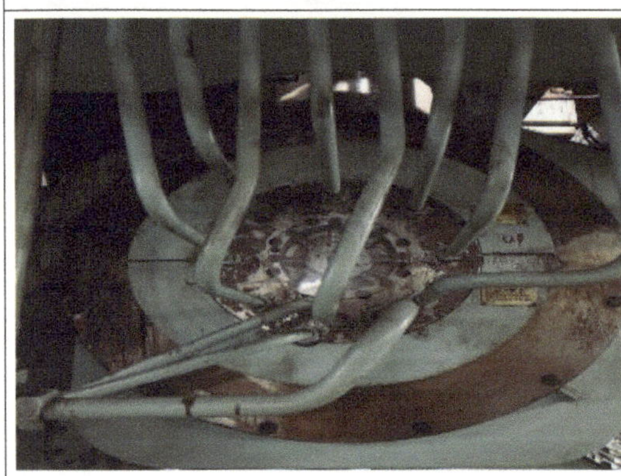

Air Distributor

Engine Side로 변경됨

9) 배기 밸브

정비 주기: 16000H

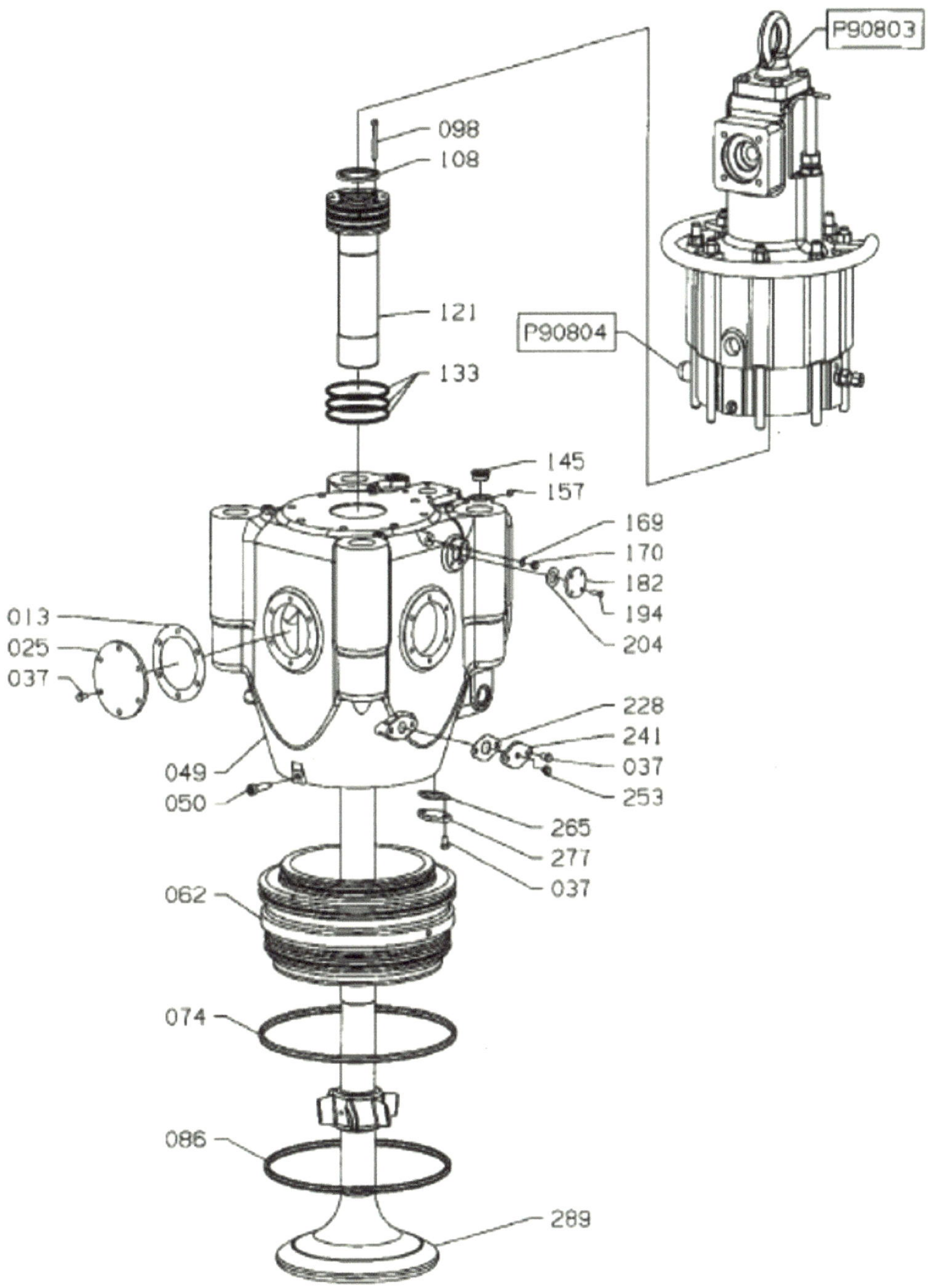

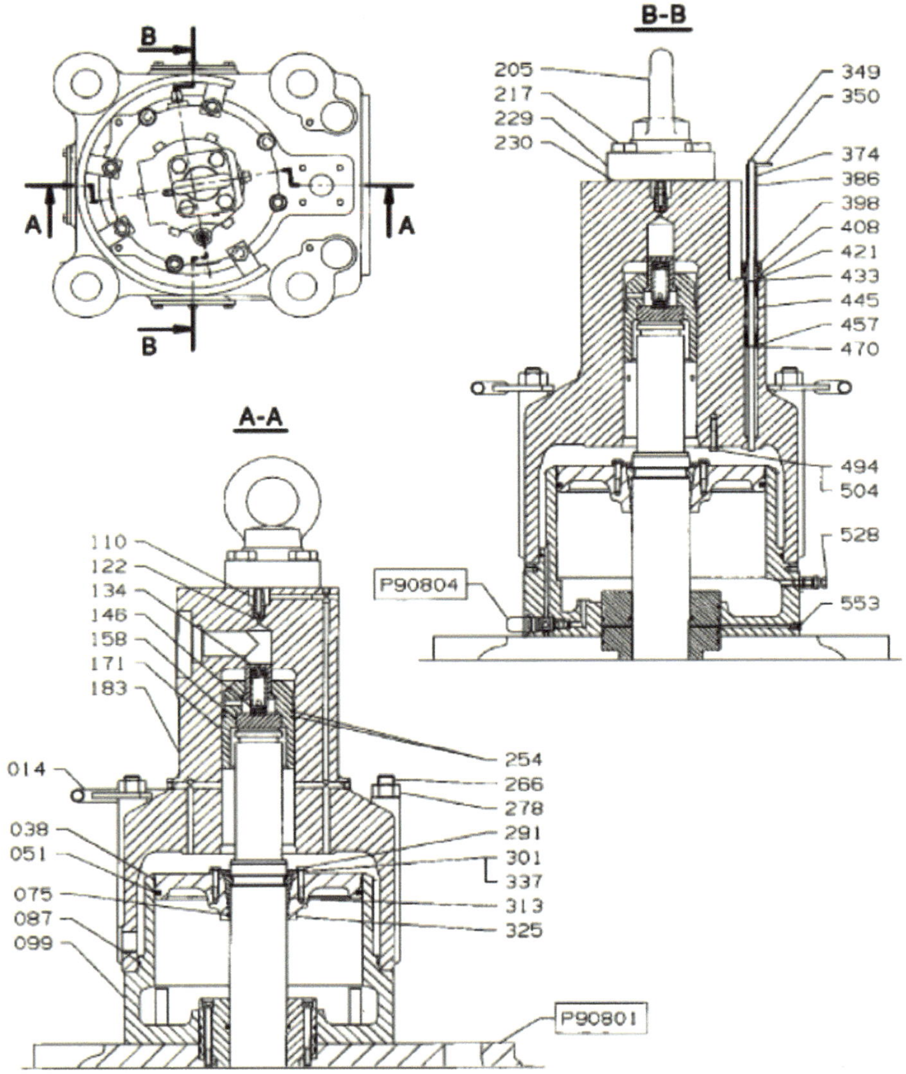
B
A
A
B
B-B
205
217
229
230
349
350
374
386
398
408
421
433
445
457
470
494
504
528
553
P90804
A-A
110
122
134
146
158
171
183
014
038
051
075
087
099
254
266
278
291
301
337
313
325
P90801

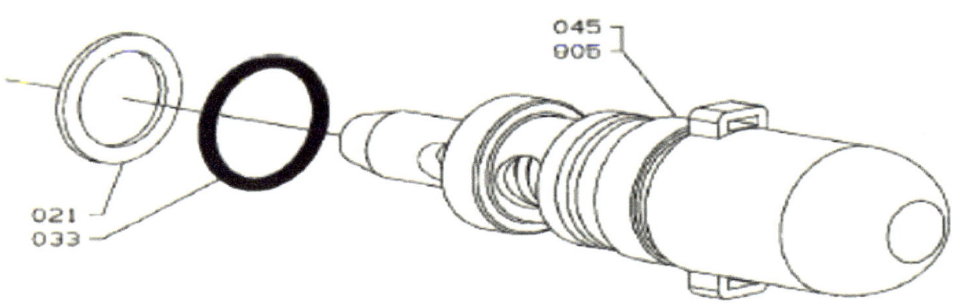
045
905
021
033

Exhaust Valve - Panel

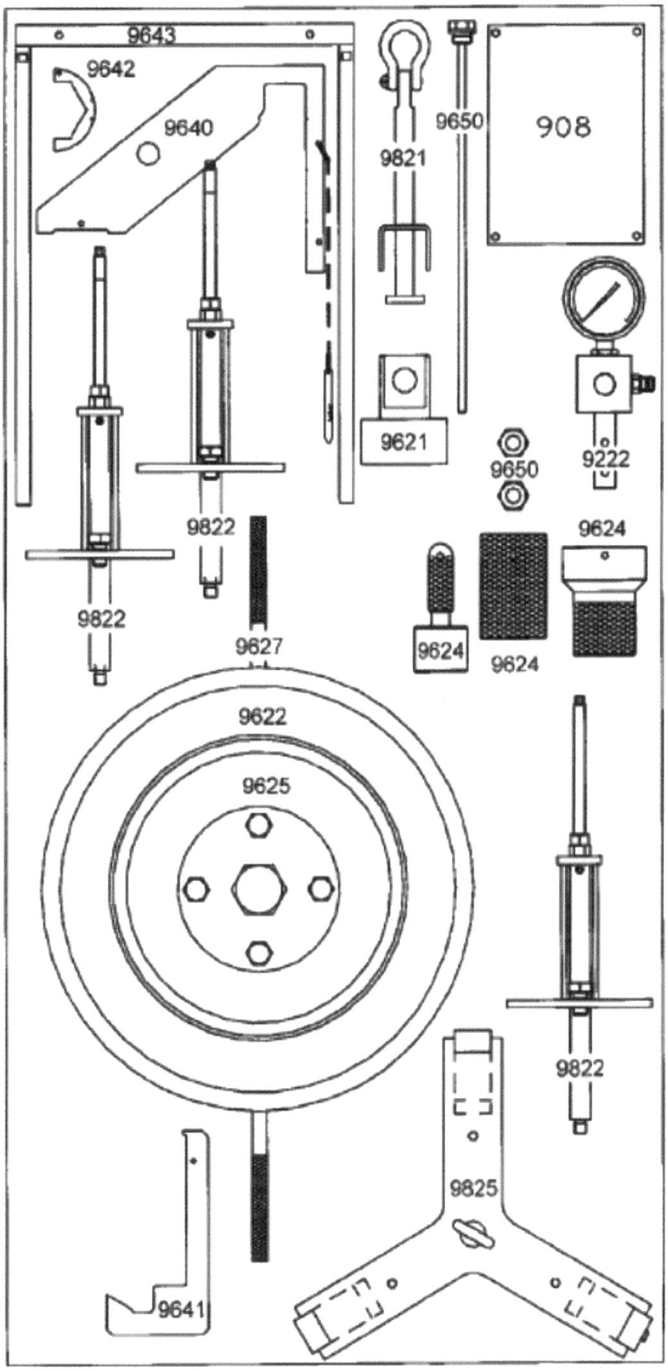

908 : Panel for tools

9222 : Pressure tester, safety valve

9621 : Lifting tool, exhaust valve

9622 : Conical ring, pneumatic piston

9624 : Grinding tool, H.P. pipe

9625 : Tool, hydraulic piston

9627 : Grinding ring, exhaust valve bottom piece

9640 : Gauge, exhaust valve spindle

9641 : Gauge, exhaust valve bottom piece

9642 : Tightening template, actuator

9643 : Bridge gauge, exhaust valve

9650 : Tool, emergency opening of exhaust valve

9821 : Lifting tool, roller guide

9822 : Lifting tool, exhaust valve roller guide

9825 : Lifting tool, exhaust valve

Exhaust Valve - Hydraulic Tools

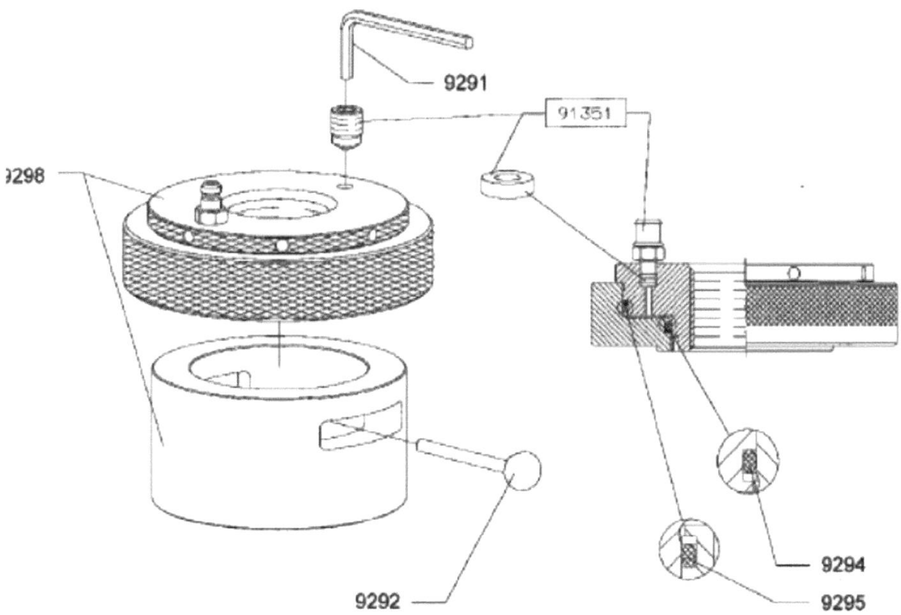

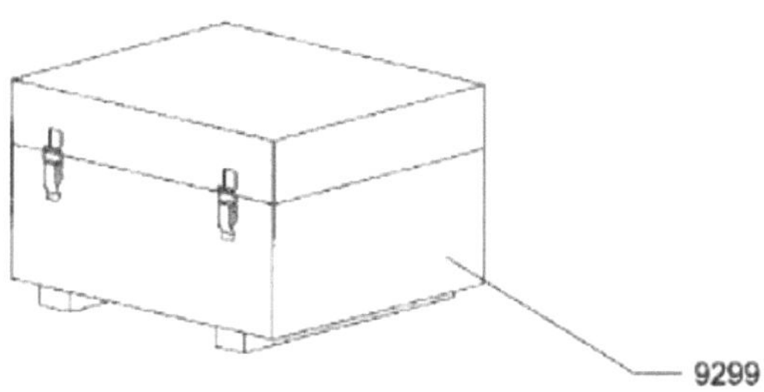

9291 : Key for hexagon socket screw

9292 : Tommy bar

9294 : O-ring with back-up ring

9295 : O-ring with back-up ring

9298 : Hydraulic jack

9299 : Hydraulic jack for exhaust valve, complete

Exhaust Valve

SAFETY PRECAUTIONS *For detailed sketch, see 900-2*

X	Stopped engine
X	Shut off starting air supply – *At starting air receiver*
X	Block the main starting valve
X	Shut off starting air distributor/distributing system supply
X	Shut off safety air supply – *Not ME engines*
X	Shut off control air supply
X	Shut off air supply to exhaust valve – *Only with stopped lubricating oil pumps*
X	Engage turning gear
X	Shut off cooling water
	Shut off fuel oil
X	Stop lubricating oil supply
X	Lock the turbocharger rotors

Data

Ref.	Description	Value	Unit
D08-04	Safety valve, opening pressure	23	bar
D08-05	Screws for lifting eye bolt, tightening torque	340	Nm
D08-06	Oil cylinder, tightening torque	340	Nm
D08-07	Safety valve, tightening torque	70 - 100	Nm
D08-09	Bottom piece seat, grinding angle	29.9 - 30.0	°
D08-10	Valve spindle seat, grinding angle	30.7 - 30.8	°
D08-11	Bottom piece seat, max. grinding	2.3	mm
D08-12	Gap of bottom piece seat	1.0	mm
D08-13	Spindle, max. burn-off	12.0	mm
D08-14	Spindle, max. grinding	2.0	mm
D08-15	Spindle stem, measuring area, min.	435	mm
D08-16	Spindle stem, measuring area, max.	610	mm
D08-17	Spindle stem, min. diameter	107.8	mm
D08-18	Bushing max. diameter, top	108.8	mm
D08-19	Bushing max. diameter, bottom	112.0	mm
D08-20	Oil cylinder max. inside diameter	125.2	mm
D08-21	Piston rings, min. thickness	4.7	mm
D08-22	Damper piston, min. distance	48,4	mm
D08-23	Damper piston, max. distance	51,4	mm
D08-24	Exhaust valve, complete	2400	kg
D08-25	Exhaust valve housing	1300	kg
D08-26	Oil cylinder	330	kg
D08-27	Air cylinder	95	kg
D08-28	Spindle	240	kg
D08-29	Bottom piece	300	kg
D08-30	Air piston	30	kg
D13-01	Hydraulic pressure, mounting	900	bar
D13-02	Hydraulic pressure, dismantling	800 -990	bar

(1) 교환 방법

① 엔진 사용 종료(Finished With Engine)후, Main L.O Pump Stop & Spring Air Valve Close 한다.

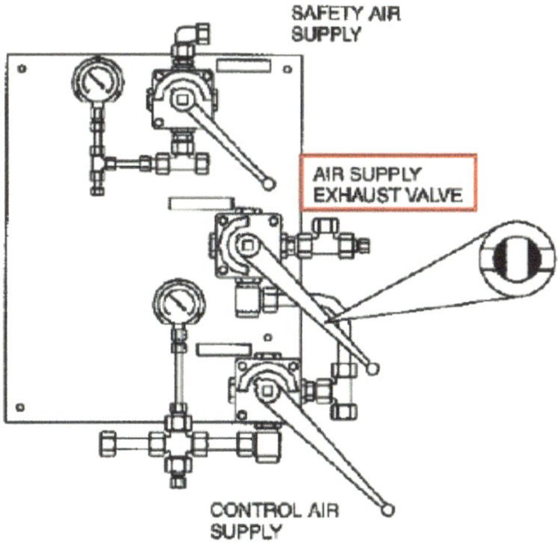

< Spring Air Valve >

* 작업 전 : Main L.O Pump Stop → Spring Air Close

작업 후 : Spring Air Open → Main L.O Pump Start

② CYL′ Head Water Drain을 시켜준다.

Inlet Valve Close & Drain Valve

Outlet Valve Close

Outlet Valve & Venting Valve

Air Vent Line 중간 Valve

Air Vent Valve

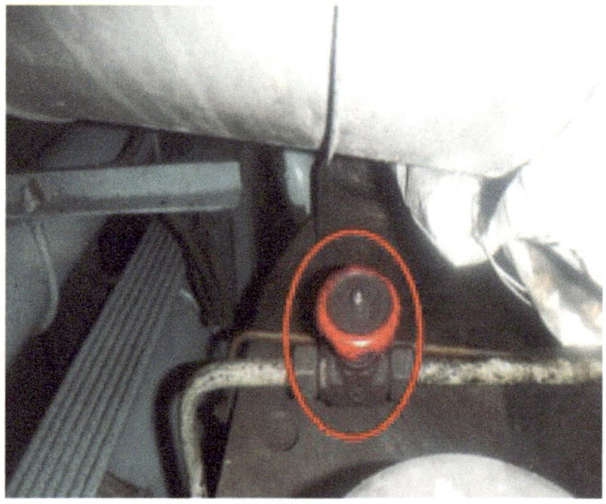

Jacket Water Drain Valve

Jacket Water Drain Position

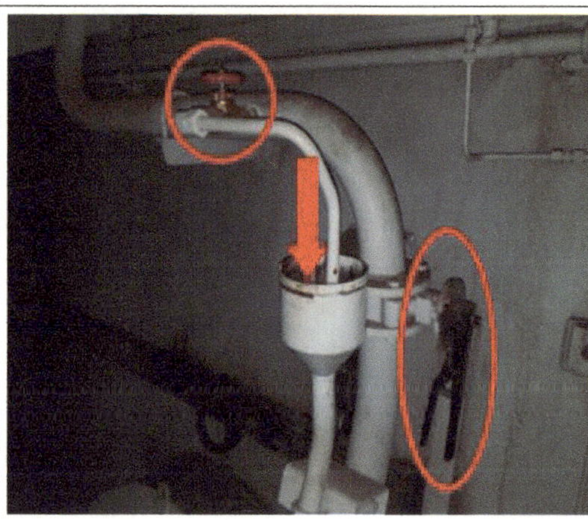

Drain Valve가 Blinding되었거나 Line상의 문제로 Drain불가 시 Inlet Valve 옆의 Drain Valve이용

㉠ EXH′ Valve의 J.C.F.W In & Out Valve(2EA) Close

㉡ Air Vent Line Valve(2EA) Close

㉢ CYL′ Water Drain Valve Open

㉣ CYL′ Venting Valve Open 하면 Drain 시작된다.

㉤ 만약에 Drain Valve가 Blinding 되었거나 Line상의 문제로 Drain불가 시 Inlet Valve 옆에 있는 Drain Valve로 Drain 한다.

(J.C.F.W In & Out Butterfly Valve Seat Ring 불량하면 Valve Close 제대로 되지 않아 전 기통 CYL′ Water Drain이 되므로 팽창 Tank Level이 줄어든다.

팽창 Tank가 HT/LT로 구분 되어있는 선박은 괜찮지만 통합된 Type의 팽창 Tank의 경우 Level 계속 줄어들면 G/E 운용에 지장을 초래하므로, Jacket Pump Stop하고 Pipe Line 재확인하여 관련 Valve들 Close 한다.

대게 팽창 Tank 출구 Valve(M/E Jacket Water로 가는 Valve,

Jacket Cooler In & Out Valve 및 By-pass Valve,

Jacket Pump Outlet Valve,

F.W.Generator(조수기) By-pass Valve Close 한 다음 다시 팽창 Tank Water 보충하여 Level을 올린 다음 CYL′ Water Drain을 한다.

팽창 Tank가 줄어 들지 않는다면 작업을 시작해도 된다.

Butterfly Valve Seat Ring 불량하면 정박 시간이 상대적으로 긴 항구에서 Line 차단 및 교환하고 예비품이 없다면 차후 선용품으로 청구한다.)

③ High Pressure Pipe Bolt를 느슨하게 풀어서 Actuator 배출구로 Pipe 속의 L.O가 배출되도록 Actuator로부터 Pipe 약간 들어올린다. (참고로 Drain Valve 없다.)

Overhead Crane과 Rope를 이용하여 옆 Cylinder High Pressure Pipe에 **'X'**모양으로 고정한다.

< HIGH PRESSURE PIPE >

< HYD′ JACK >

EXH´ Valve 연결된 Pipe들 전부 풀고 EXH´ Valve Expansion Joint에 위/아래 방향을 표시 및 취외하고 Expansion Joint는 Overhead Crane에 Chain Block과 Rope를 이용하여 취외한다.

(Expansion Joint를 반대로 조립 하면 배기 가스로 인해서 Expansion Joint 끝단이 손상 되거나 파손될 수 있고 파손된 부품이 T/C로 들어가서 T/C 손상 시킨다.

Expansion Joint 내부 확인하면 EXH´ Valve에서 나가는 부분이 더 넓고 EXH´ Manifold로 들어가는 부분이 더 좁다.)

④ E/R Overhead Crane 이용하여 HYD´ Jack을 EXH´ Valve Stud Nut에 설치한다.

HYD´ Jack 과 HYD´ Hose를 Pump에 연결한다.

(HYD´ Jack 을 Stud Nut에 체결하고 HYD´ Jack Venting Valve를 Open 상태에서 Round Nut가 Stud에 밀착(Clearance = 'X')이 되도록 Tight 한다.

Close 할 때 Venting Valve로 L.O가 비산하지 않도록 천으로 막아준 상태에서 Round Nut를 시계 방향으로 Tight하여 Stud에 완전히 밀착 시킨다.

상기 방법처럼 하지 않고 유압을 걸어주면 HYD´ Hose, HYD´ Jack에 문제 발생할 수 있다.)

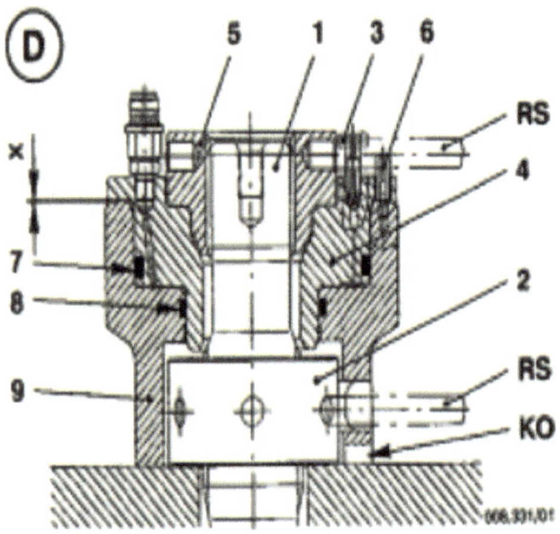

⇒ Clean the threads of waisted studs 1 of the valve cage as well as the area around nuts 2 (Fig. 'D').

⇒ Place hydr. tensioning device 94252 consisting of two pre-tensioning jacks on the two waisted studs 1 (Fig. 'D') which have to be loosened (arrangement according to Fig. 'E').

⇒ Open vent screw 3 of each pre-tensioning jack.

⇒ Screw round nuts 5 onto waisted studs 1 until there is no or very little clearance at 'x'.

⇒ Subsequently slacken round nuts by about 1½ turns (for loosening only).

< 이해 안되면 상기 그림 설명 참조 >

(간혹 유압을 걸어주는 것이 실린더를 눌러주는 것으로 생각할 수 있으나, 인장 Stud이기 때문에 Stud를 인장력으로 잡아 당겨주는 것이다. 그로 인하여 간극이 생기고 Nut를 풀 수가 있다.)

HYD´ Jack Round Nut를 Stud에 밀착 후 Venting 하였으면 Round Nut를 Full Tight에서 반 시계 방향으로 한 바퀴 반을 풀어준다.

(유압 원리를 이해하면 되며, **많이 풀거나 안 풀면 HYD′ Nut와 HYD′ Jack의 Piston이 서로 붙을 수 있다.)**

HYD′ Jack의 Venting Valve를 살짝 Open하고 HYD′ Pump에 Hose와 Jack을 연결시키고 유압을 올리면서 Venting Valve로 L.O가 나오는 것이 육안으로 확인되면 Venting Valve Close 한다.

유압을 **900~1000Bar**를 걸어준 다음 Stud Nut가 움직이면 한 바퀴 반 Open 한다.

그 다음 유압을 **500Bar**까지 낮추어 주고 Stud Nut를 조금 더 풀어준 다음에 유압 해제한다.

HYD′ Jack을 Overhead Crane으로 빼내고 Stud Nut를 다 풀어준다.

다시 말해, **두 번에 걸쳐 900~1000Bar, 500Bar로 유압을 걸고 해제**를 하면 HYD′ Jack과 Head Stud Nut가 달라붙는 일이 없다.

(유압을 걸 때 HYD′ Hose 에서 L.O 누설하지 않는지 확인한다.
만약에 L.O가 누설하면 HYD′ Hose 제대로 체결되지 않았거나 호스 불량한 것이니 교환한다.

참고로 원래 HYD′ Stud Nut 풀 때 압력이 체결 할 때 보다 더 크다.)

⑤ HYD′ Nut를 다 풀고 E/R Overhead Crane 이용하여 EXH′ Valve 상부에 장착된 Lifting Tool에 걸어 천천히 취외한다.

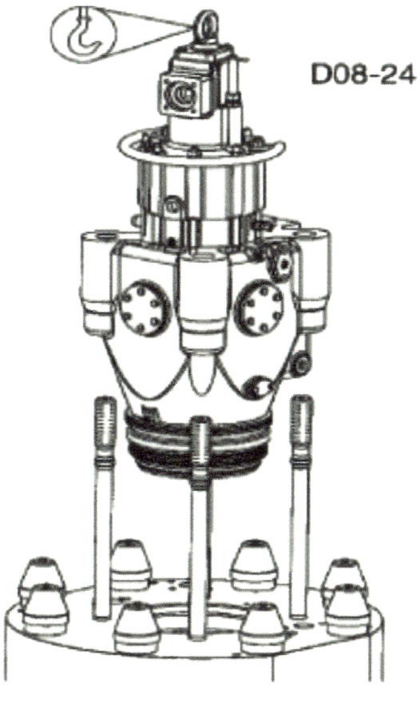

〈 EXH′ Valve Lifting 〉

취외하여 Engine Plate Side에 내려놓을 때는 보다 단단한 Frame인 장소에 내려 놓는다.

EXH′ Valve 장착된 안쪽(Cylinder Head) 소제한다.

취외한 Expansion Joint의 **Metal Gasket(91003-0152-325)** 빼내고 내부 소제한다.

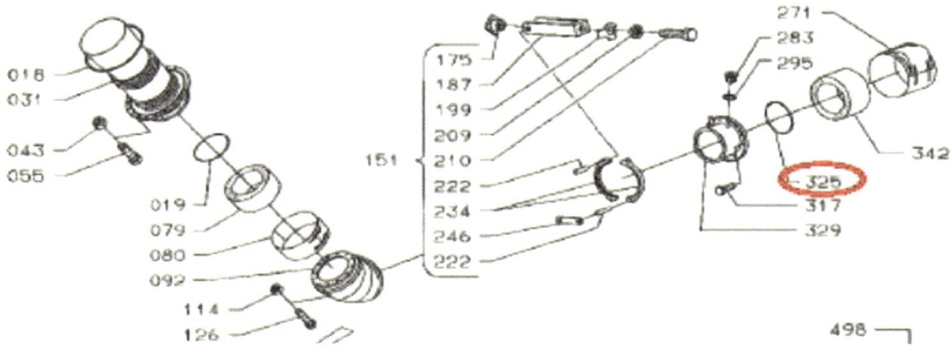

< Expansion Joint Metal Gasket >

소제 완료 후, Piston에 올라가서 Engine Turning 한다.

Engine Turning하면서 Cylinder 내부 Liner 상태 및 Cylinder Head Crack 유무를 확인한다.

(Cylinder Head Crack부 의심될 경우 Color Checker를 이용하여 Crack 유무를 확인한다.)

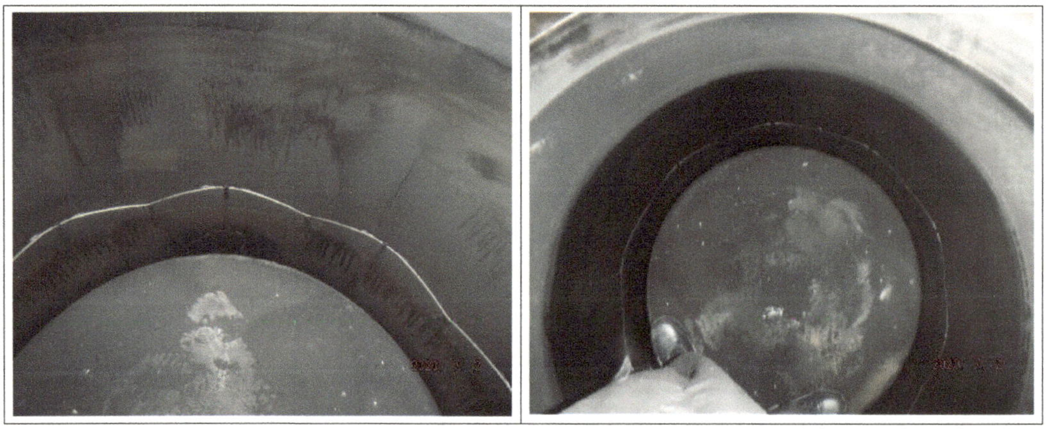

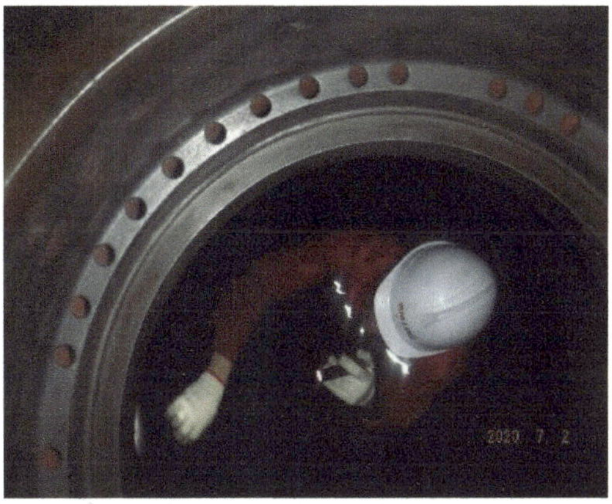

Engine Turning을 하면서 Piston 상부에서부터 내려가면서 Cylinder Head에 부착된 F.I.V 또는 Starting Air Valve의 상태 및 Cylinder Liner의 상태를 확인한다.

Cylinder Cover Tools
Grinding Tool : 90151-0204-9821

Grinding Tool 장착

⑥ Cylinder head & liner 상태를 확인했다면 상기 그림과 같이 Grinding Tool을 조립하여 Lapping을 한다.

Compound를 Cylinder Head 측 EXH' Valve 장착 부에(또는 Tool 하단 부에) 골고루 도포하고 Grinding Tool을 안착한다.

무게가 상당하니 반드시 Overhead Crane과 Rope를 이용하여 장착한다.

Lapping이 끝났다면 Tool을 취외하고 정비된 EXH' Valve 조립 준비를 한다.

⑦ 정비된 EXH' Valve을 조립하기 전에 O-ring(**90801-0214-074 & 086**)를 조립한다.

E/R Overhead Crane을 이용하여 EXH' Valve 정확히 Center를 맞춰서 조립한다. **(이때 주의할 점은 정확히 Center를 맞추는 것이 중요하며 Center 맞추면 서서히 조립한다.**
만약에 정확하게 Center를 맞추지 않고 조립하면 배기가스 및 LO 누설한다.)

EXH' Valve 제대로 조립하였다면Overhead Crane 이용하여 HYD' Jack을 Stud Nut에 조립하고 Venting Valve Open한다.

Round Nut 이용하여 HYD' Jack의 Piston을 Cylinder에 밀착되도록 체결하면서 Venting 한다.

HYD' Jack Venting Valve Open하고 Round Nut **Full Tight**한 뒤 HYD' Pump로 유압 형성한다.

Venting Valve로 L.O가 나오는 것을 확인한 뒤 Venting Valve Close 한다.

유압을 500BAR까지 올리면서 Stud Nut를 계속 천천히 시계방향으로 Full Tight 한다.

HYD' Pump 유압을 해제하고 900BAR까지 올려 Stud Nut Full Tight 한다.
(상기 언급한 유압 원리를 이해했다면 이때는 Full Tight 상태에서 유압을 걸어준다.)

EXH' Valve Thread Insert(**90801-0214-157**)상태 확인하여 불량 시 교환하고 E/R Overhead Crane 및 Rope이용하여 Expansion Joint 및 Metal Gasket(**91003-0152-325**) 조립한다.

< EXH' Valve Thread Insert : 90801-0214-157 >

* 배기 밸브 THREAD INSERT 교환 방법

Thread Insert & Tool

Thread Insert & Tool

Tool 사용하여 Thread Insert 교환

Tool 사용하여 Thread Insert 교환

Thread Insert 상태 불량 부분 반드시 교환한다.

만약에 불량 부분 교환하지 않는다면 Expansion Joint Bolt 조립이 제대로 되지 않아서 배기가스를 누설할 수 있다.

Tool이 없으면 상기의 사진 참조하여 제작한다.

(Expansion Joint 조립 시 원래 방향대로 조립하고 Metal Gasket 누락하면 Engine 시동 중에 EXH′ Gas 누설한다.
또한 Expansion Joint Bolt 조립 전에 Thread Hole과 정확히 Center 맞추고 Tight 한다.
일부 Bolt 제대로 조립이 안되면 Chain Block 이용하여 조금씩 Expansion Joint 좌우로 움직여 Bolt를 Thread Hole에 맞춘 다음 조립한다.)

⑧ 소제한 High Pressure Pipe Disc(**90806-0090-071**)의 O-ring(**90806-0090-083**) 교환한다.
(O-ring은 Disc에 2개씩 조립되며 Hydraulic Pipe Disc도 예비품 있다면 교환 정비한다.)

High Pressure Pipe 조립할 때 Pipe 아래 부분 먼저 Bolt 걸어주고, 남은 위 부분을 조립할 때는 Chain Block, Rope이용하여 아래로 당겨서 Bolt 조립하고 **210Nm**으로 골고루 Tight 한다.

취외한 Jacket Water Pipe Gasket 바꿔주고 관련 액세서리 전부 조립한 다음에 J.C.F.W Valve 원상 복귀하여 Head Water를 채운다.
(팽창 Tank가 Venting 역할 하지만 개인적으로는 J.C.F.W 입구 Valve Open, Venting Valve를 Open하여 Venting을 한번 더 확인한다.)

⑨ Spring Air Valve Open하고 Spring Pressure 복구 되면 Main L.O Pump를 기동한다.

고정된 EXH′ Valve Indicator 풀어내고 Main Engine Air-Blow 준비한다.

Main Engine Air-Blow하면서 EXH′ Valve Indicator가 위아래로 제대로 움직이는지 확인한다.

출항 후에 조립한 부분의 누설 유무, 작동상태 및 이상 소음 재확인한다.

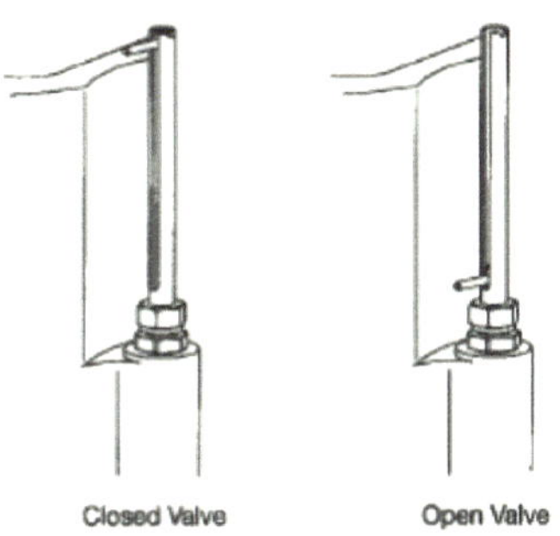

〈 EXH′ Valve Indicator 〉

1.

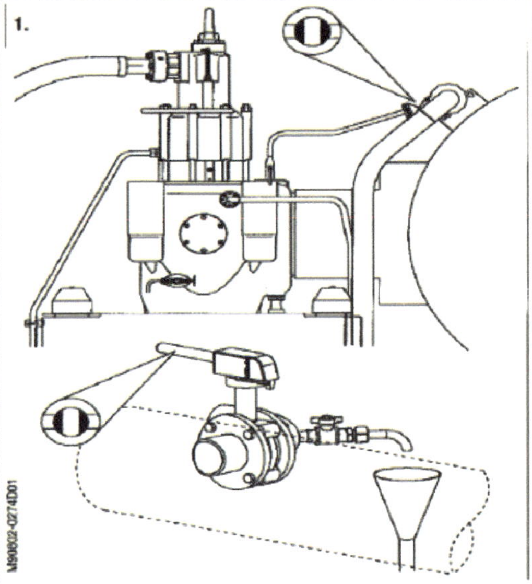

L.O Pump Stop
Cooling Water Inlet & Outlet Valve Close
Cooling Water Drain Valve Open
Spring Air Close

2.

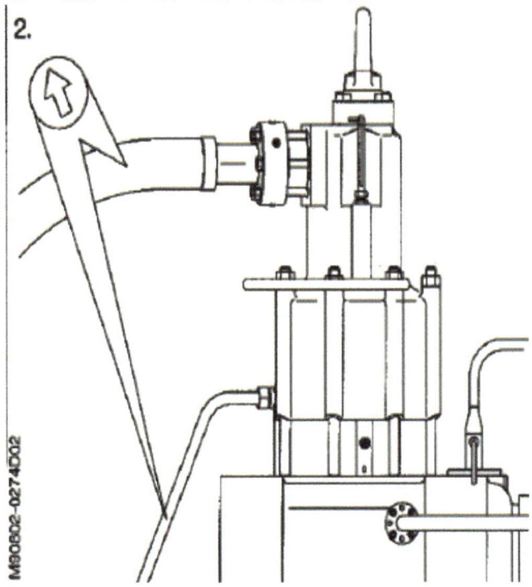

High Pressure Pipe & Return Oil Pipe 분해

3.

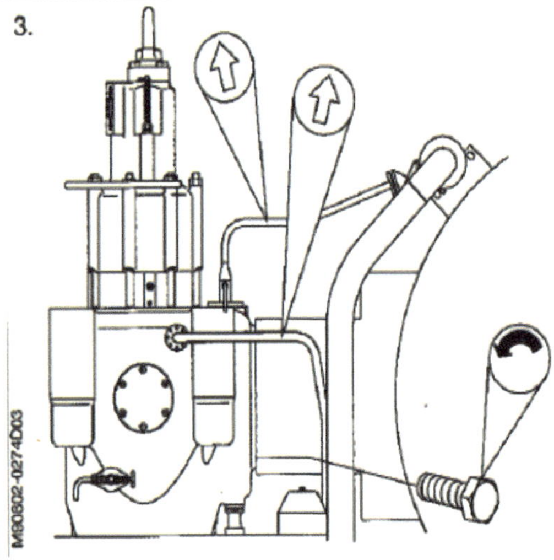

Cooling Water & Spring Air Pipe 분해하고
Overhead Crane, Chain block & Rope
이용하여 Expansion Joint 잡아준 상태에서
Expansion Joint Bolt 분해한다.
Metal Gasket 취외한다.

4.

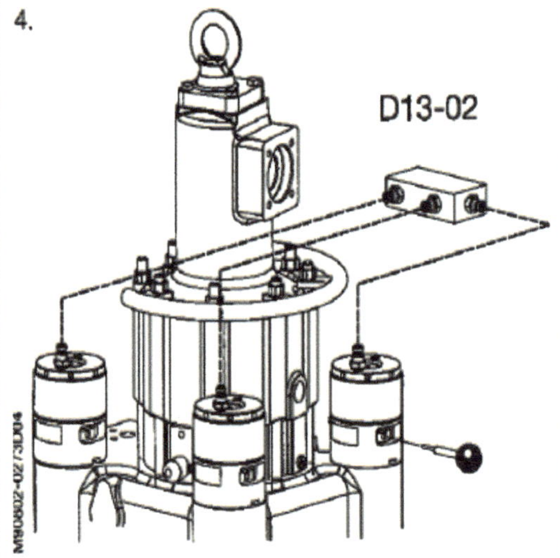

유압을 걸어 HYD′ Nut 분헤한다.
(950~1000Bar)

5.

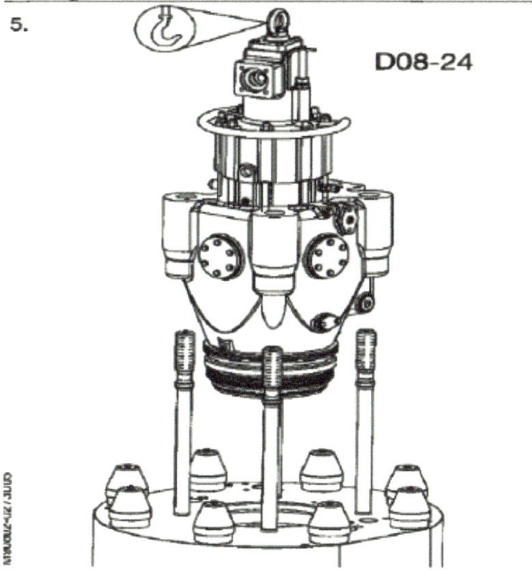

Overhead Crane을 Oil Cylinder 상부에 달린
Lifting Tool에 걸어 EXH´ Valve를 올린다.
Valve 분해하기 전에 주변 연결된 액세서리가
있는지 한번 더 확인한다.
Cylinder 내부 Brush소제한다.

2.

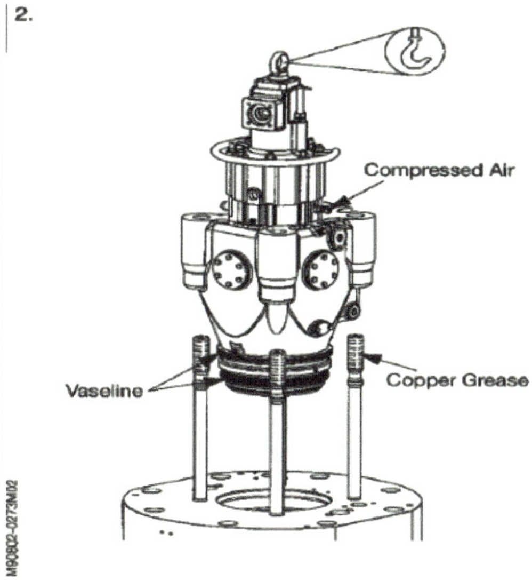

정비된 EXH´ Valve 올린 상태에서
Seat O-ring 교환하고 Compress Air 공급하여
Valve Spindle을 올린다.

3.

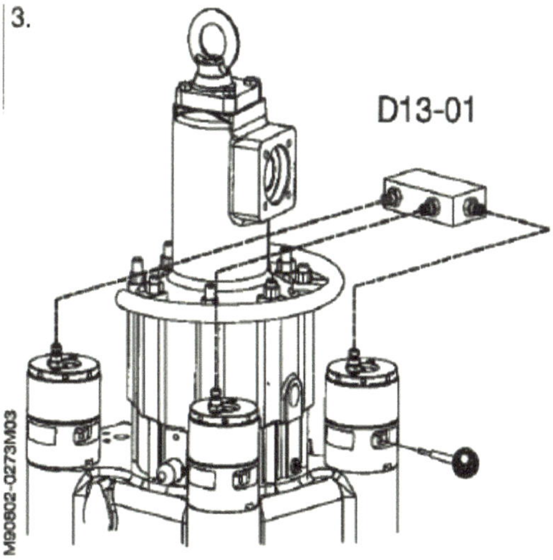

유압을 걸어 Nut Tight 한다.
(900Bar)

4.

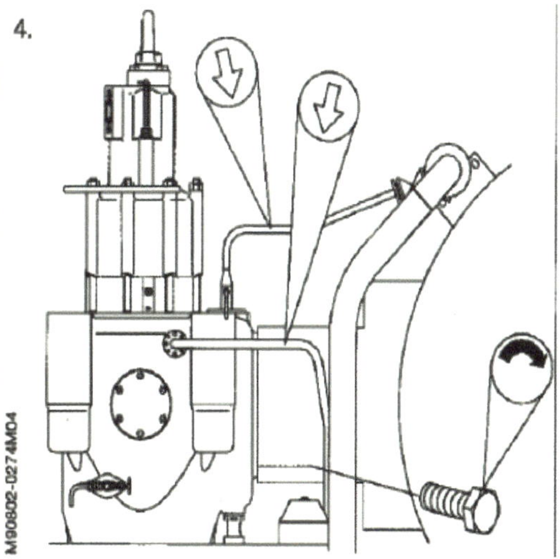

Cooling Water & Spring Air Pipe조립한다.
Overhead Crane, Chain block & Rope
이용하여 Expansion Joint 잡아준 상태에서
Metal Gasket 반드시 교환한다.
Expansion Joint Bolt Tight 하기 전
Thread Insert 확인하여 불량 시 교환한다.

5.

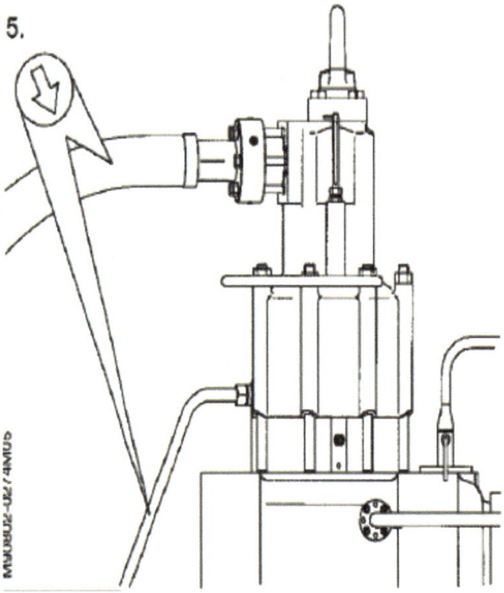

High Pressure Pipe Disc O-ring 교환 후
Pipe & Bolt 조립하고 210Nm Tight 한다.
Return Oil Pipe 조립한다.
(Disc 상태 불량 시 교환한다.)

6.

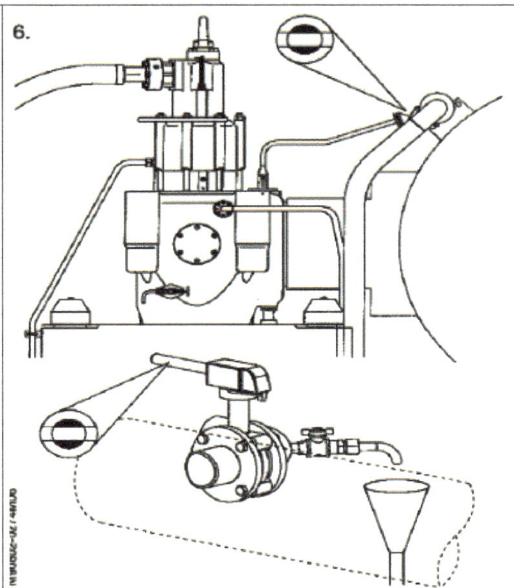

Cooling Water Drain Close
Cooling Water Inlet & Outlet Valve Open
Spring Air Open
L.O Pump Start

7.

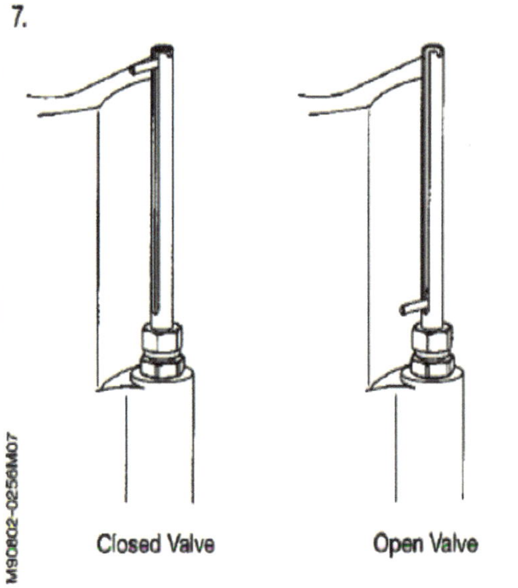

EXH′ Valve Indicator 고정 위치에서 분해하고
Main Engine Air Blow하여 움직임 확인한다.

특별하게 어려운 작업은 아니지만
1기사의 Overhead Crane 조작이 중요하며,
조립 시 정확히 Center를 맞추는 게
매우 중요하다. Expansion Joint 조립 전
반드시 Metal Gasket 교환하고
Thread Insert 불량 시 교환한다.
Expansion Joint와Thread Insert Hole이
정확히 맞은 상태에서 Bolt Tight 한다.
Bolt Tight 시 뻑뻑한 느낌이 드는데
억지로 조립 시 Bolt가 손상되면 일이 커지므로
다시 Center를 맞춘다.
Cooling Water Inlet & Outlet Valve Open하면
팽창 탱크로 자동 Venting 되지만,
EXH′ Valve Venting Valve 이용하여
Venting 재확인한다.

*** 참고**

① Metal Gasket(91003-0152-325) 장기간 교환하지 않은 경우 Exhaust gas누설

→ 누설 심하지 않으면 가까운 항구에서 정박 후 교환, 누설 심하면 Engine Stop 후 교환한다.

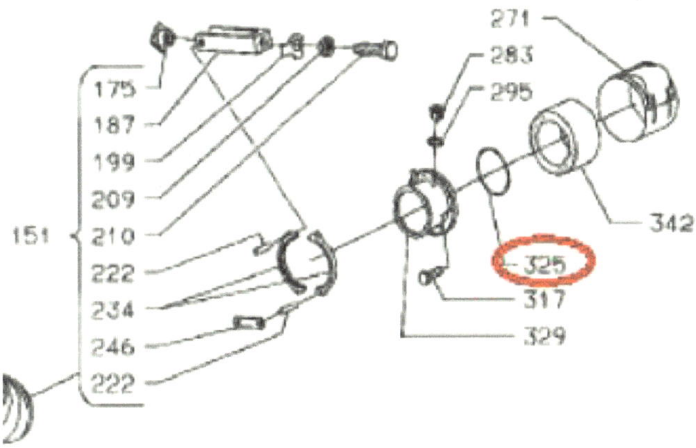

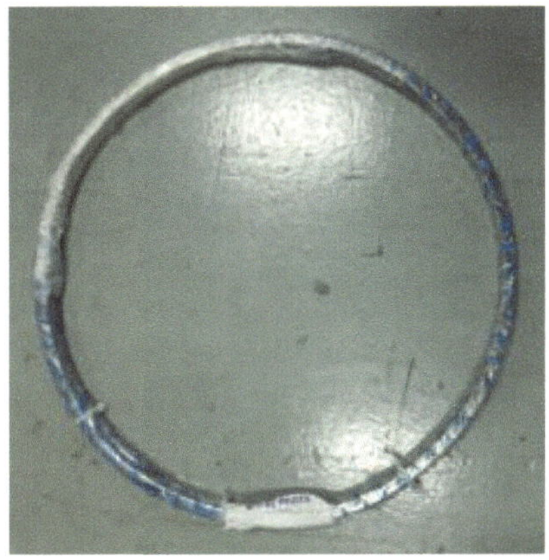

② Jacket Water Line Gasket 손상으로 Jacket Water 누설

→ Engine Stop 후 교환한다.

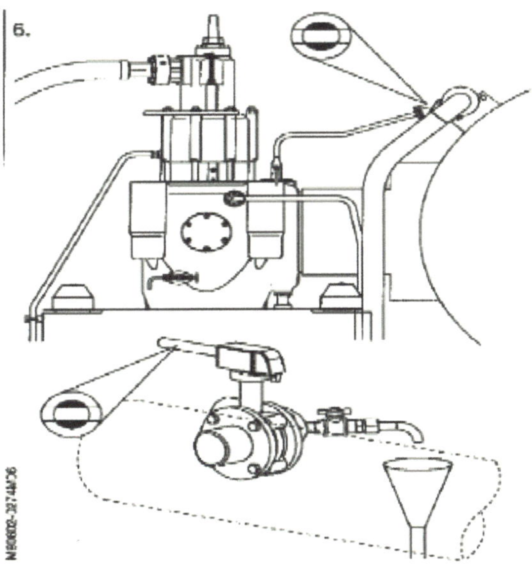

③ **EXH′ Valve Drain Nipple 손상으로 누설**

→ Engine Stop 후 교환한다.

(2) 소제 방법

① 취외한 EXH′ Valve는 분해하기에 앞서 소제를 먼저 시행한다.

Air Cylinder**(90803-0053-099)**하부 Non-return Valve**(99803-0053-528)**에 Screwdriver 이용하여 Air Drop 한다.

② EXH′ Valve Lifting Tool에 Overhead Crane을 걸어 들어 올리면 Spindle이 아래로 내려온다.

Spindle과 Seat 접촉면을 밀봉할 수 있는 Rubber Packing을 놓고 Seat Tray에 안착한다.

③ Water Line을 준비하여 Valve Housing 내부를 Spindle Guide 아래까지 물을 채운다.

약 하루 정도 물을 채우면 Valve Housing 내부의 카본이 쉽게 떨어지고 소제도 원활하다.

④ 내부 Water Drain후 Brush로 Valve Housing 내부 깨끗이 소제한다.

Spindle Guide 내부는 Brush 소제하지 않는다.

EXH′ Valve Seat 안착시킬 수 있는 Tray Spindle과 Seat사이 Rubber Packing을 넣고 Valve를 Tray에 안착한다.	EXH′ Valve 안착시킨 다음 Water Line 준비한다.

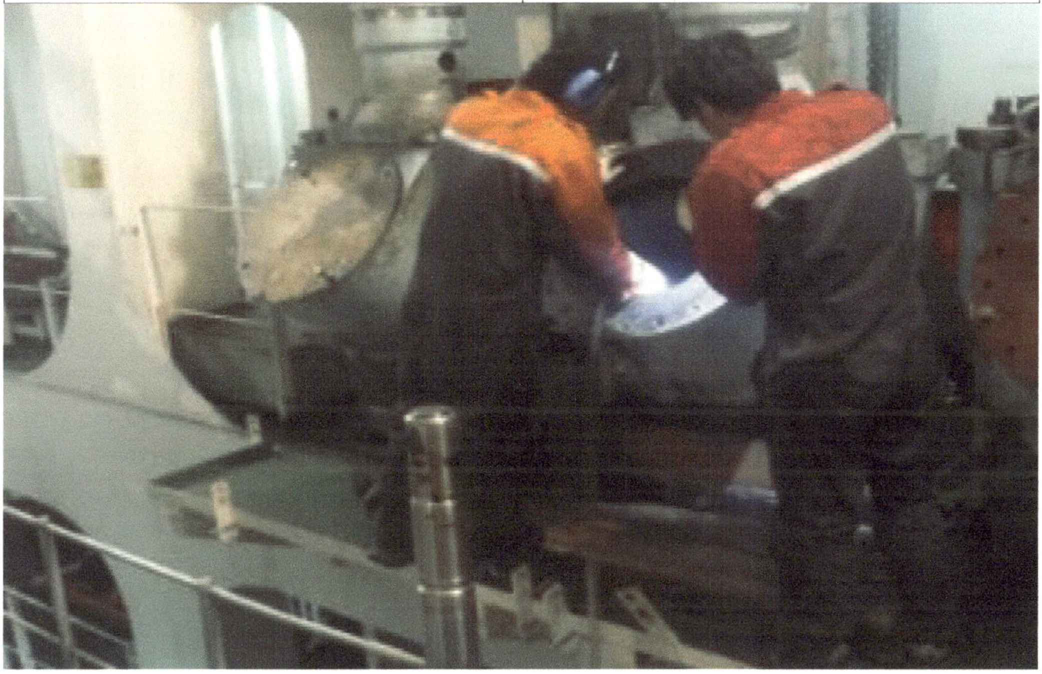

Water를 넘치지 않게 채운 다음 덮어준다.

(3) 분해 방법

① Oil Cylinder(**90803-0053-183**)의 Safety Handrail(**90803-0053-014**) & Nut(**90803-0053-278**) 제거한 뒤 Oil Cylinder를 취외한다.

② Air Cylinder(**90803-0053-099**)하부 Non-return Valve(**99803-0053-528**)에 Screwdriver

이용하여 Air Drop 한다.

③ Air Cylinder의 Flange(**90803-0053-301**), Screw(**90803-0053-291**)을 취외하고 Rubber Hammer, Zinc이용하여 Air Piston(**90803-0053-325**)를 아래방향으로 내린다.

Conical Locking Ring(**90803-0053-313**) 취외한다.

④ 취외한 Oil Cylinder를 다시 조립하고 EXH′ Valve Housing 올리면 Spindle (**90803-0214-289**)과 분리된다.

Seat Locking Screw(**90801-0214-050**) 풀고 Hammer, Zinc이용하여 Seat(**90801-0214-062**)를 분해한다.

⑤ Oil Cylinder, Air Piston, Air Cylinder순서로 취외한다.

⑥ Spindle Guide(**90801-0214-121**)내부 상태 확인 후 필요 시 교환한다.

⑦ 취외한 Spindle, Seat 마모도 계측하고 초과 및 불량하면 재생 수리 요청한다.

마모도가 초과하지 않거나 전반적으로 상태 양호 시 소제하여 사용한다.
(D08-13 Spindle, max. burn-off : 12.0mm
D08-11 Seat, max. grinding : 2.3mm / D08-12 Seat gap : 1.0mm)

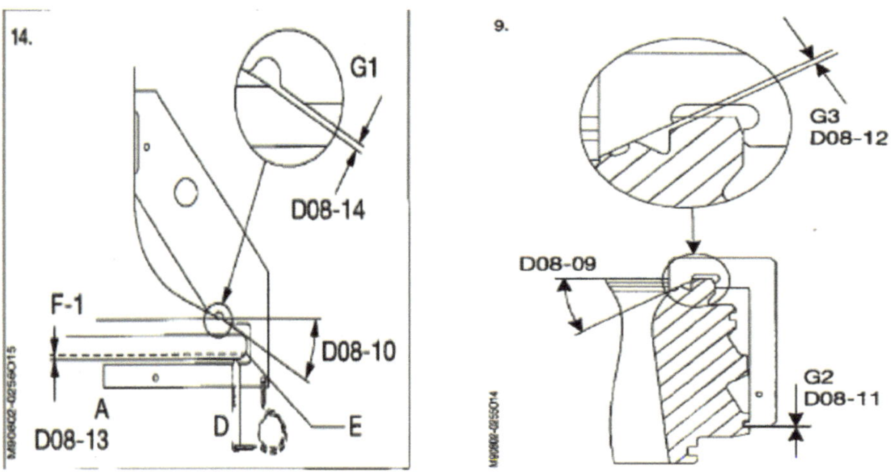

〈 Spindle & Seat 마모도 계측 〉

* 취외한 Seat 및 Spindle은 Lashing 한다.

⑧ 필요 시 Oil Cylinder 상부 Gasket(**90803-0053-230**) 교환한다.

Oil Cylinder를 눕혀놓고 Flange를 제거하고 Piston(**90803-0053-134**)를 분해한다.

Oil Cylinder 내부에 긁힌 자국 확인 하고 계측한다.

Piston Ring(**90803-0053-254**) 2EA 상태 확인하고 필요 시 교환한다.

⑨ 취외한 Part O-ring 및 Sealing Ring 전부 빼내고 내부 소제하여 조립 준비한다.

(Spindle Guide 내부는 브러쉬 소제하지 않는다.)

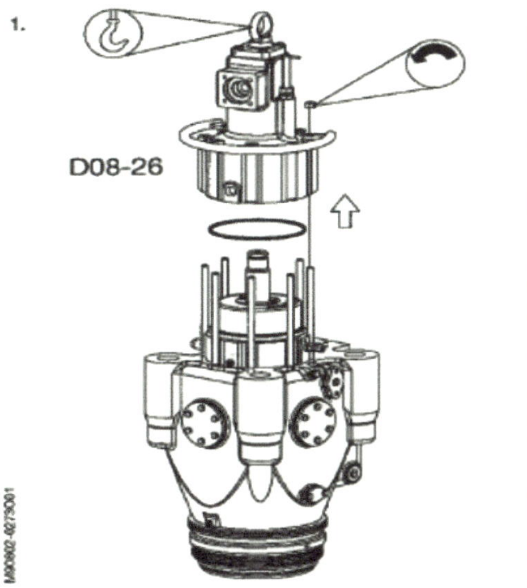

Oil Cylinder의 Safety Strap과 Nut 제거한 뒤 Oil Cylinder 취외한다.

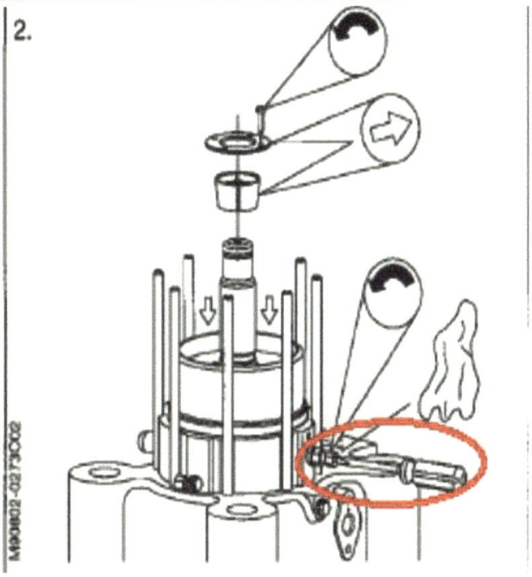

Non-return Valve에서 Screw Driver 이용하여 Pressure Drop한다.

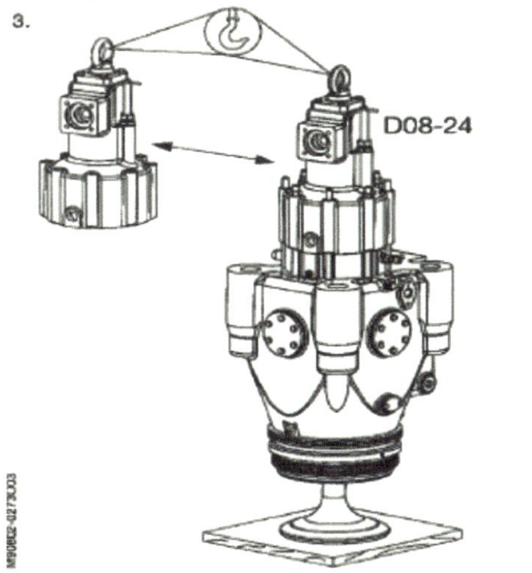

Air Cylinder Flange, Screw 분해 후 Air Piston을 아래로 내려 Conical Locking Ring 분해한다. Oil Cylinder 다시 조립하고 Valve Housing 올리면 Spindle이 분해된다.

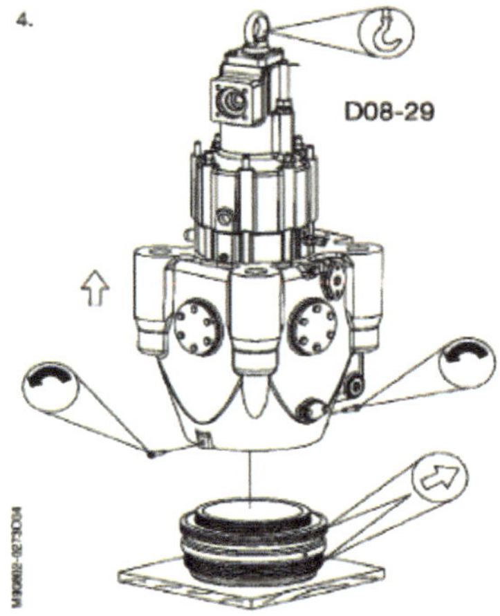

Seat Locking Screw 풀고 Hammer, Zinc이용하여 Seat 분해한다.

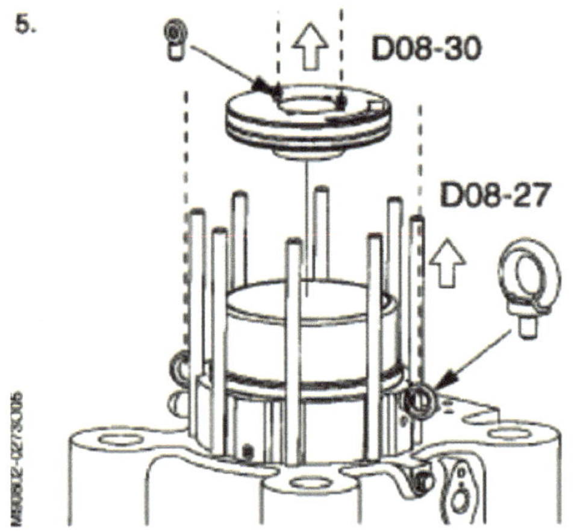

Oil Cylinder, Air Piston, Air Cylinder순서로 취외한다.

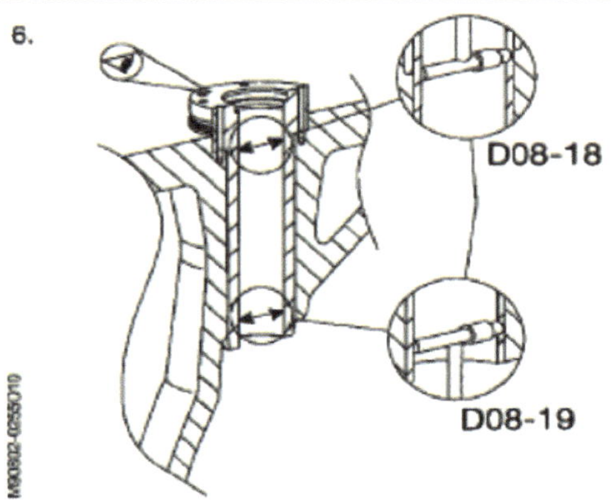

Spindle Guide(90801-0214-121)내부 확인하고 필요 시 교환한다.

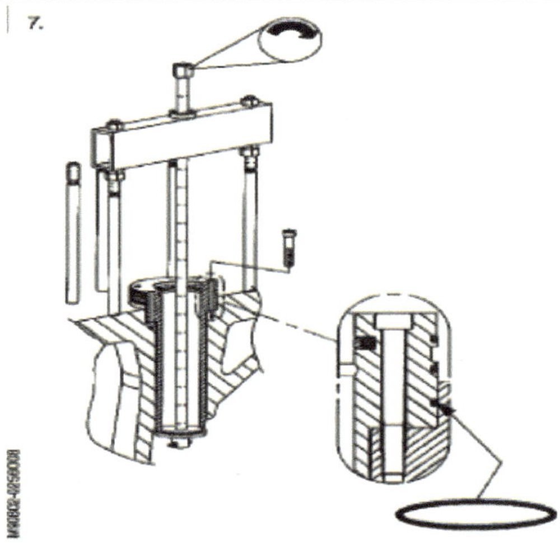

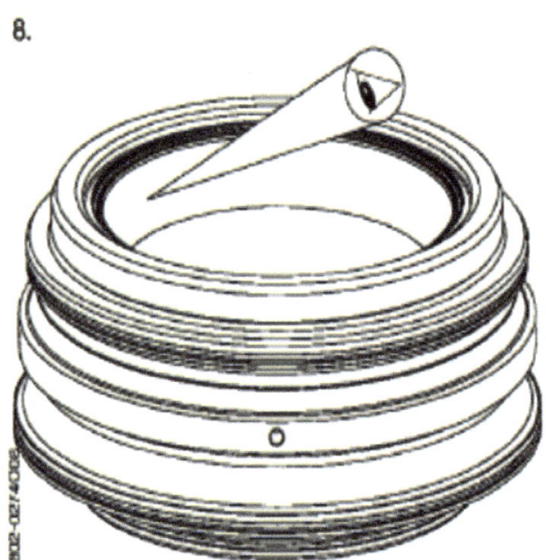

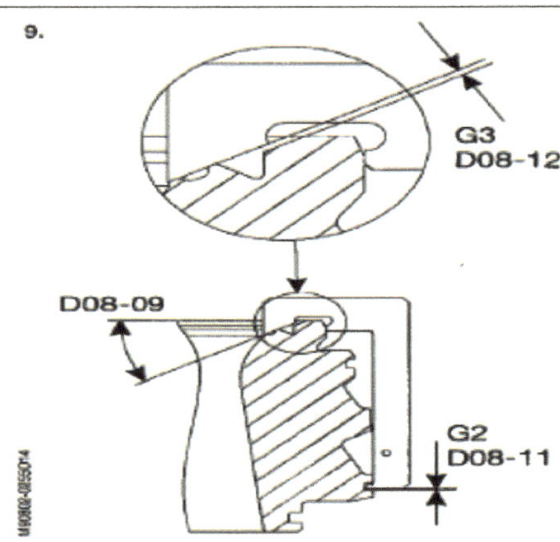

D08-11 : 2.3mm, D08-12: 1.0mm

Seat Measuring Tool : 90851-0196-9641

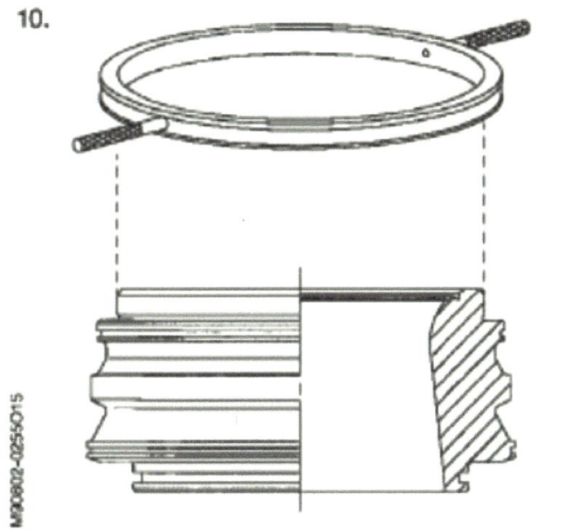

Spindle Lifting Tool : 90851-0196-9621

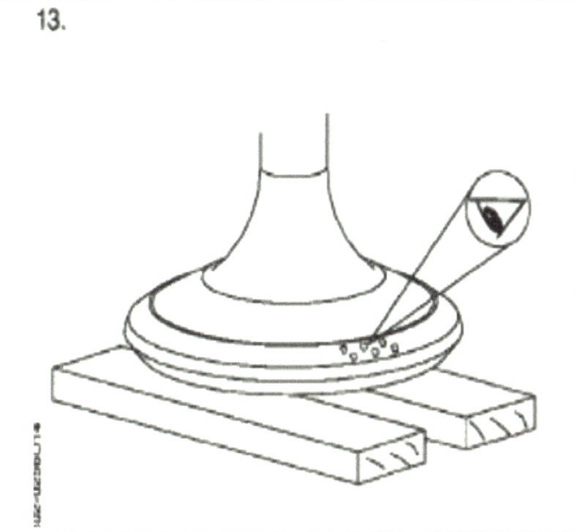

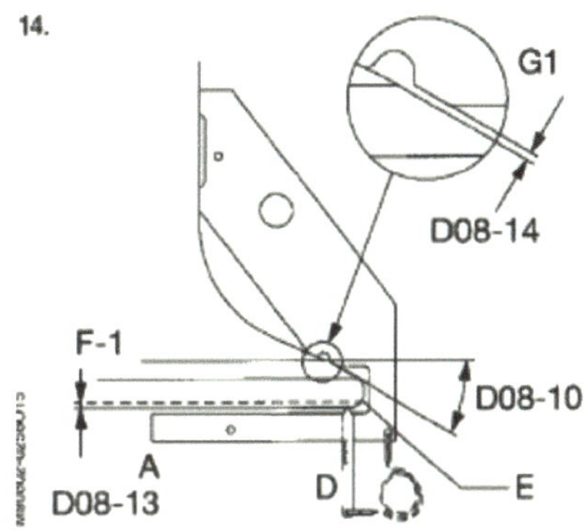

D08-13 : 12.0mm

Spindle Measuring Tool : 90851-0196-9640

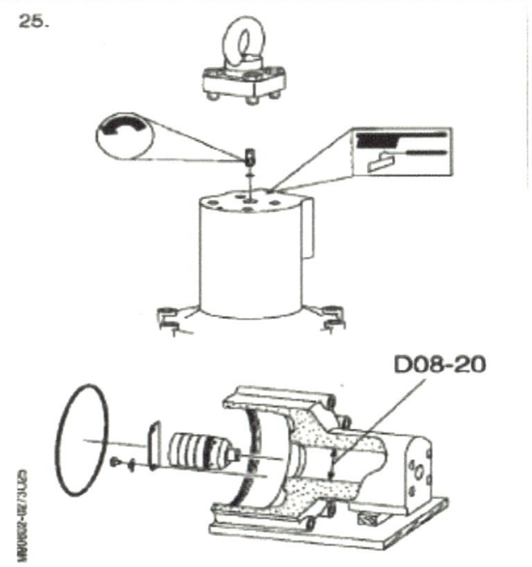

Oil Cylinder를 눕혀 Flange 제거 및 Piston 분해
Oil Cylinder 내부 상태 확인 및 계측한다.

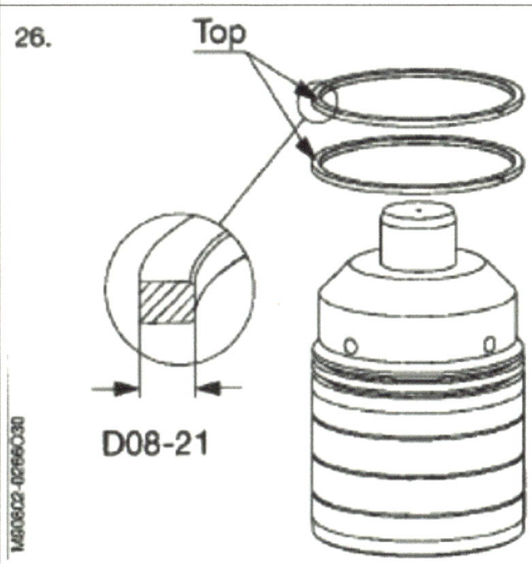

Piston Ring 상태 불량 시 신품 교환하고
Piston Ring Top Mark 위가 되도록 조립한다.

(4) 조립 방법

① Spindle Guide(**90801-0214-121**)의 O-ring(**90801-0214-133**) 및 Sealing Ring (**90801-0214-108**) 조립한다.

(Spindle Guide 안쪽 Sealing Ring 조립할 때 손상되지 않도록 주의하고 Spindle Guide 바깥쪽 L.O Line으로 Compressor Air를 넣어서 Spindle Guide 안쪽으로 Air 나오는지 확인한다.

그 부분이 Valve Spindle L.O Hole이다.)

② Housing에 조립되는 Packing(**90801-0214-013 & 169 & 204 & 228 & 265**) 교환한다.

또한 Thread Insert(**90801-0214-157**) 상태 불량 부분 교환한다.

③ Oil Cylinder(**90803-0053-183**)를 Overhead Crane 이용하여 Housing(**90801-0214-049**)에 조립 하여 들어 올리고 조립할 Seat(**90801-0214-062**)에 Valve Housing 안착한다.

그 후 Seat Screw(**90801-0214-050**) Tight 한다.
(Seat에 적힌 Number 기록하여 이력 관리한다.)

④ Valve Spindle(**90801-0214-289**)와 Valve Housing Spindle Guide에 L.O를 골고루 도포하여 부드럽게 조립될 수 있도록 한다.

Seat 조립된 Housing을 Overhead Crane 이용하여 다시 들어올리고 Valve Spindle 조립한다.

그 후 Oil Cylinder 다시 취외한다.
(Spindle Guide 안으로 Valve Spindle 조립 시 주의하고 Spindle Guide 내부 Sealing Ring이 손상되지 않았는지 확인한다. 또한Valve Spindle에 적힌 Number 기록하여 이력 관리한다.)

⑤ Air Cylinder(**90803-0053-099**) 아래에 있는 Safety Valve(**90804-0018-045**) 분해한 다음 내부의 O-ring(**90803-0053-033**) 및 Gasket(**90803-0053-021**) 교환한다.

압력 시험 장치에 Safety Valve를 놓고 시험 장치와 고압 펌프를 Hose로 연결한다.

Data를 참고하여 Opening Pressure 점검하고 Opening Pressure 맞지 않으면 Screwdriver 사용하여 잠겨있는 Nut를 풀고 Pressure 값을 Data참고하여 조정한다.

Safety Valve를 Air Cylinder에 단단히 조립하고 보호용 캡을 조립한다.

⑥ Air Cylinder 내부 상태 양호 시 Spindle Guide O-ring 유의하면서 Air Cylinder를 조립한다.

Air Cylinder Sealing Ring(**90803-0053-087**) 조립한다.

⑦ Air Piston(**90803-0053-325**)의 Sealing Ring(**90803-0053-051 & 075**)과 Guide Bearing(**90803-0053-038**)을 조립한다.

Instruction Book에 100℃ 물에 5분 동안 넣거나 Tool(**90851-0196-908**)과 Conical Locking Ring을 사용하라 적혀 있으나, 일반적으로 사용하는 노란색 노끈을 이용하면 쉽게 조립할 수 있다.

Sealing Ring에 노끈을 연결하여 상하좌우 방향으로 당기면서 조립하거나, 한 쪽 노끈을 연결하여 집어넣고 나머지 부분에 노끈을 하나 더 연결하여 시계 또는 반 시계방향으로 당기면서 조립하면 쉽게 조립할 수 있다.
(Sealing Ring 조립 시 손상되지 않도록 주의!)

⑧ Air Cylinder Drain Hole까지 L.O를 채우고 Air Cylinder 내부 벽면에 L.O를 도포한다.

Air Cylinder 상부에 Conical Guide Ring 조립 후 내부 벽면에 L.O를 도포하고 나무 또는 Zinc로 Air Piston을 부드럽게 조립한다.
(Air Piston 부드럽게 조립 되지 않으면 Screwdriver 이용하여 Non-return Valve를 눌러서 내부 압축 공기를 빼면서 조립한다.)

Conical Locking Ring(**90803-0053-313**)을 넣고(**Spindle을 잡아주는 방향으로**) Compress air 공급하여 Air Piston과 Conical Locking Ring을 올린다.

Conical Locking Ring을 Valve Spindle 홈에 끼우고 Air Piston를 Air Cylinder와 수평 되게 하고 Conical Guide Ring을 취외한다.

Flange Screw(**90803-0053-301 & 291**)을 조립하고 Tight 한다.

⑨ 정비된 Oil Cylinder를 조립한다.

Safety Handrail Nut(**90803-0053-014 & 278**)를 조립하고 340Nm으로Tight 한다.

*** Copper Packing 예비품이 없다면 Gas Torch로 가열 후 상온 냉각하면 재사용 가능하다. 또한 Thread Insert 교환은 앞서 언급한 교환 방법을 참조하여 상태 불량 시 교환한다.**

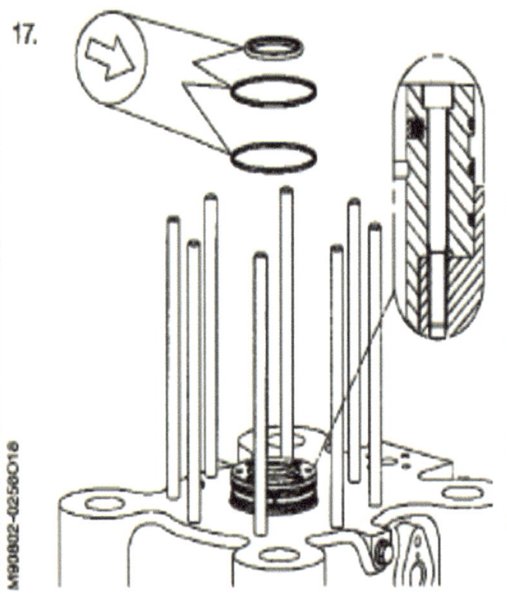

Valve Housing의 Spindle Guide O-ring 및 Sealing Ring을 조립한다.

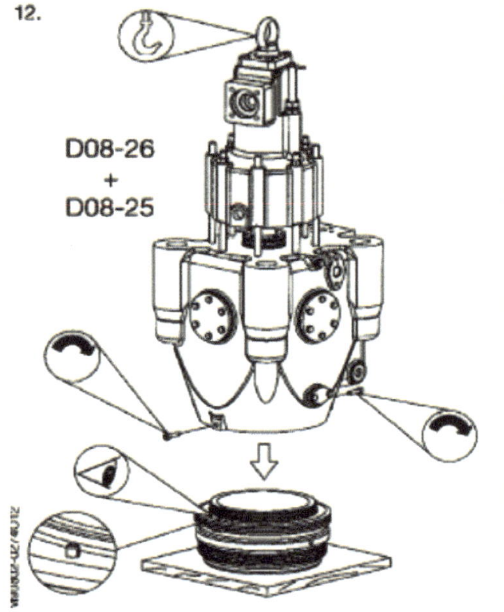

Oil Cylinder 조립하고 Valve Housing을 Valve Seat에 안착 시킨다.

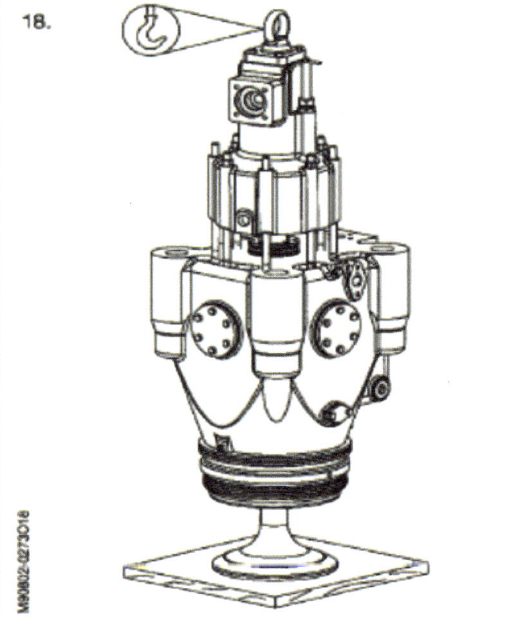

Valve Spindle과 Spindle Guide에 L.O를 충분히 도포하여 부드럽게 조립되도록 한다.

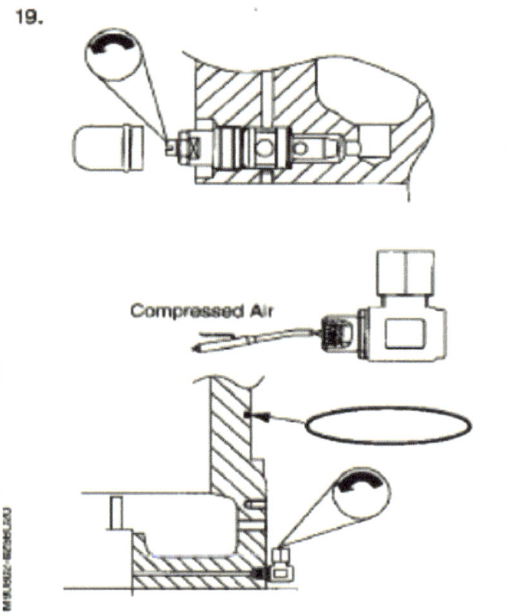

Safety Valve 분해하여 O-ring & Gasket 교환한다.

20.

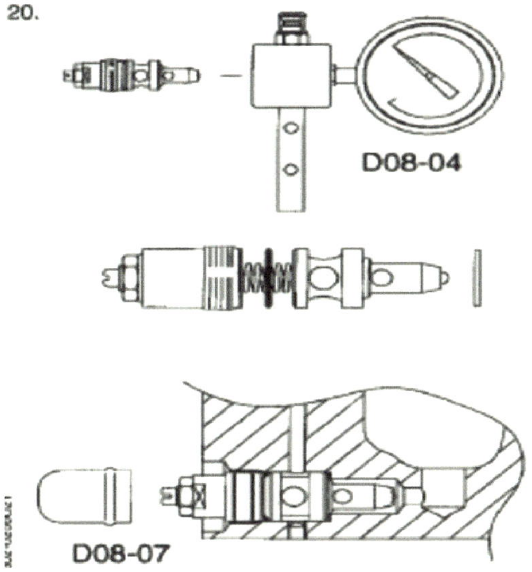

분해한 Safety Valve Test 및 압력 조정 후 조립한다.

21.

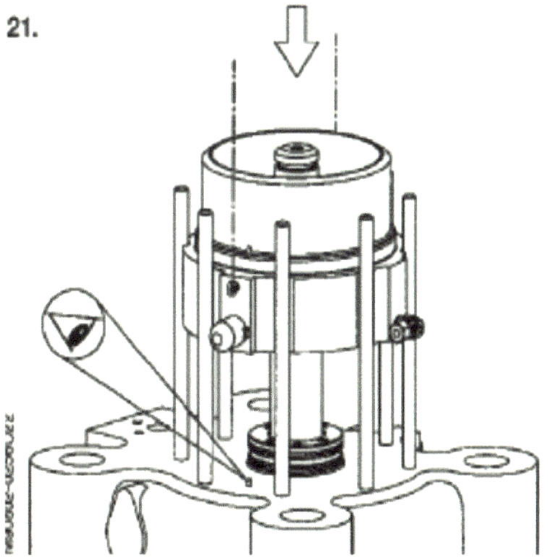

Oil Cylinder 제거하고 Air Cylinder 조립한다.
(Engine에 따라 Guide Pin 없는 경우도 있음)
Air Cylinder O-ring 조립한다.

22.

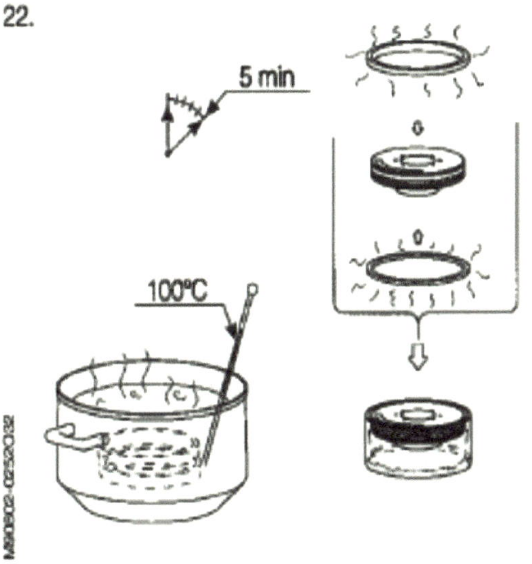

Air Piston Sealing Ring을 조립한다.
(노란색 노끈을 이용하는 것이 훨씬 수월)

23.

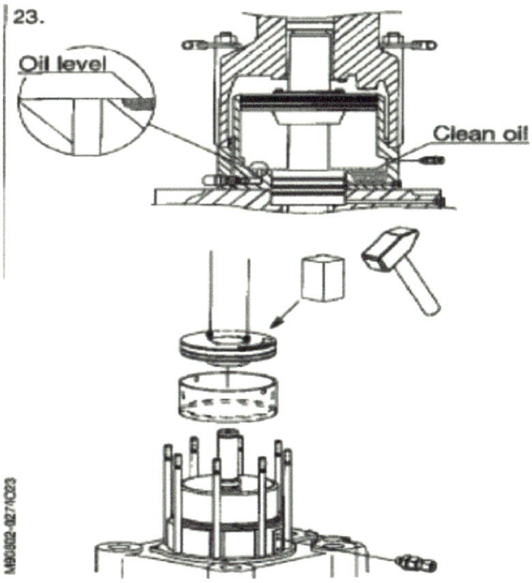

Air Cylinder Drain Hole까지 L.O를 채우고 내부 벽면에도 L.O를 도포한다.
Air Cylinder 상부에 Conical Guide Ring을 조립하고 내부 벽면에 L.O를 도포한다.
나무 또는 Zinc 이용하여 Air Piston 조립한다.

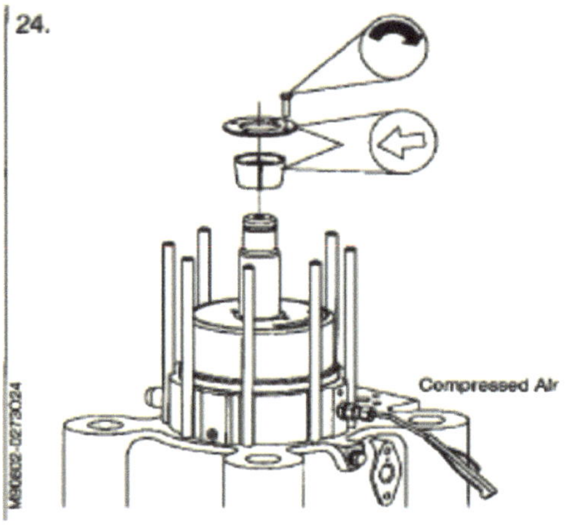 Non-return Valve에Compress Air 공급하여 Air Piston과 Conical Locking Ring을 올려 Conical Locking Ring은 Valve Spindle홈에, Air Piston은 Air Cylinder 상부와 수평이 되도록 하고 Conical Guide Ring 취외한다. Spindle, Flange, Screw 조립 및 Tight 한다.	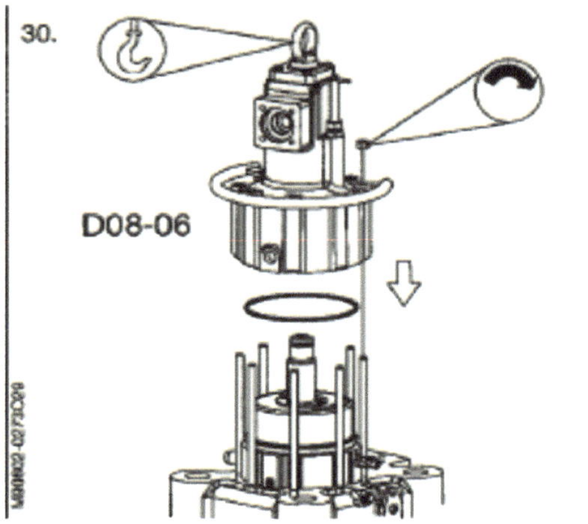 정비된 Oil Cylinder 조립하고 Safety Handrail & Nut를 조립하고 340Nm으로Tight 한다.

(5) TEST 작업

① Overhead Crane과 EXH′ Valve Lifting Tool이용하여 EXH′ Valve 올린 다음 Air Line 준비한다.

② 아래 그림과 같이 Gauge가 장착된 Test Tool이 있는 경우가 많으나 없다면 그림과 같이 제작한다.

< EXH′ VALVE TEST TOOL >

< TEST SPINDLE >

Compress Air를 Non-return Valve로 공급하여 Spindle이 제대로 작동하는지 확인한다.

Compress Air를 넣으면 Spindle이 올라가고 Air를 빼면 Spindle이 천천히 내려온다.

EXH' Valve 올린 상태에서 Air를 넣어 Gauge 상으로 압력이 형성되면 Air Valve를 Close한 상태에서 최소 15분 이상 Spindle이 내려오지 않는지 확인한다.

Spindle이 만약에 내려온다면 내부 O-ring 및 Piston Ring의 조립이 잘못 된 것이다.

(중요한 점은 Spindle과 Valve Cage 터치 부분이나 안쪽을 확인한다고 손을 넣는 행동은 절대 하지 않도록 한다. 육안으로만 확인하면 되며 손으로 확인하는 도중에 누군가 실수 또는 의사소통의 문제로 인해 Air Valve를 작동시켜서 Spindle이 움직이면 큰 인명사고가 발생할 수 있다. 이와 같은 사고 사례가 빈번하니 반드시 주의한다.)

③ Test가 제대로 되었다면 보관하는 곳에 단단히 안착한다.

10) 윤활유 고압관

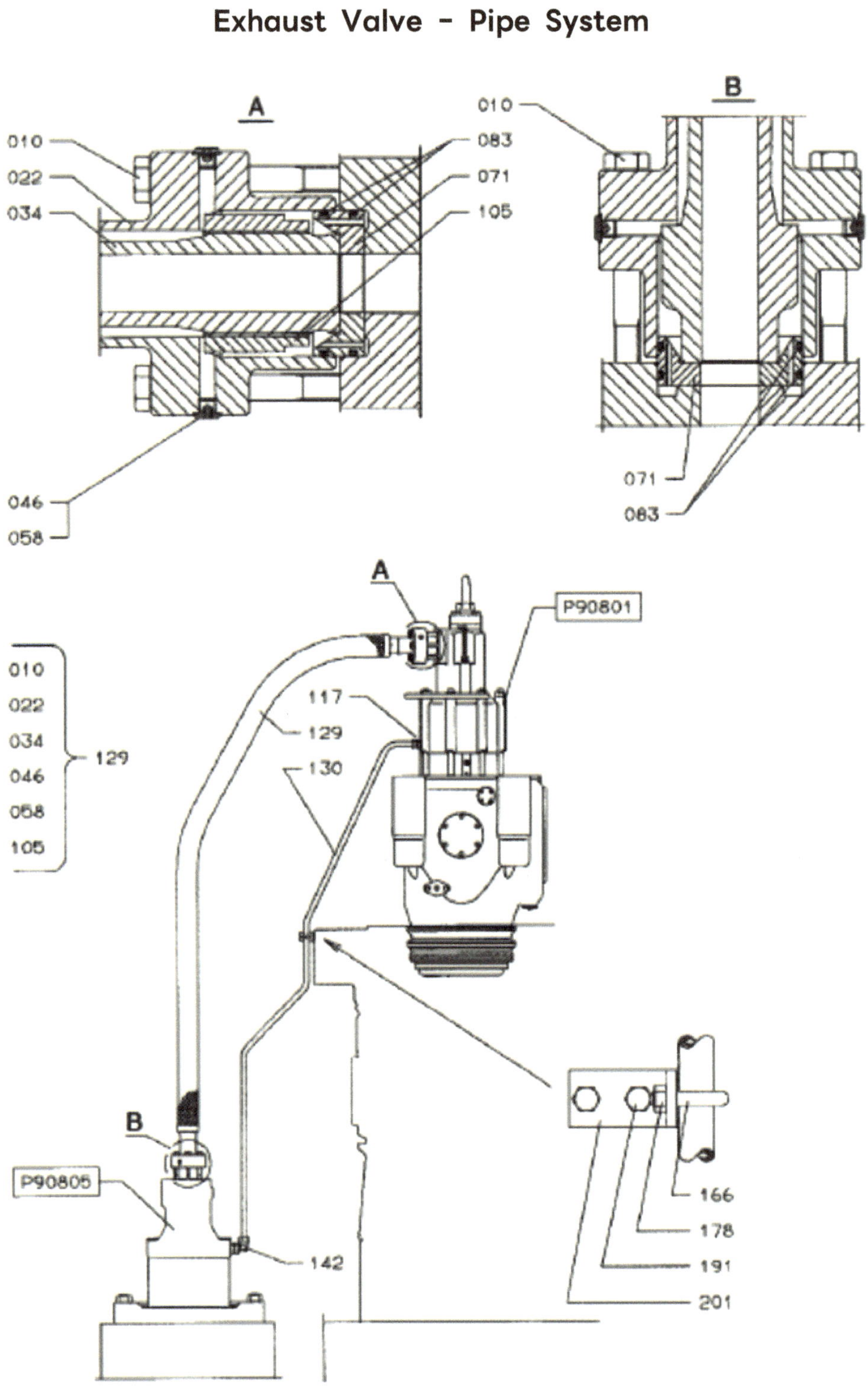

Exhaust Valve - Pipe System
A
010
022
034
083
071
105
046
058
B
010
071
083
A
P90801
010
022
034
046
058
105
129
117
129
130
B
P90805
142
166
178
191
201

High - Pressure Pipe

SAFETY PRECAUTIONS *For detailed sketch, see 900-2*

X	Stopped engine
X	Shut off starting air supply – *At starting air receiver*
X	Block the main starting valve
X	Shut off starting air distributor/distributing system supply
X	Shut off safety air supply – *Not ME engines*
X	Shut off control air supply
X	Shut off air supply to exhaust valve – *Only with stopped lubricating oil pumps*
X	Engage turning gear
	Shut off cooling water
	Shut off fuel oil
X	Stop lubricating oil supply
	Lock the turbocharger rotors

Data

Ref.	Description	Value	Unit
D08-01	High-pressure pipe, tightening torque	210	Nm
D08-02	Distance of high-pressure pipe lower edge to pressure flange	24	mm
D08-03	High-pressure pipe	105	kg

① 교환 방법은 앞서 설명한 EXH' Valve에 포함된다.

High Pressure Pipe Bolt를 느슨하게 풀어서 Actuator 배출구로 Pipe 속 L.O가 배출되도록 Actuator로부터 Pipe 약간 들어올린다.**(참고로 Drain Valve 없다.)**

② High Pressure Pipe 접촉면 점검하여 이상 있으면 핸드 그라인더로 Pipe의 끝 단 연마한다.

처음엔 Carborundum No.200 과 같이 거친 연마제로 연마하고 마지막엔 No.500 으로 세밀하게 마무리한다.

③ Trust Bushing(**90806-0090-105**) 모서리에서 Pipe 상부 끝 단의 접촉면까지 거리를 측정한다.

(D08-02 : 24mm)

④ High Pressure Pipe 위 아래 Disc(**90806-0090-071**)의 O-ring(**90806-0090-083**)을 교환한다.
만약 Disc의 상태가 불량하다면 Disc도 교환한다.

⑤ High Pressure Pipe 조립할 때 Pipe 아래 부분 먼저 Bolt 걸어주고 남은 위 부분을 조립할 땐 Chain Block, Rope이용하여 아래로 당겨서 Bolt 조립하고 210Nm

으로 Tight 한다.

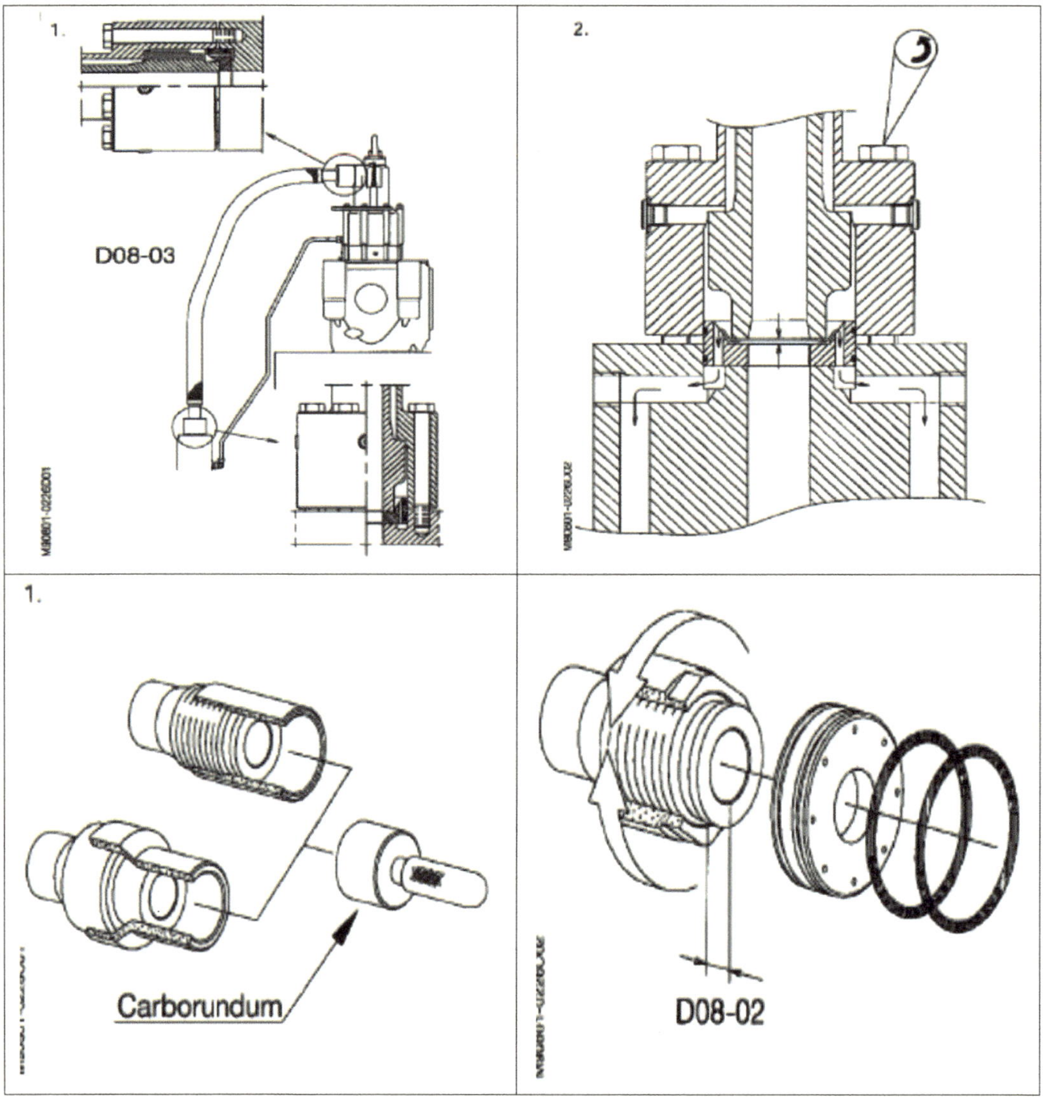

11) 연료유 오토 바이패스 인디케이터 필터

* 좌측 바이패스 필터 / 우측 오토 백워싱 필터

< Ratchet Pawl, Spring, Split Pin & Air Cylinder Joint >

참고로 Ratchet Pawl는 양측에 존재하며 사진상 반대편 아래쪽에 하나 더 있다.

Cylinder Joint 이용하여 Stud 작동길이를 조절할 수 있다.

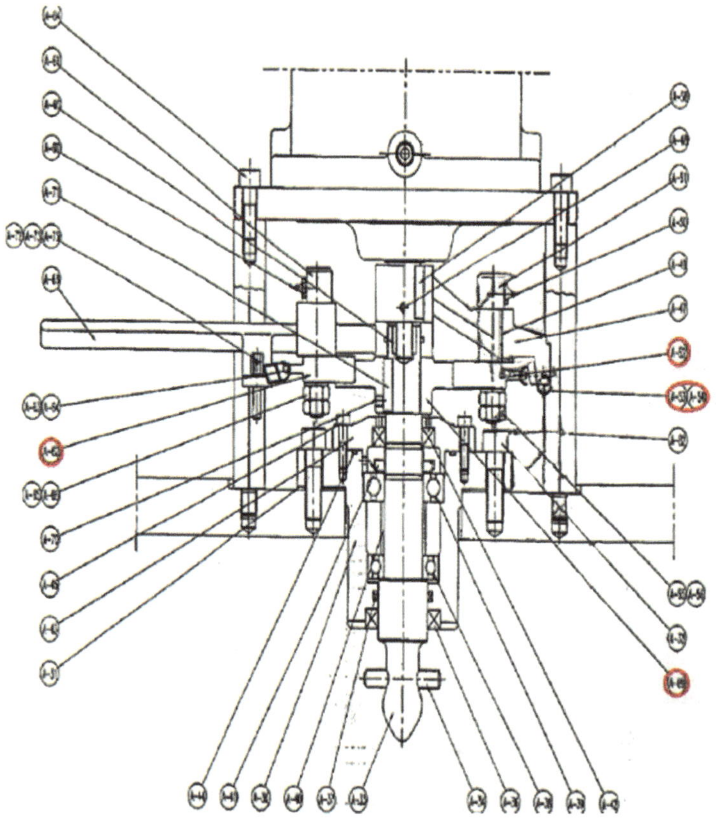

A-84	SOCKET BOLT	SCM435	4	
A-82	ACT. STAND	SS400	4	
A-75	STOPPER PIN	SS400	1	
A-73	CHAIN	SS400	1	
A-72	SPLIT PIN	SWRM17	3	
A-71	KEY	S45C	1	
A-70	SET SCREW	SCM435	1	
A-69	RATCHET WHEEL	S45C	1	
A-68	MANUAL HANDLE	FCD400	1	
A-66	CASTLE NUT	SS400	2	
A-65	PLAIN WASHER	SS400	1	
A-64	SPLIT PIN	SWRM17	1	
A-63	SPRING	SUS304	1	
A-62	RATCHET PAWL	S45C	1	
A-61	SET PIN	SCM435	1	
A-60	GREASE NIPPLE	BsBM2	2	
A-58	KEY	S45C	1	
A-56	CASTLE NUT	SS400	2	
A-55	PLAIN WASHER	SS400	1	
A-54	SPLIT PIN	SWRM17	1	
A-53	SPRING	SUS304	1	
A-52	RATCHET PAWL	S45C	1	
A-51	SET PIN	SCM435	1	
A-50	GREASE NIPPLE	BsBM2	1	
A-49	NEEDLE BEARING	SUJ2	3	
A-48	SET SCREW	SCM435	1	
A-47	AUTO LEVER	S45C	1	
A-46	OILLESS METAL	[illegible]	1	
A-45	SOCKET BOLT	SCM435	4	
A-44	O-RING	NBR	1	
A-43	COVER	FCD400	1	
A-42	OIL SEAL	NBR	1	
A-41	U-NUT	S45C	1	
A-40	COLLAR	SGP	1	
A-39	BEARING	SUJ2	1	
A-38	BEARING	SUJ2	1	
A-37	X-RING	NBR	1	
A-36	OIL SEAL	NBR	1	
A-34	PARALLEL PIN	S45C	1	
A-33	SHAFT	SCM435	1	
A-32	SOCKET BOLT	SCM435	4	
A-31	GESKET	ASB.FREE	1	
A-30	BEARING BOX	FC250	1	

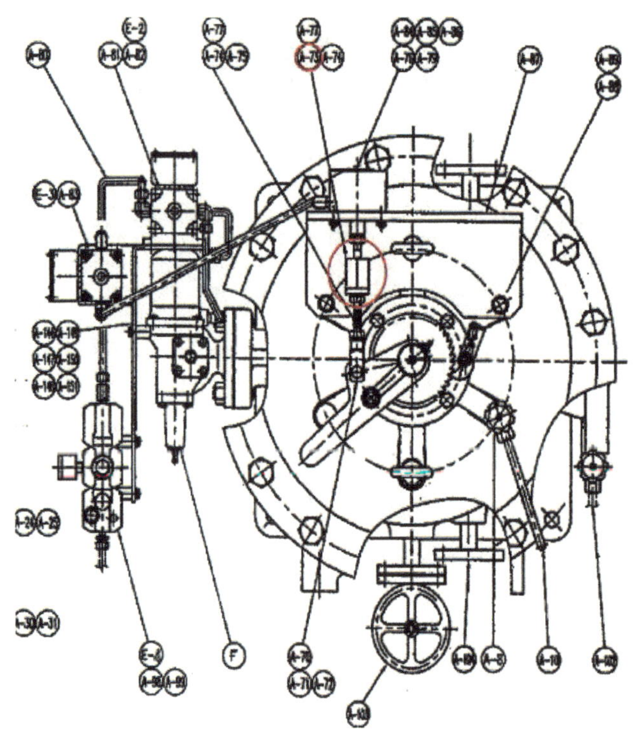

A-80	TUBE	COPPER	–	
A-79	TUBE FITTING	BsBM	1	
A-77	SPRING WASHER	SWRH62	3	
A-76	AIR CYLINDER	-----	1Set	
A-75	STUD BOLT	SS400	1	
A-74	HEX. NUT	SS400	3	
A-73	CYLINDER JOINT	S45C	1Set	
A-72	C-RING	SK-5	1	
A-71	PIN	SS400	1	
A-70	KNUCKLE	SS400	1	
	DETAIL "A"			
A-31	SPRING WASHER	SWRH62	8	
A-30	HEX. NUT	SS400	8	
A-29	STUD BOLT	SS400	8	
A-28	BEARING	SUJ2	1	
A-27	O-RING	VITON	1	
A-26	ELEMENT STAND	FC250	1	
A-25	SPRING WASHER	SWRH62	1	
A-24	HEX. NUT	SS400	1	
A-23	SOCKET BOLT	SCM435	1	
A-22	SPRING	SUS304-WPA	1	
A-21	COLLAR	STPG370	1	
A-20	SPRING HOLDER	STPG370	1	
A-19	SHAFT	SS400	1	
A-18	PLUG	S25C	2	
A-17	BACK-WASHING ARM	FCD400	1	
A-16	SOCKET BOLT	SCM435	4	

Air Cylinder Joint로 Stud 작동길이 조절

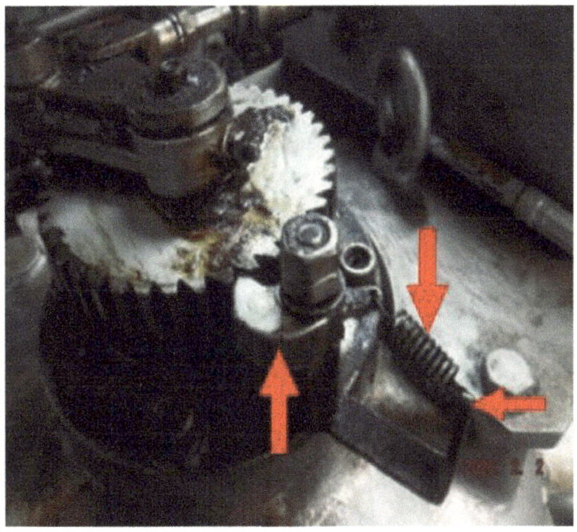

Ratchet Pawl, Spring, Split Pin

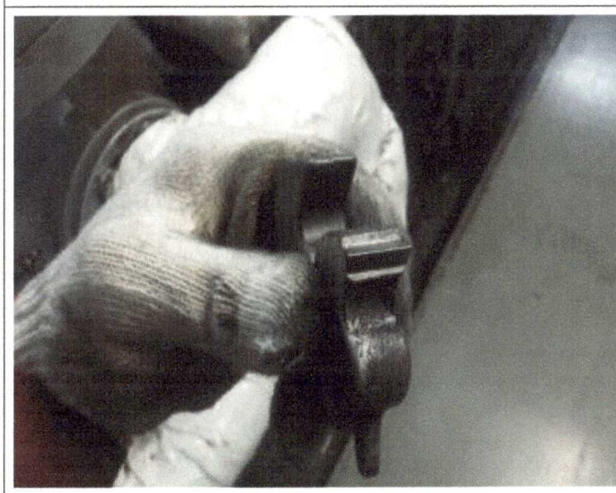

신품 Ratchet Pawl과 취외품 Ratchet Pawl

취외품 Ratchet Pawl(양측)과 Spring

승선하는 선박에서 Filter Auto Backwashing함에도 불구하고 Filter 차압 발생 시 By-pass Filter로 변경하고 Filter 분해하여 Element 소제한다.

누설하는 경우는 Lower(하단부) Gasket 또는 내부 O-ring교환하면 된다.

By-pass Filter 또한 마찬가지이며 필요 시 좌, 우측 Filter Position 변경하여 사용한다.

F.O & L.O Auto Filter Ratchet Pawl 마모 및 손상되어 Ratchet Wheel과 맞물리지 않고 헛돌아가는 경우가 발생할 수 있다.**(Backwashing이 되지 않는 다는 것을 의미)**

마모가 심하지 않다면 Air Cylinder Joint를 이용하여 Stud 작동길이를 조절하여 Ratchet Wheel과 Ratchet Pawl 맞물리게 할 수 있다.

마모가 아래와 같이 심할 경우 Ratchet Pawl, Spring, Split Pin 전부 교환한다.

주기적으로 Filter Element 상태 확인하여 필요 시 소제한다.

* 연료유 인디케이터 필터

매 UMA순찰 때마다 차압 확인한다.

차압이 심하면 3-Way Valve 변경하여 Filter Change, Steam Line 차단 후 Filter Element 소제한다.

3-Way Valve Sealing Ring이 정상이라면 운항 중에 취외 및 소제하고, Sealing Ring 불량할 경우 정박 중에 Pump Stop후 작업 진행한다.

3-Way Valve 변경하여 Filter Change한 다음에 소제할 Filter Cover 취외했는데 누설하면 Sealing Ring이 불량한 것이다.

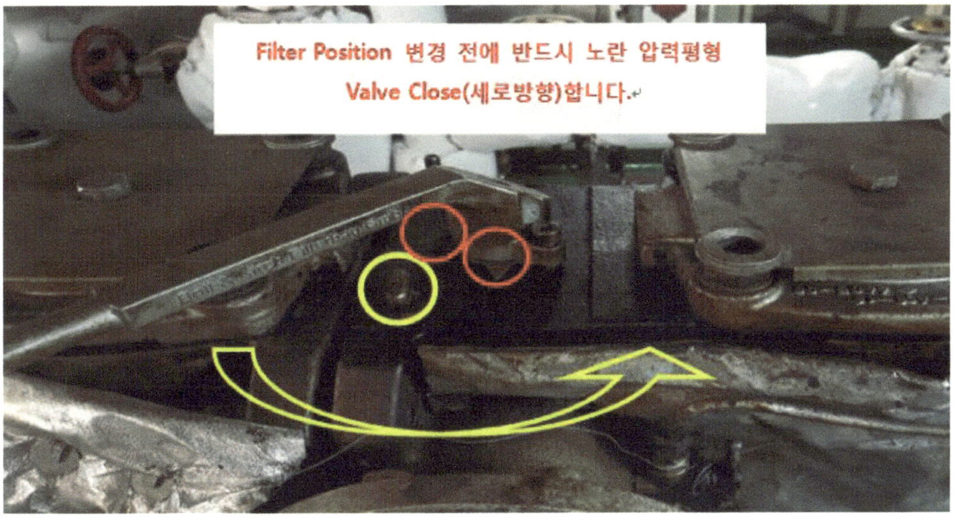

ver safety devices

A Open position
B Closed position

〈 Pressure Compensation Valve 〉

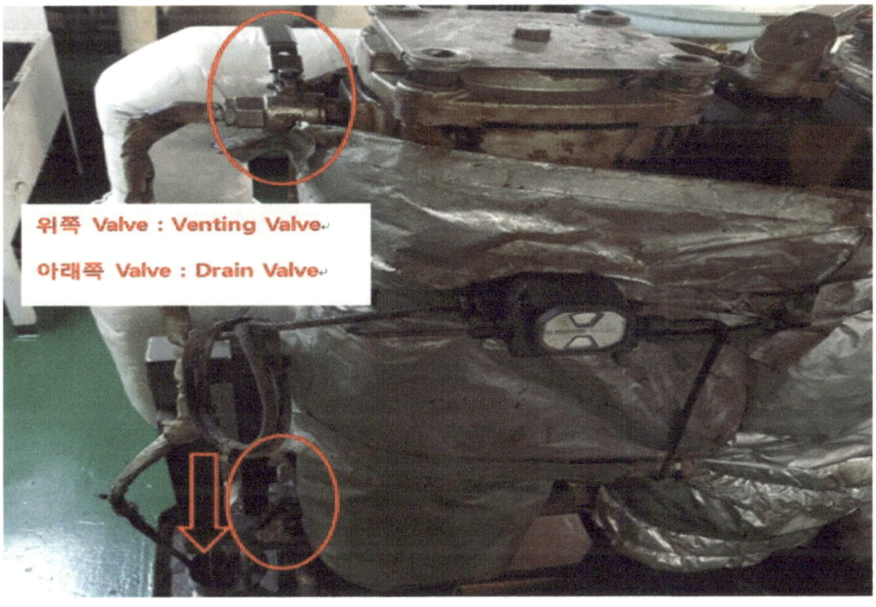

아래와 같이 인디케이터 필터 하부에서 Steam이 누설하는 경우가 있다.

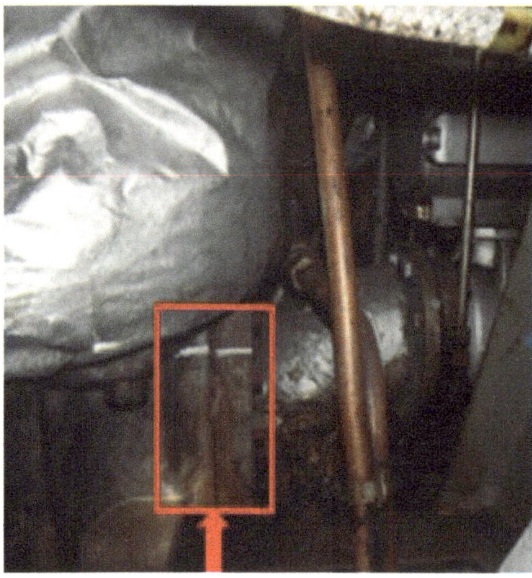

상기 표시된 부분은 Steam만 들어가는 부분이며 Steam 누설 시 Gasket이 손상된 경우이다.

Indicator Filter Steam Inlet Valve Close 후 연결된 Steam Line 분해하여 Gasket 교환한다.

Indicator Filter Steam Inlet Valve는 양측 Filter Common으로 연결되어 있으므로 Valve Close 시 양측 Filter에 Steam 공급되지 않는다.

연료유가 계속 순환하면서 사용하고 있고 다른 부분에서도 Heating하고 있기 때문에 운항 중에 잠시 Steam Valve Close하여도 무방하다.

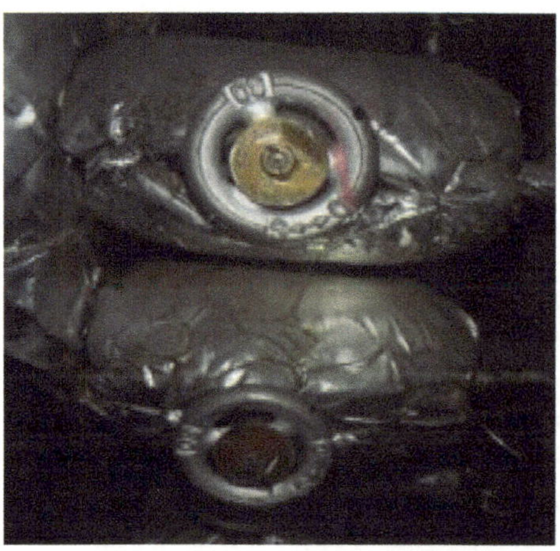

〈 Filter Steam Inlet Valve 〉

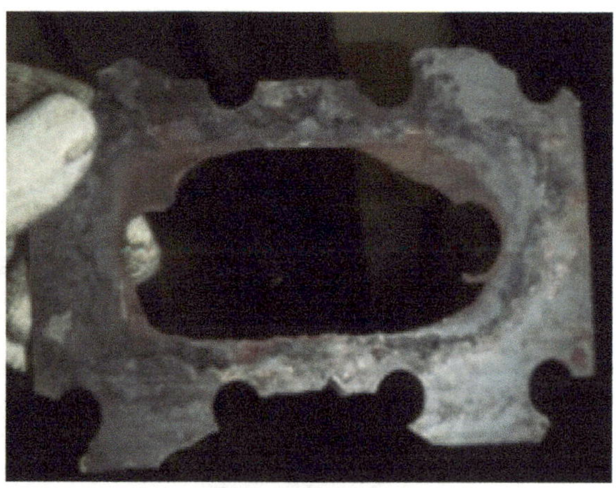

〈 손상 된 Gasket 〉

Steam Gasket으로 교환하고 조립 후에 다시 Steam Valve를 열어 준다.

FO By-pass Filter와 같이 차압이 발생하면 Filter Position 변경 후 분해하여 Element 소제한다.

주기적으로 Filter Element 상태 확인하여 필요 시 소제한다.

12) 윤활유 오토 바이패스 필터

< L.O Auto Filter(좌) / L.O By-pass Filter(우) >

승선하는 선박에서 Filter Auto Backwashing함에도 불구하고 Filter 차압 발생 시 By-pass Filter로 변경하고 Filter 분해하여 Element 소제한다.

누설하는 경우는 Lower(하단부) Gasket 또는 내부 O-ring교환하면 된다.

By-pass Filter 또한 마찬가지이며 필요 시 좌,우측 Filter Position 변경하여 사용한다.

F.O & L.O Auto Filter Ratchet Pawl 마모 및 손상되어 Ratchet Wheel과 맞물리지 않고 헛돌아가는 경우가 발생할 수 있다.**(Backwashing이 되지 않는 다는 것을 의미)**

마모가 심하지 않다면 Air Cylinder Joint를 이용하여 Stud 작동길이를 조절하여 Ratchet Wheel과 Ratchet Pawl 맞물리게 할 수 있다.

마모가 아래와 같이 심할 경우 Ratchet Pawl, Spring, Split Pin 전부 교환한다.

주기적으로 Filter Element 상태 확인하여 필요 시 소제한다.

13) 스타팅 에어 밸브

정비 주기 : 12000H

Starting Valve

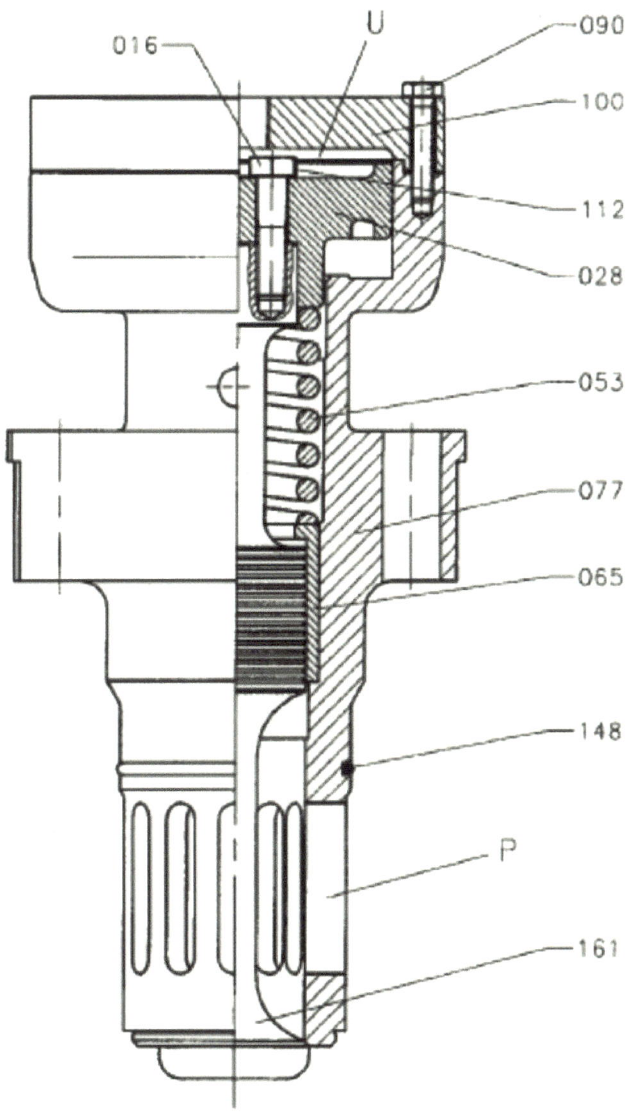

Starting Air Valve

SAFETY PRECAUTIONS *For detailed sketch, see 900-2*

X	Stopped engine
X	Shut off starting air supply – *At starting air receiver*
X	Block the main starting valve
X	Shut off starting air distributor/distributing system supply
X	Shut off safety air supply – *Not ME engines*
X	Shut off control air supply
	Shut off air supply to exhaust valve – *Only with stopped lubricating oil pumps*
	Engage turning gear
	Shut off cooling water
	Shut off fuel oil
	Stop lubricating oil supply
	Lock the turbocharger rotors

Data

Ref.	Description	Value	Unit
D07-09	Piston tightening torque	140	Nm
D07-11	Valve tightening angle	60	°
D07-12	Starting air valve	75	kg

RT-FLEX Engine과 달리 5-2Way Solenoid Valve가 없으며 Starting Air Valve 상부에 Control Air Pipe가 Direct연결된다.

(1) 교환 작업

① F.W.E(Finished With Engine)후 작업을 하되 Control Safety Air Valve Close 할 필요 없다.

② Starting Air Valve에 연결된 Air Pipe ,Nipple Copper Packing을 분해하고 Eye-bolt를 조립한다.
또는 Starting Air Valve Cover(90704-0043-100) 양쪽에 Eye-bolt를 조립해도 된다.

③ Air Pressure Drop을 위해 Starting Air Valve 양쪽 Nut 풀고 Stud 끝단에 걸쳐 둔다.

④ Overhead Crane에 Rope와 Chain Block을 걸고 Chain Block을 Eye-Bolt에 연결한다.
Chain Block 이용해서 Starting Air Valve 올려 Pressure Drop한 다음 다시 내

려준다.

Starting Air Valve 양쪽 Nut 풀어서 취외하고 Overhead Crane에 연결된 Chain Block 이용하여 Starting Air Valve 분해한다.

⑤ Starting Air Valve 분해된 Cylinder Head 내부를 깨끗이 소제한다.

⑥ 정비품 Starting Air Valve에 Eye-Bolt 조립하고 E/R Overhead Crane, Rope Chain Block을 Eye-Bolt에 연결하여 분해 때와 같이 조립한다.

(조립 시 Housing Seat에 Molykote 바른다.)

그 후 Starting Air Valve Nut Tight 한다.

⑦ 취외한 Starting Air Valve Nipple(Air Pipe와 연결됨)을 소제하고 신품 Copper Packing과 같이 정비품 Starting Air Valve에 조립한다.

(본선 Copper Packing 예비품 없다면 Gas Torch로 가열 후 상온 냉각하면 재사용 가능하다.)

⑧ Air Pipe 조립하고 Engine 시동 Test 한다.

Starting Air Valve 상부에 연결된
Air Pipe, Nipple, Copper Packing

Starting Air Valve Nut 살짝 풀어서
Pressure Drop 한다.

(2) 분해 및 소제 방법

① 취외한 Starting Air Valve 수평으로 놓은 다음, Sealing Ring(**90704-0043-148**) 분해한다.

② Cover Screw(**90704-0043-090**) 분해하여 Cover(**90704-0043-100**) 취외한다.

③ 고정 Bolt의 Locking Plate(**90704-0043-112**) 풀어서 취외한다.

Spindle(**90704-0043-161**)에 Piston(**90704-0043-0028**)을 고정시키는 Bolt(**90704-0043-016**)는 Spindle 아래 측에 Spindle 고정 Tool을 조립하여 Spindle이 움직이지 않게 한 뒤 분해한다.

Bolt 분해 시 Spring(**90704-0043-053**)의 힘으로 인해 Piston이 Valve Housing (**90704-0043-077**) 밖으로 나온다.

④ Housing 위쪽으로 Piston Spring 취외한다.

Housing 아래쪽으로 Spindle 취외한다.

⑤ Grinding Ring(Valve Housing Lapping Tool)으로 Housing Seat 연마한다.

Housing Spindle Seat 잘 접촉되도록 연마할 때는 Spindle 고정 Tool로 Spindle을 돌리면서 연마한다.

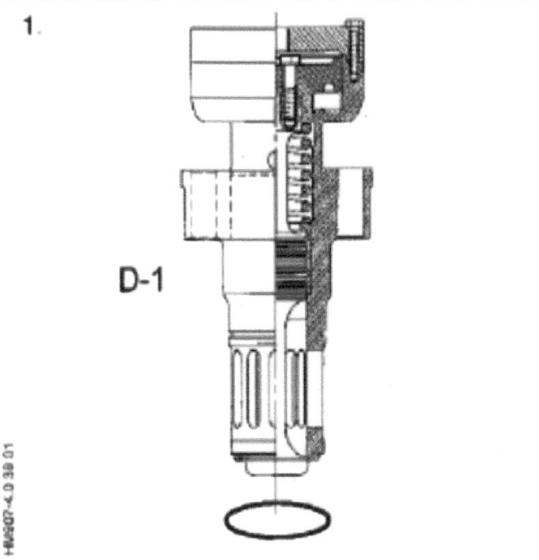

Sealing Ring 취외한다.

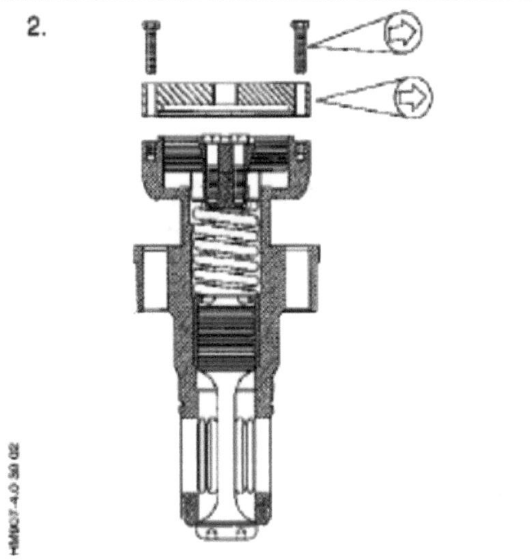

Cover Screw 분해하여 Cover 취외한다.

3.

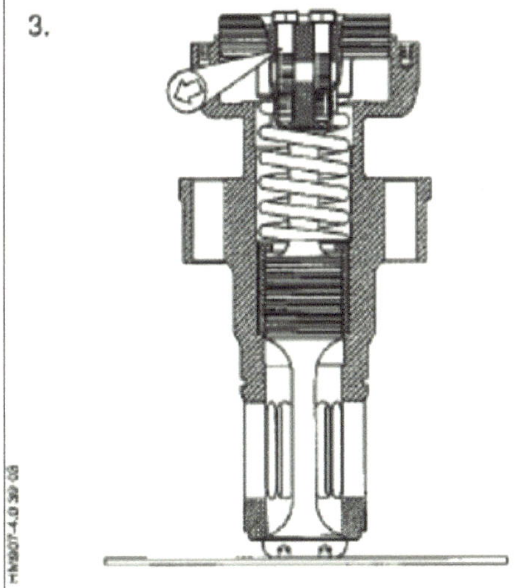

Spindle에 Piston을 고정시키는 두 개 Bolt는 Spindle 고정 Tool로 반대측 잡아주면서 푼다. Bolt를 풀 때는 Spring 힘 때문에 Piston이 조금씩 Valve Housing 밖으로 나온다.

Spindle 고정 Tool (1)

Spindle 고정 Tool (2)

4.

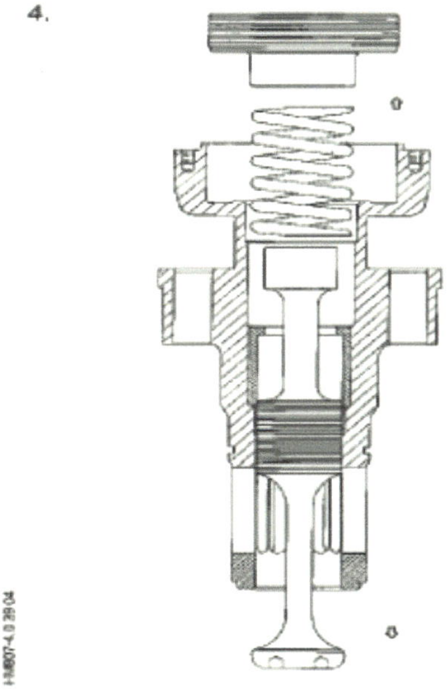

Housing 위쪽으로 Piston & Spring 취외한다. Housing 아래쪽으로 Spindle 취외한다.

Starting Air Valve 분해

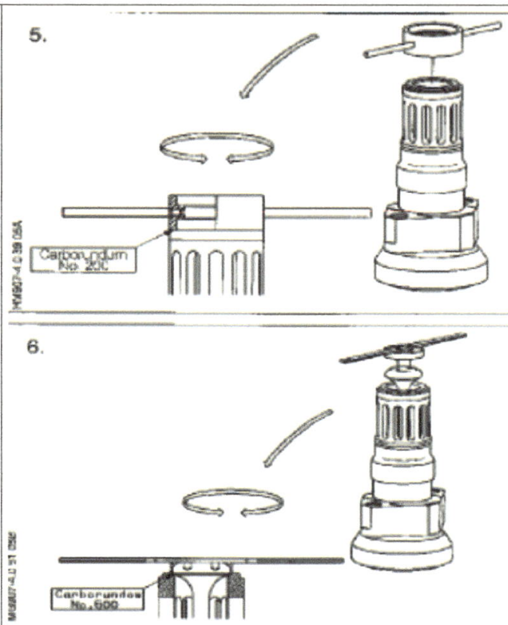

Grinding Ring으로 Housing Seat 연마한다.
Housing과 Spindle Seat가 잘 접촉되도록
Spindle 고정 Tool로 Spindle 돌려준다.

Housing & Spindle Seat Grinding Tool
(Valve Housing & Spindle Lapping Tool)

Housing & Spindle Seat Grinding Tool
(Valve Housing & Spindle Lapping Tool)

(3) 조립 방법

① Valve Housing 아래로 Spindle을 조립하고 Housing 위로 Spring과 Piston을 조립한다.

(Piston을 조립할 때 Housing 내부와 Piston에 L.O를 도포하여 조립한다.

또한 Spindle 조립 시 좌우로 움직이면 Housing 내부 Bushing(90704-0043-065) 불량이므로 교환한다.)

Locking Plate와 Bolt를 조립하고 Bolt Tight하면 Piston이 Spring을 눌러준다.

Spindle 고정 Tool로 Spindle을 고정시키고 Piston 접촉면이 Spindle에 완전히 밀착될 때까지 Bolt(Torque : **140Nm**) Tight 한다.

② Locking Plate로 Bolt 고정한다.

③ Cover를 Housing에 얹히고 Cover Screw를 Tight 한다.

④ Housing Sealing Ring을 조립하고 이물질이 들어가지 않게 밀봉한다.

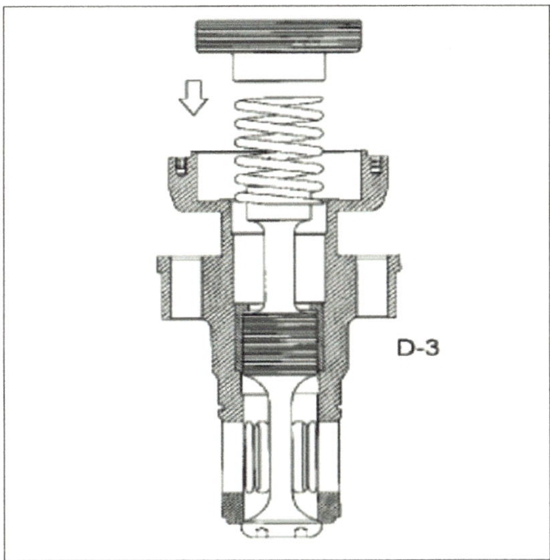

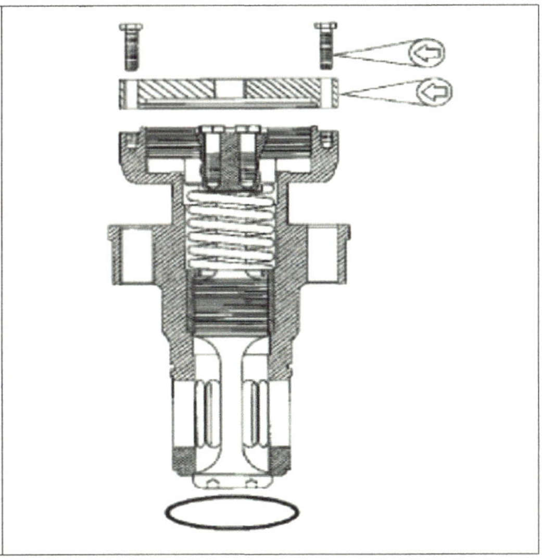

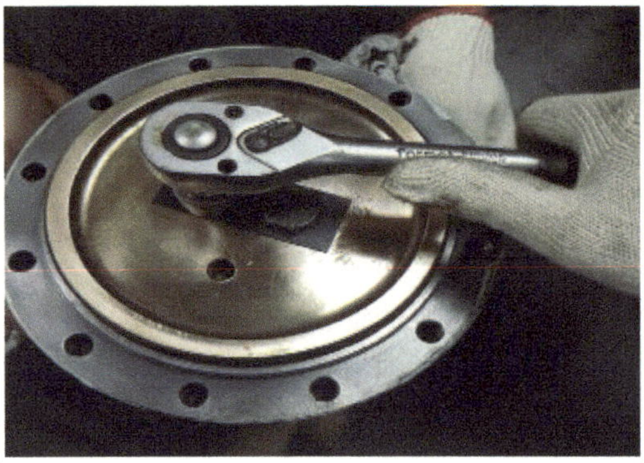

Bolt Tight

Spindle 고정 후 Torque Tight (140Nm)

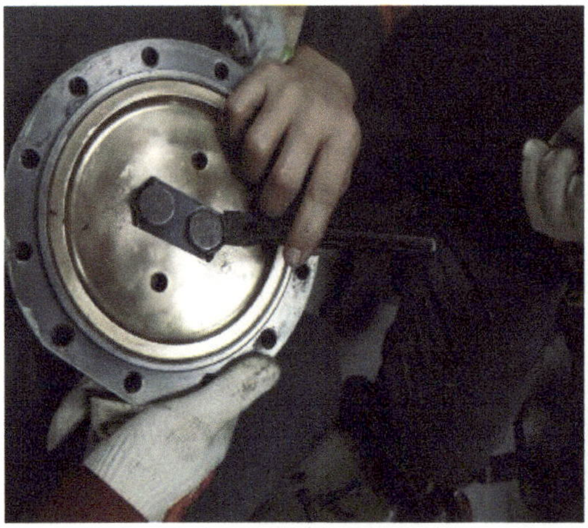

Locking Plate 고정

Locking Plate 고정

Cover Bolt Tight

정비품 밀봉

Starting Air Valve 내부 Bushing

Valve Spindle

Spring

내부 L.O 도포하여 조립

Piston

Piston Bolt(140Nm) Locking Plate

(4) TEST 방법

Test 방법은 간단하다.

Cover 상부에 Size 맞는 Ball Valve를 조립하고 Air Hose를 연결한다.

Air를 공급하면 Piston과 Spring을 눌러 Spindle 아래로 Open되어 Air Drain되는지 확인한다.

Air 차단 시 Spring으로 인해 Spindle Close 된다.

14) 에어 쿨러

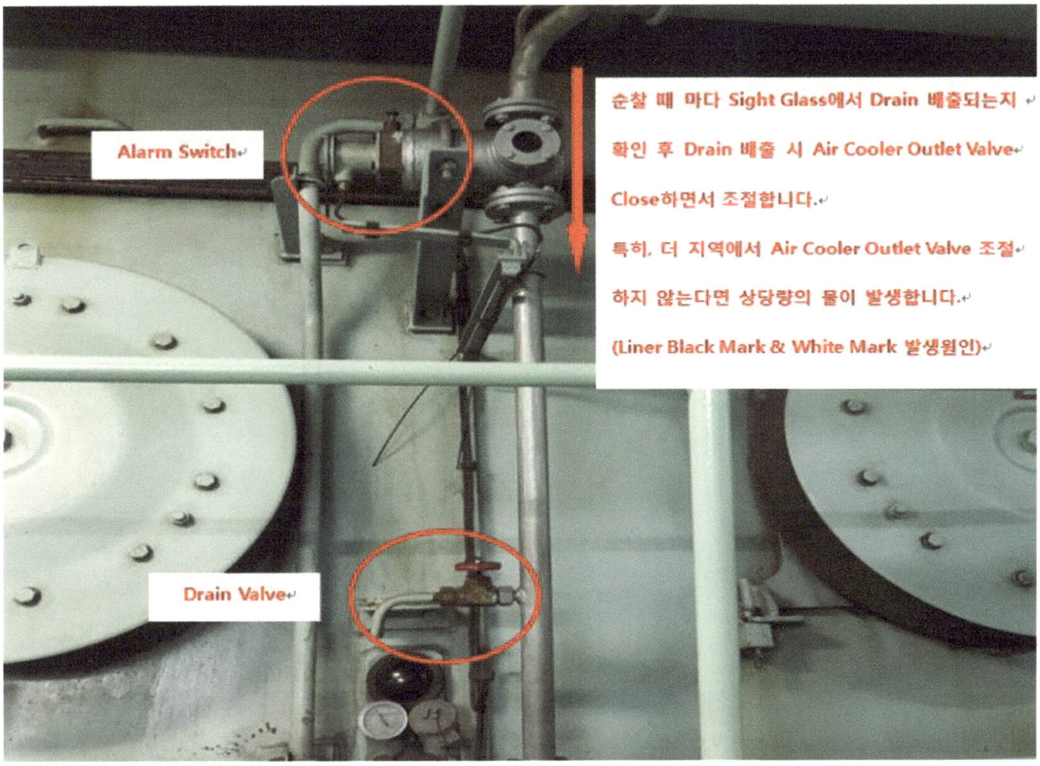

또한 운항 중에 Cooler F.W 온도를 평소에 주의 깊게 관찰한다.

필요 시 Chemical Cleaning을 시행한다.

오랜 사용으로 또는 Dry Dock에서 Frame Ring(91005-0119-056) 교환하지 않을 경우 누설한다.

아래 도면 참조하여 UMA 순찰 때마다 Air Cooler 양측의 Frame에서 누설 유무를 확인한다.

Air Cooler

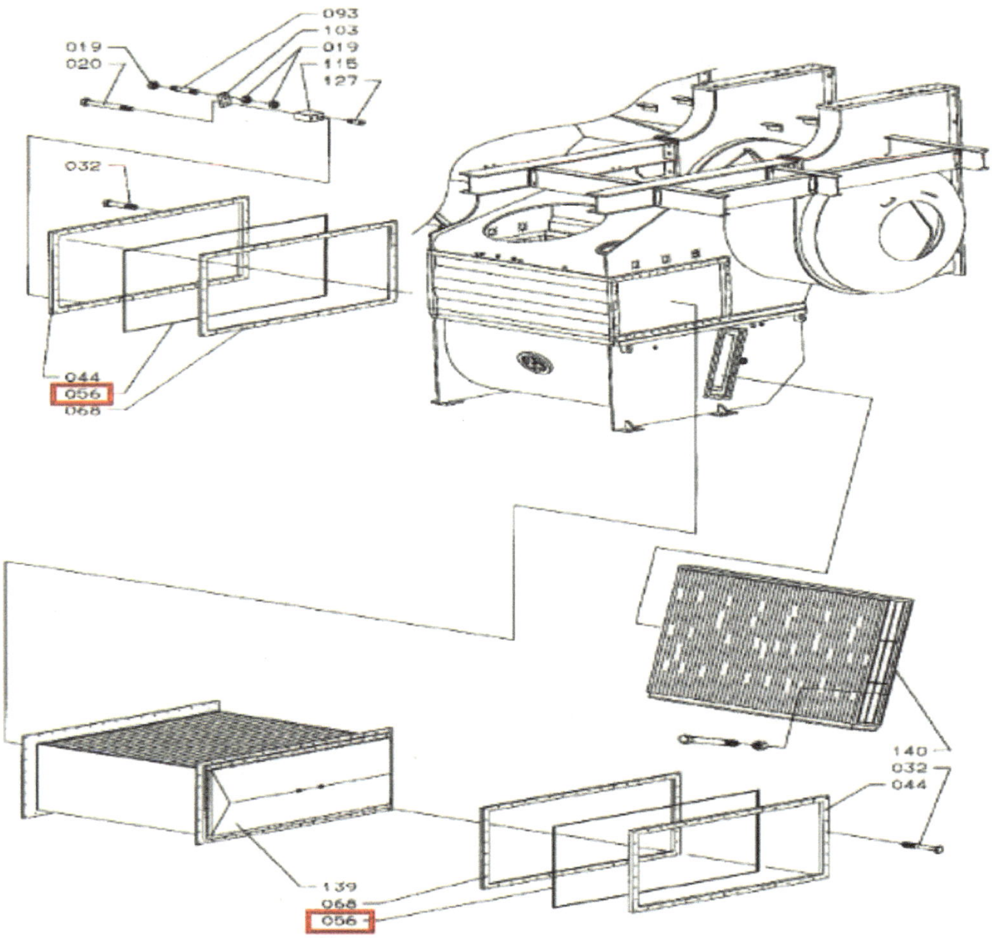

< Air Cooler Frame Ring : 91005-0119-056 >

15) 보조 블로워

정비 주기 : 3000H

AUX. Blower교환 주기는 3000H이며 F.W.E(Finished With Engine)후 Power Off하고 작업하며, 고온 다습한 구역을 항해하는 선박은 교환 주기가 일반적으로 더 짧다.

Local Start Panel에서 NO.1 & 2 Ampere 비교 시 Ampere 높은 쪽이 교환 주기 도래한 것이다.

취외한 AUX. Blower 본선에서 정비(Bearing, Gas Lock seal, Seal Ring, Sealing Bush, Gasket 교환)를 하거나 육상 재생 수리 신청한다.

당장 본선에서 급하게 정비를 요하는 긴급 상황 또는 부득이한 상황 제외 본선 감독과 협의하여 육상수리 진행한다.(육상수리 시 Balancing Test 포함)

Bearing의 경우 Open Type은 주기적으로 Grease 주입 필요하고 Seal Type은 Grease 주입 시 Motor로 들어가 Motor 손상되니 주입하지 않는다.

경험상 Open Type Bearing이 OV'HL 주기가 더 빠르며, 주기적으로 Bearing 검진기로 Bearing의 상태를 확인한다.

Drain Pipe에서 Air가 나오면 Air Lock-seal 불량 또는 교환 주기가 도래했다는 것을 의미한다.

Power Off 후 항상 Paint Marker로 표시를 하고 Power Cable 취외 및 조립을 시행한다.

Junction Box까지 교환하게 되면 시간이 오래 소요되며 AUX. Blower Cable만 취외하고 조립하면 작업 시간을 단축할 수 있다.

본선 정비 시 Stuffing Box, Gas Lock-seal의 방향 확인하여 조립하고 Gasket 또한 교환한다.

예전 Model Motor의 경우 Power Cable J/Box 내부에 Heater Cable 연결할 수 있도록 되어있으나 있으나 최근에 보급되는 신품들은 Heater Cable을 연결하지 않도록 되어 있다.

Stop했을 때 Space Heater의 작동원리는 아래와 같다.

Stop중 시퀀스 회로를 보면 UVW 3상 전원에 Heater 선 연결하여 공급 Power에서 Transformer 거쳐 변압을 시키고 Power Cable 자체에 전압을 공급하여 Heater를 작동시킨다.

조립 완성 후 Local Start Panel에서 Blower Start 한다.

Panel 측 **Ampere**와 Local 측 **AUX.Blower Impeller 회전방향** 및 **소음, 진동**을 점검한다.

*** 보급된 재생품 AUX.Blower의 Impeller 회전방향이 반대라면 Control Panel Power Off하고 Panel 내부의 Power Cable 3상 중에서 2상만 바꾸어 조립한다.**

상기와 같이 진행하여 Impeller 회전방향을 조절하였음에도 Ampere가 높게 유지되고 Motor의 소음 발생하거나 Trip된다면 앞서 Impeller 회전방향을 조절한 Cable 옆에 위치한 Power Cable 3상 중에 2상만 바꾸고 조립하면 된다.

(간혹 재생품의 경우 이와 같은 문제 발생하며 Y-△기동을 생각하면 이해하기 수월하다.)

Power Cable 변경 전

Power Cable 변경 후

16) 컨트롤 에어 드라이어

순찰 때마다 Air Drain 및 Pressure Gauge 확인하고 냉매(R-404a) 부족 시 보충하되 보충 주기가 짧아지면 냉매 라인, Gauge연결부 및 Compressor 연 결부 Pipe 거품 Test하여 냉매 누설 부를 확인한다.

또한 Cooling Fan Motor 정상적으로 작동하는지 확인한다.

Cooling Fan Motor 자체가 불량하거나 Power Cable이 제대로 연결되지 않으면 작동하지 않으니 순찰 때마다 확인한다.

Pressure Gauge, Lamp Switch, Compressor, Air Filter(0.01 MICRON, 오른쪽 그림)은 본선에 예비품을 보유하는 것이 좋다.

장착된 Air Filter 주기적으로 상태 확인하여 오손 시 교환한다.

17) 터보차저 컷-오프 시스템

Turbocharger Cut-off System은 요즘 대부분의 선박에 설치 되어있는 경우가 많다.

설치 목적은 유류절감 목적이다.

7~8기통 Engine 장착된 선박은 2대 Turbocharger가 있으며, 보통 1번 Turbocharger에 Cut-off System을 장착하였다.

12기통 Engine 장착된 선박은 3대 Turbocharger가 있으며, 보통 2번 Turbocharger에 Cut-off System을 장착하였다.

평소 2대 또는 3대 Turbocharger 기동 중에 1대를 Cut-off하면 그만큼 나머지 Turbocharger의 회전속도가 올라가면서 소기 압력이 올라가고, 결국에는 적은 기름으로 완전연소 유도할 수 있어 설치한 것이다.

실제로 각종 Data들이 Turbocharger Cut-in과 Cut-off했을 때 기름소모량의 차이를 보여준다.

하지만 Turbocharger도 정해진 Spec이 있기에 무작정 RPM 올릴 수 없으며, 승선하는 선박 마다 Turbocharger Cut-off시 정해진 Limit RPM이 있다.

Turbocharger Cut-off시 과도하게 높은 RPM으로 운용 시 Turbocharger High

RPM Alarm 발생하겠지만, 소기 압력이 너무 높아지면 폭발 압력이 높아져서 Engine Condition에 전반적으로 악영향을 미칠 수 있다.

하지만 특정 부하 이상에서는 VIT가 작동하여 일정하게 폭발 압력이 유지되지만, 무엇보다도 소기 압력이 높아지면 소기 계통 손상, 파손 및 부식과 같은 문제가 발생할 수 있다.

또한 Cut-off 상태에서 Cut-in으로 변경할 때는 Close 된 Swing-gate Valve를 Open 하기 때문에, Engine운전 중 Dead-slow Slow RPM 에서 변경해도 된다.

반대로 Cut-in 상태에서 Cut-off로 변경할 때는 Open된 Swing-gate Valve를 Close하기 때문에 가급적 Engine Stop & Dead-slow RPM에서 변경한다.

Cut-in 상태에서 Cut-off로 변경할 때 Exhaust gas 압력으로 인하여 Swing-gate Valve가 제대로 작동하지 않을 수 있기 때문이다.

결론적으로 Turbocharger Cut-off System은 기름 절감 차원에서 설치한 것이며, High RPM 운전 필요할 때는 Turbocharger Cut-in 해야 한다.

Cut-off 상태에서 Cut-in으로 변경 시 Engine 운전 중 Dead-slow & Slow RPM 변경, Cut-in 상태에서 Cut-off로 변경 시 Engine Stop & Dead-slow RPM에서 변경한다.

* 터보차저 컷-오프 방법

① Turbine & Compressor측 Swing-gate Valve에 조립된 Safety Pin을 제거한다.

② Air Supply Valve Open하고 Position Valve 중립에서 TC Cut-off로 변경한다.

③ Swing-gate Valve 변경된 것을 확인 후 Safety Pin을 기존에 조립된 반대측으로 조립한다.
(조립되면 Turbocharger Cut-off System Panel에서 Valve Close 점등 확인 가능.)

④ Air Supply Valve Close하고 Position Valve TC Cut-off에서 중립으로 변경한다.
(Air Supply Valve Close하고 Position Valve 중립에 놓으면 Air Drain된다.)

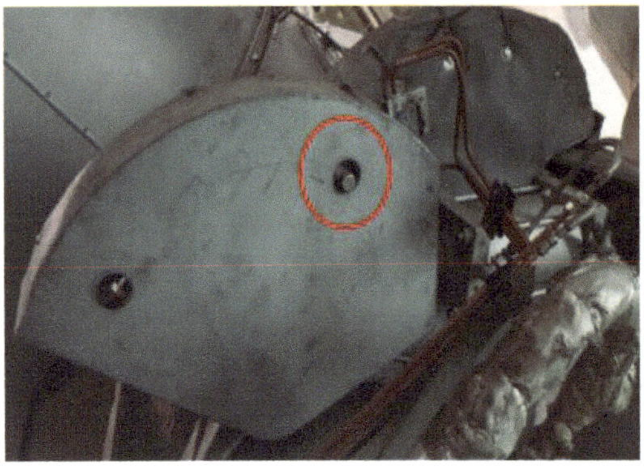

Turbine 측 Safety Pin 제거

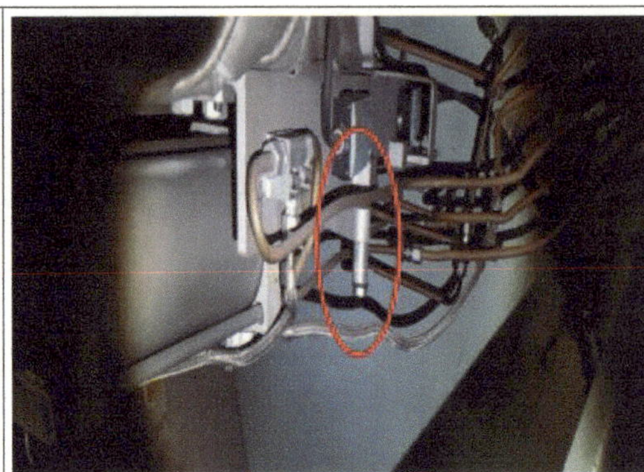

Compressor 측 Safety Pin 제거

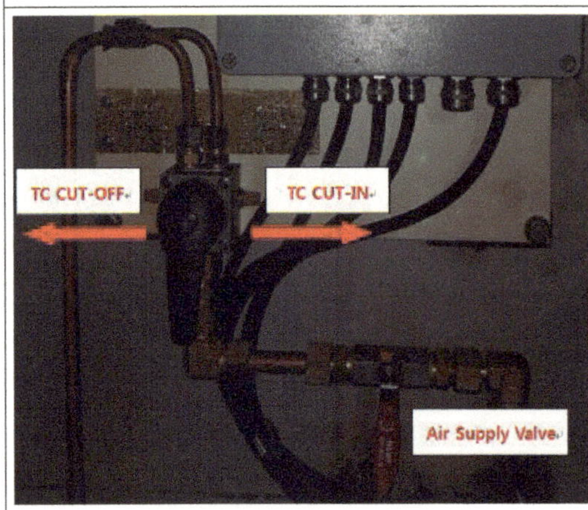

현 상태가 Normal 상태
(Air Valve Close, Position Valve 중립)

Air Valve Open,
Position Valve TC Cut-off 변경한다.

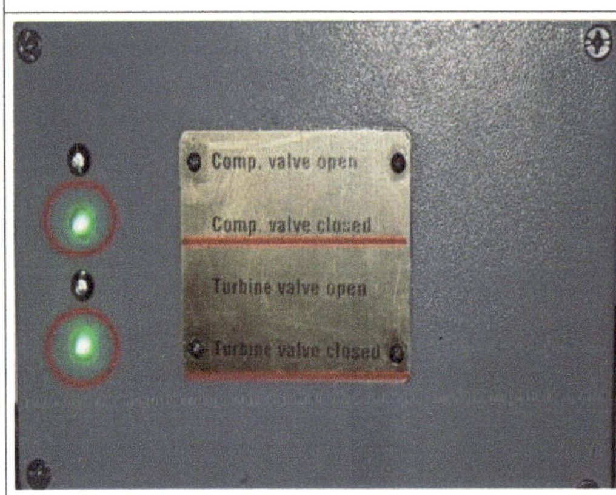

점등 상태 확인

Air Valve Close, Position Valve 중립하면
Air Drain 된다.

* Swing-gate Valve 제대로 움직이고 Safety Pin이 제대로 조립되었는데도 Lamp 점등되지 않을 경우 Limit Switch를 점검한다.

Swing-gate Valve 제대로 움직이지 않는다고 Valve 작동을 위해 해머 사용은 하지 않도록 한다.

해머로 Valve를 쳐서 움직인다면 기존에 장착될 때 맞추었던 Position에서 달라지게 되고 결국 제대로 동작이 안된다.

Valve가 Full Close 또는 Full Open 되지 않아서 조정이 필요할 때는 아래 사진을 참조한다.

18) 터보차저 드라이 클리닝 & 드레인 밸브

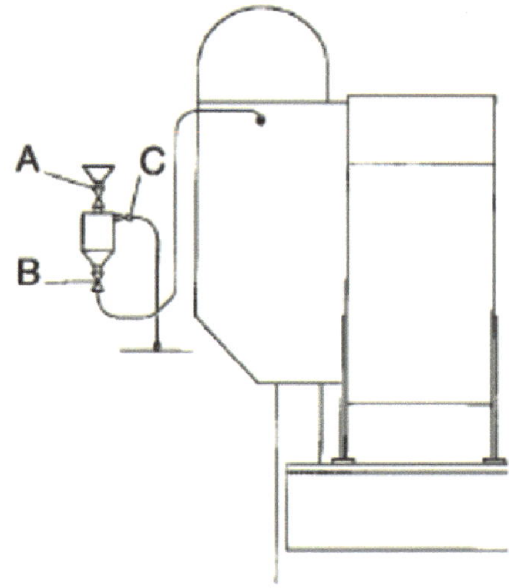

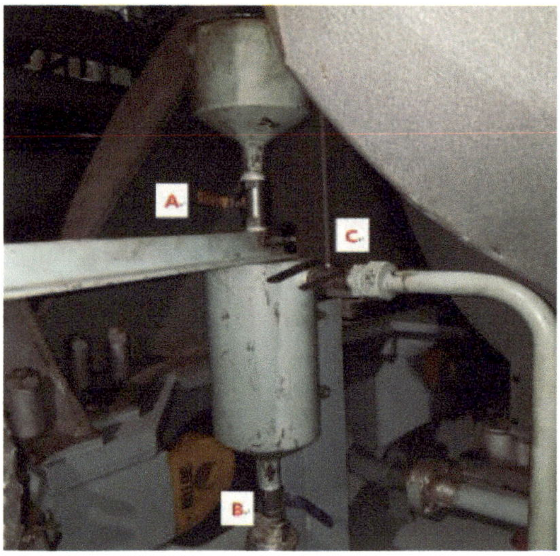

Cleaning Procedure:

1. Carry out cleaning for every 24 to 50 hours of operation, based on observations.
2. Preferably clean the turbocharger at full load. Do not clean below half load.
3. Close valve **A**.
4. Open valves **B** and **C**, to blow out any deposits and/or condensate in the connecting pipe.

   After about two minutes, close valves **B** and **C**.
5. Slowly open valve **A** to relieve the pressure in the tank.
6. Fill the tank with the quantity of granules specified in the table.
7. Close valve **A**.
8. Open valves **B** and **C**, to blow-in the granules. After one to two minutes, close valves **B** and **C**.
9. Slowly open valve **A** to relieve the pressure in the tank.

Caution!
1. The drain openings in the gas casings must remain closed while dry cleaning the turbine.
2. It is possible that during dry cleaning of the turbine, some of the blown-in solid particles or sparks will escape through the funnel.

매 **100H** 마다 본선 상황에 맞게 Engine RPM 증속하여 T/C Dry Cleaning 시행한다.

사전에 Instruction Book을 참조하여 T/C Max. RPM및 T/C Outlet Temp Limit 값을 확인한다.

(1) 터보차저 드라이 클리닝

① (A) Valve Close 한다.

② (B) & (C) Valve Open하여 Tank Blow out하고 (B) & (C) Close 한다.

③ (A) Valve 천천히 Open하여 Tank내 Pressure Drop 한다.

④ 활성탄을 2/3 정도 채우고 (A) Valve Close 한다.

⑤ (B) & (C) Valve 5분정도 Open한 다음 (B) & (C) Valve Close 한다.

⑥ (A) Valve 천천히 Open하여 Tank내 Pressure Drop 한다.

< 활성탄 >

(2) 터보차저 드레인 밸브

Turbocharger마다 Drain Pipe가 있으며 Economizer
소제 시 Drain Valve Open합니다.
(만약 Turbocharger Drain Valve로 Water가 나온다면
상부에 설치한 Canvas 또는 Drain Line 재 확인합니다.)

19) 피스톤 분해 및 정비 도구

Cylinder Cover Panel

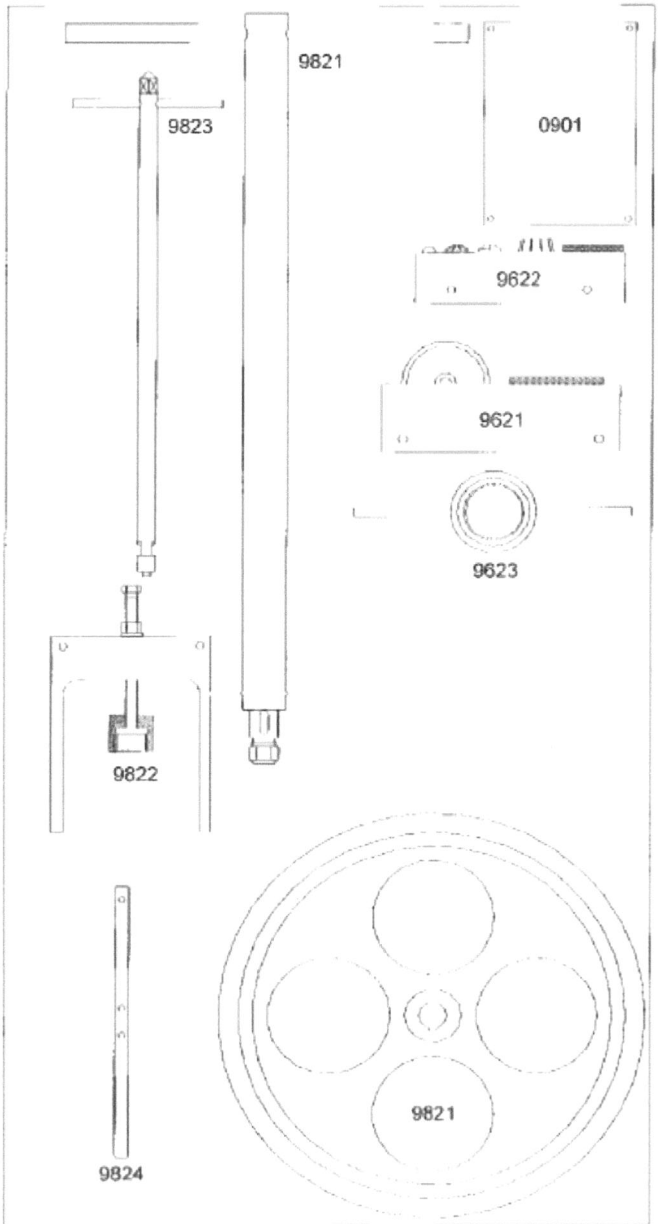

9621 : Milling and grinding tool for starting valve seat

9622 : Milling and grinding tool for fuel valve seat

9623 : Grinding ring for starting valve

9821 : Grinding tool

9822 : Dismantling tool for fuel valve

9823 : Handle

9824 : Pin wrench for starting valve

Cylinder Cover - Tools

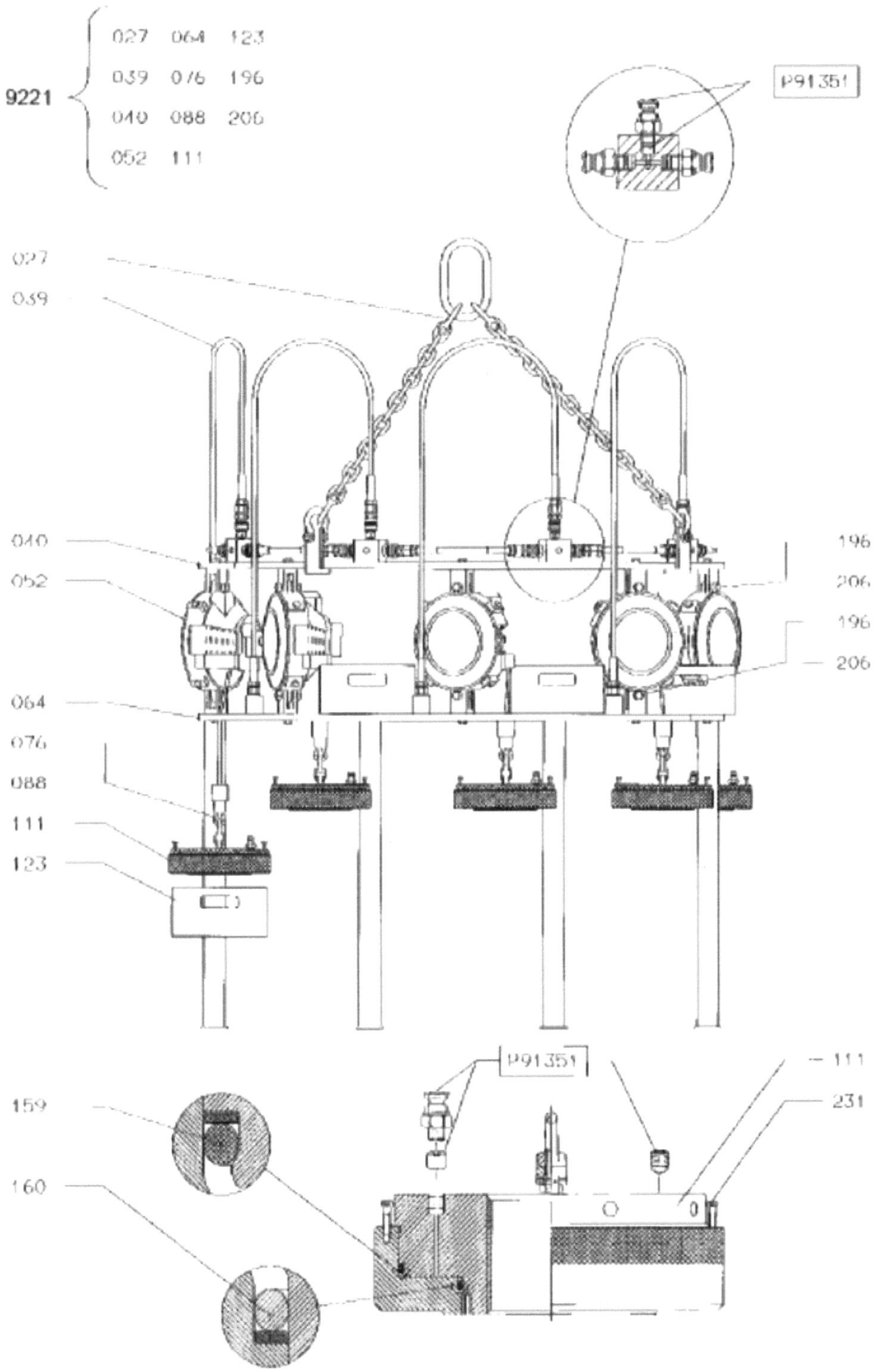

027 : Lifting device

039 : Hose with unions

040 : Balancing ring assembly, upper

052 : Spring balance

064 : Balancing ring assembly, lower

076 : Heart-shaped thimble

088 : Wire lock

111 : Jack-hydraulic, complete

123 : Jack-hydraulic, support

159, 160 : O-ring with back-up ring

196 : Head screw

206 : Nut

231 : Screw

9221 : Cylinder cover tightening tool, complete

Cylinder Cover - Support Tool

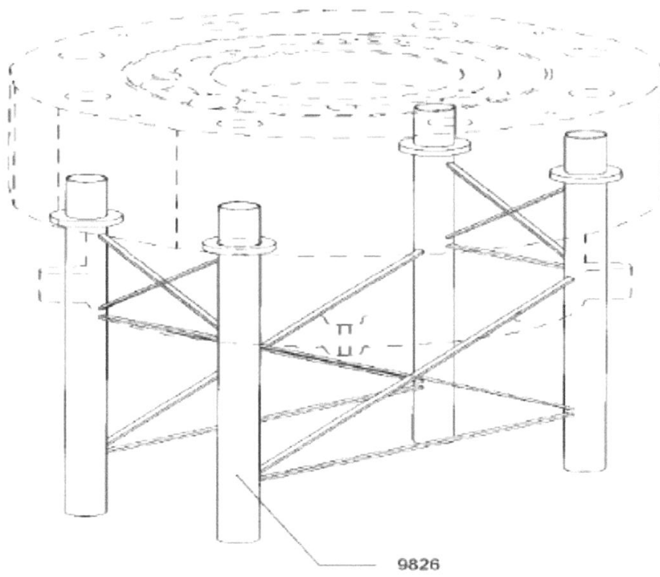

9826 : Cylinder cover rack

Cylinder Cover - Lifting Tool

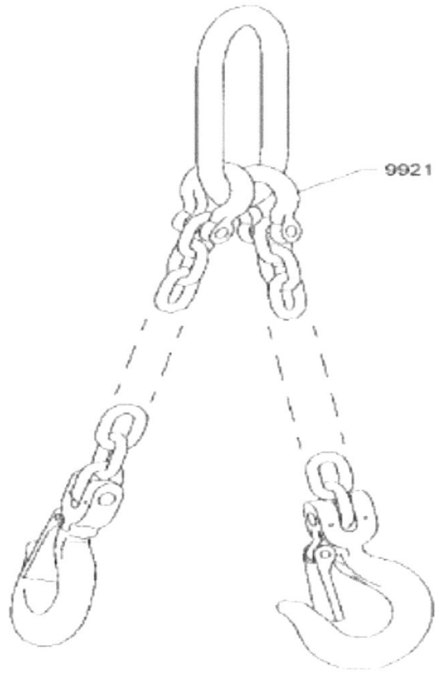

9921 : Lifting chains for cylinder cover

Piston and Piston Rod Panel

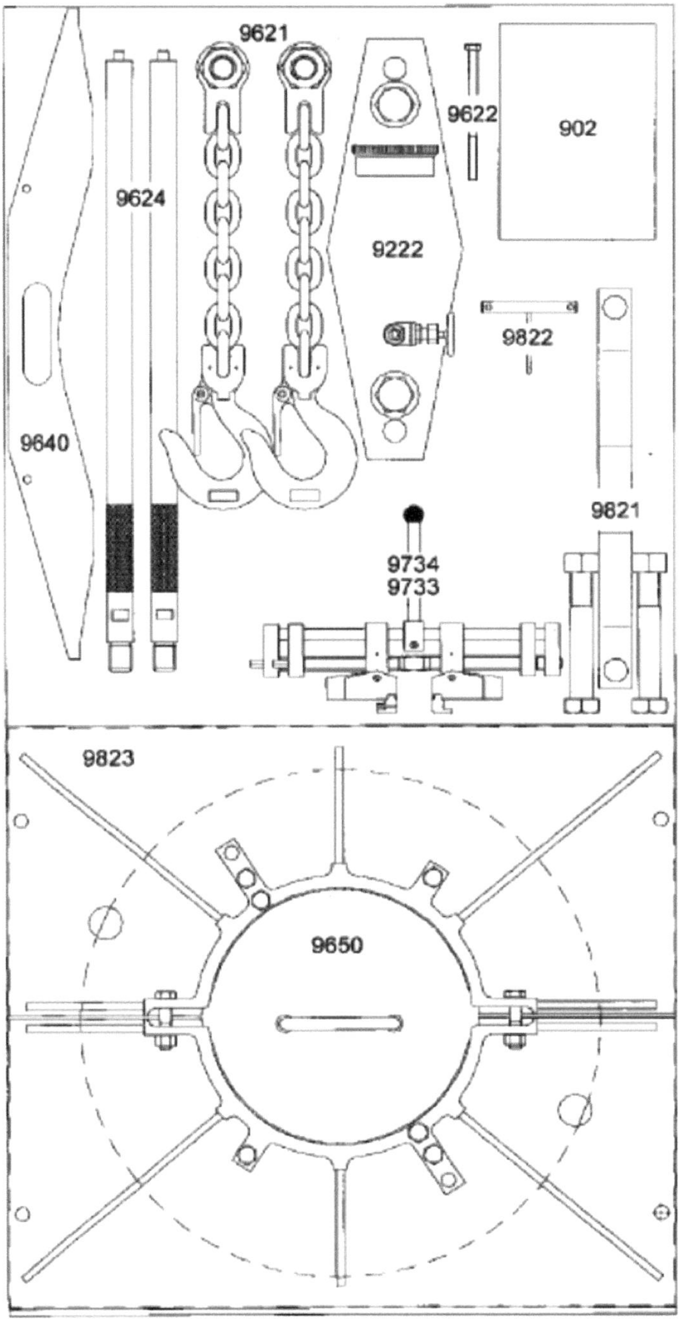

902 : Panel for tools

9222 : Pressure test tool for piston

9621 : Lifting tool for cylinder liner

9622 : Dismantling screw

9624 : Distance piece

9640 : Template for piston top

9650 : Cover for stuffing box hole

9733, 9734 : Piston ring expander, set

9821 : Lifting tool for piston rod foot

9822 : Mounting tool for stuffing box spring

9823 : Worktable for stuffing box

Piston and Piston Rod - Tools

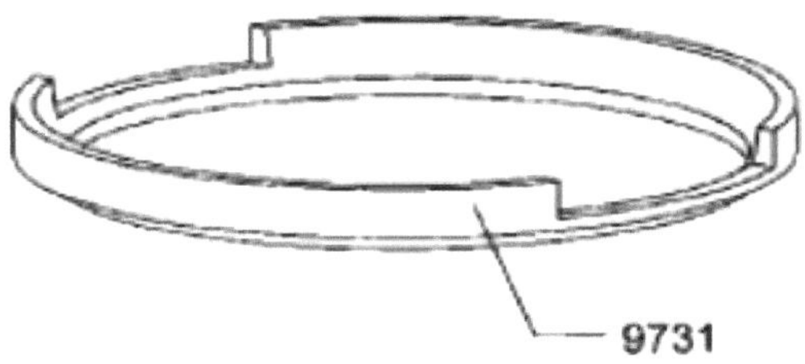

9731 : Guide ring for piston

Piston and Piston Rod - Support Tools

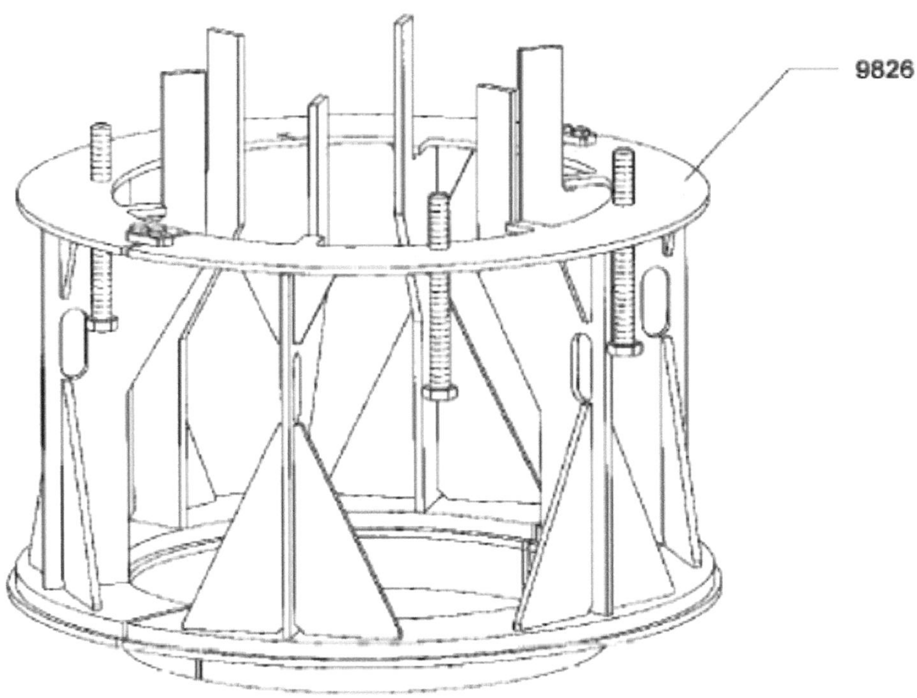

9826 : Support iron for piston

Piston and Piston Rod - Hydraulic Tools

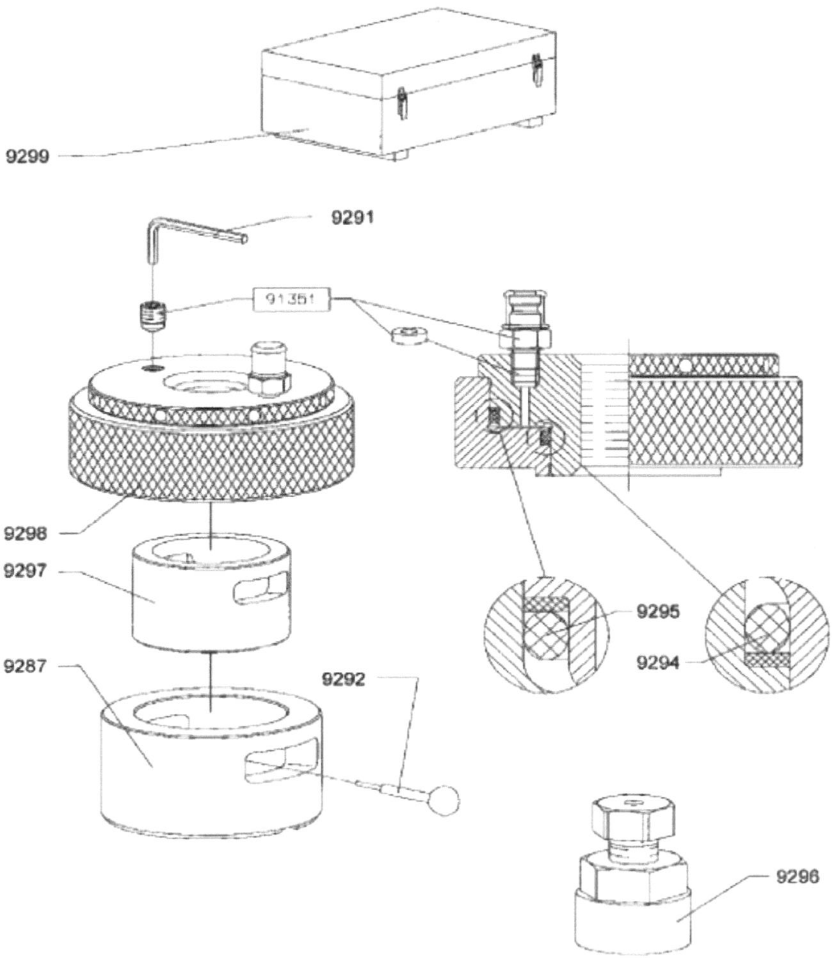

9287 : Support

9291 : Key, hexagon socket screw

9292 : Tommy bar

9294 : Sealing ring with back-up ring

9295 : Sealing ring with back-up ring

9296 : Stud setter

9297 : Support

9298 : Jack - hydraulic, complete

9299 : Steel box for hydraulic jack

Piston Crown - Hydraulic Tools

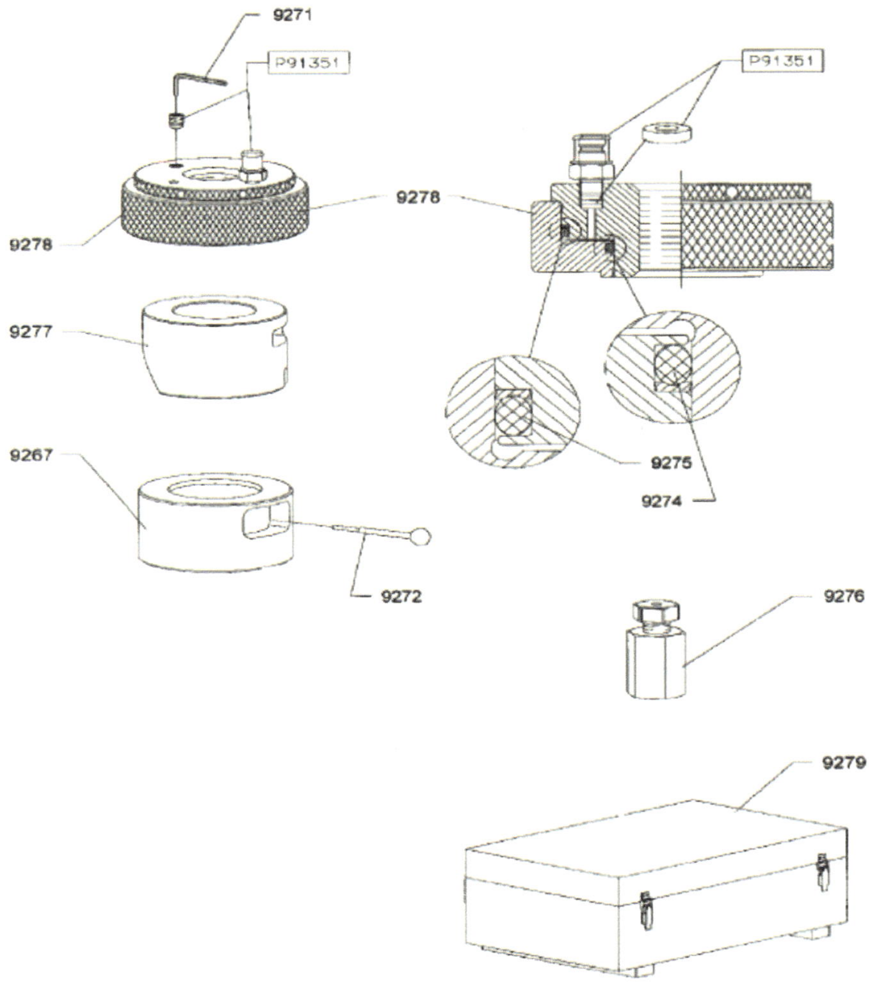

9267 : Support

9271 : Key, hexagon socket screw

9272 : Tommy bar

9274 : Sealing ring with back-up ring

9275 : Sealing ring with back-up ring

9276 : Stud setter

9277 : Support

9278 : Jack-hydraulic, complete

9279 : Steel box for hydraulic jack

Piston - Lifting Tools

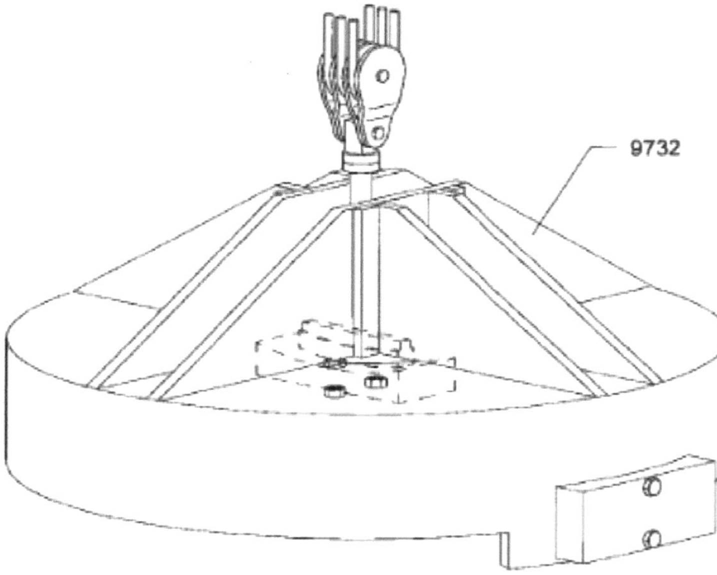

9732 : Lifting tool for piston

Cylinder Liner - Lifting Tools

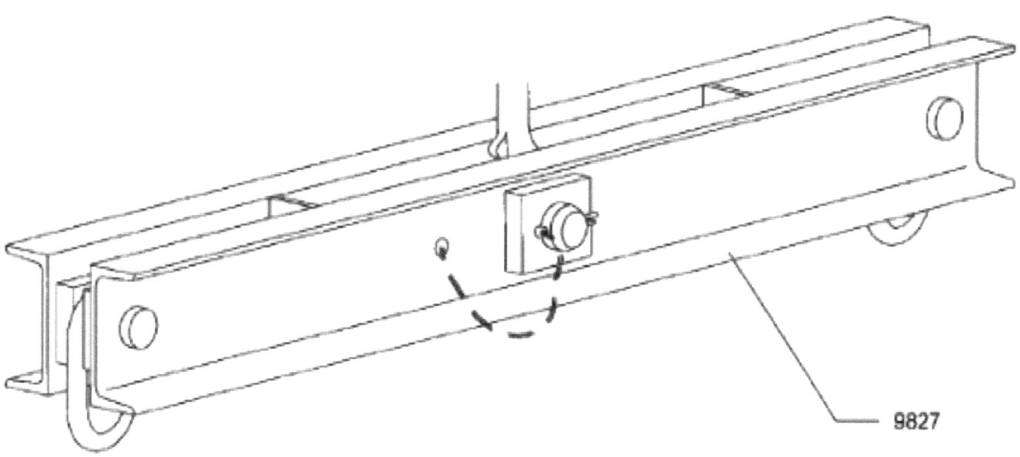

9827 : Cross bar for cylinder liner

Cylinder Liner - Tools

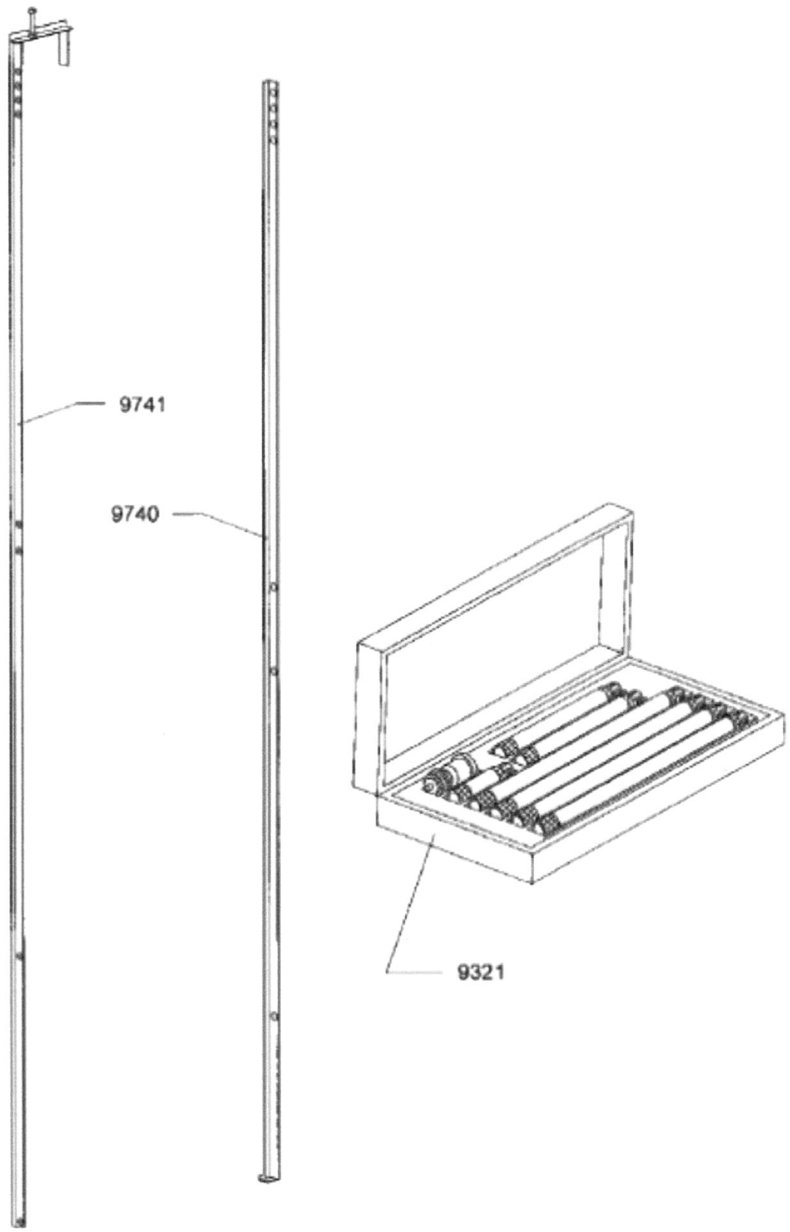

9321 : Measuring tool for cylinder liner

9740 : Measuring rod for cylinder liner, short

9741 : Measuring rod for cylinder liner, long

20) 인디케이터 테스터

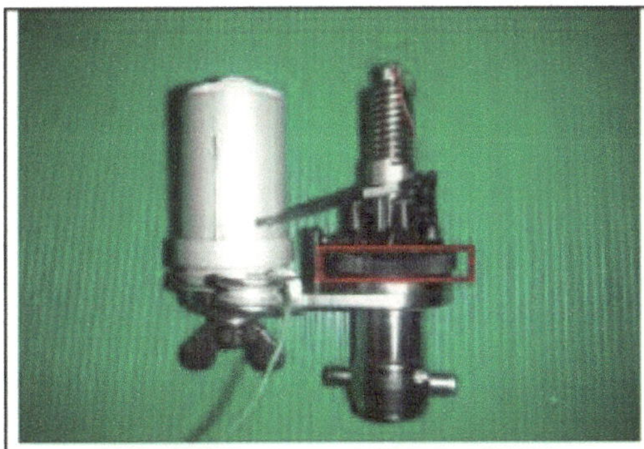

상기 표시된 Ring Nut를 Open합니다.

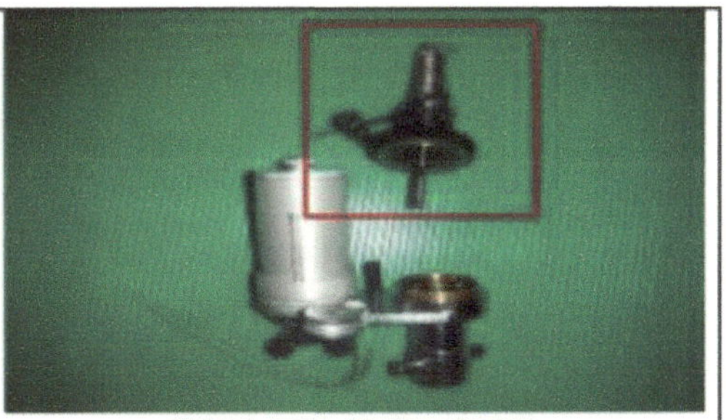

그림과 같이 분해가 됩니다.
(Piston 손상되지 않게 조심하고, Carbon이 많이 물어 있다면 잘 빠지지 않을 수 있습니다.)

Piston & Piston Rod를 소제 합니다.

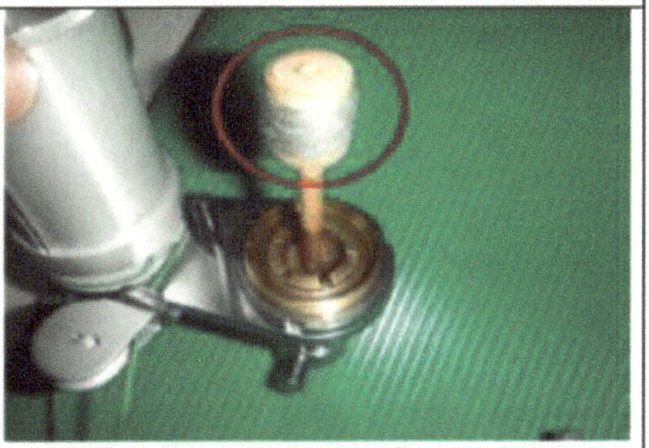

상기 Tool로 Cylinder 내부를 소제 합니다.

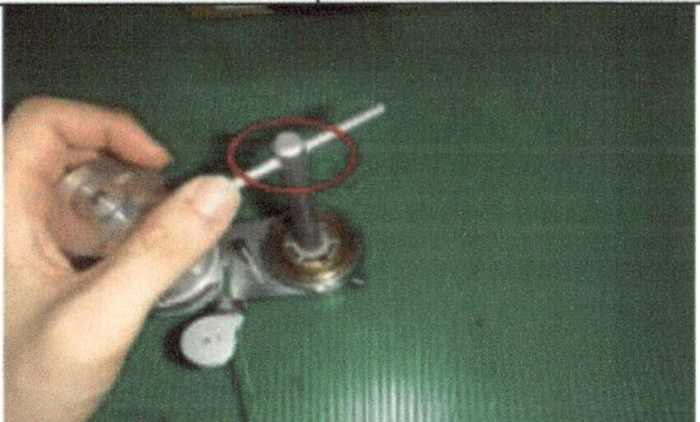

상기 Tool로 Cylinder 분해 후 내부 소제하면 됩니다.
조립은 역순으로 하시면 됩니다.

< Cylinder 내부 소제 Tool / Cylinder 분해 Tool >

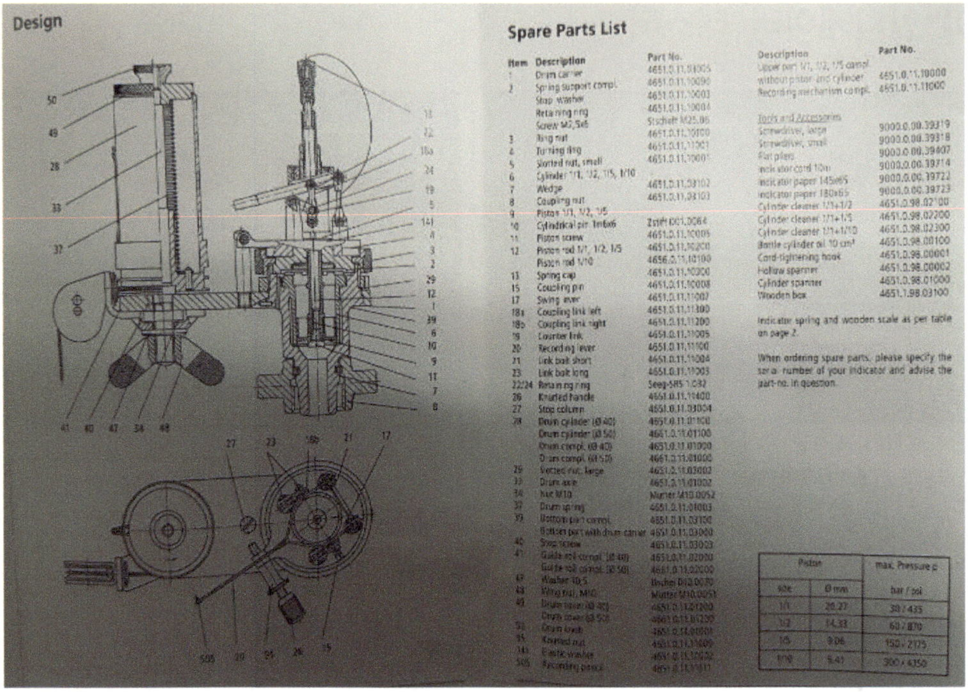

< Indicator Tester >

< Recording Pencil >

(1) 인디케이터 테스터로 P-COMP & P-MAX 계측 방법

① Test Cock 조금 Open하여 Test Cock에 남아있는 잔류 가스 및 이물질을 불어낸다.

② Indicator Tester를 Test Cock에 조립하고 측정 용지를 Indicator Tester에 장착한다.

③ 용지에 Base Line을 그린 다음 Test Cock을 Open하여 리듬을 맞춰 계측한다.

Indicator Tester 정비 시 연필심(Recording Pencil)이 불량하면 교환한다.

④ 계측한 수인선도 우측 상부에 해당 Cylinder 번호를 적고 나머지 Cylinder 계측한다.

⑤ 계측 자를 이용하여 수인선도 용지의 P-max & P-comp를 확인한다.

Test Cock 조금 Open하여 이물질 Blow

Indicator Tester 설치

Indicator Tester에 용지 장착

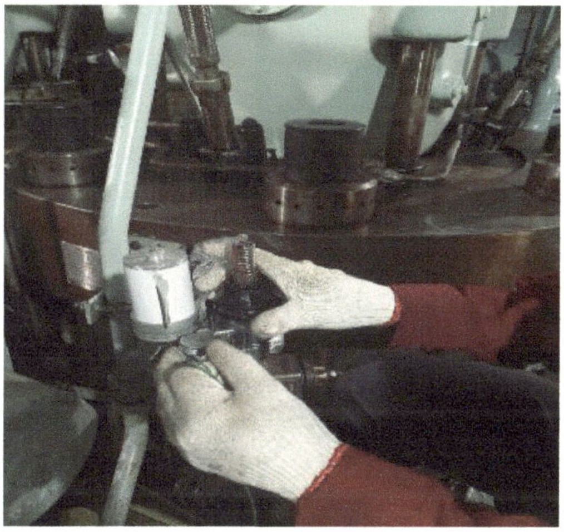

용지에 Base Line 그리기

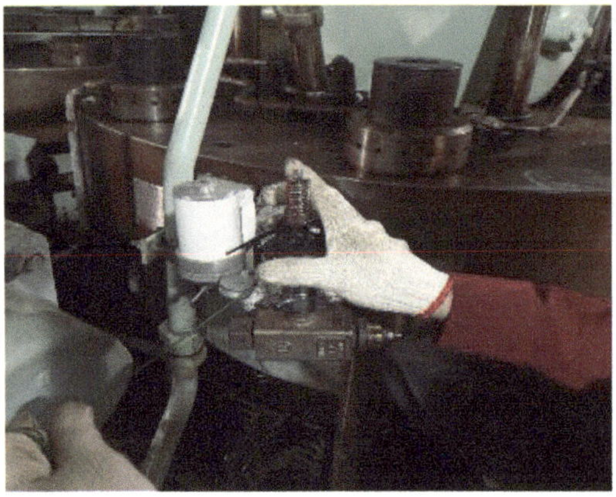

용지에 Base Line 그리기

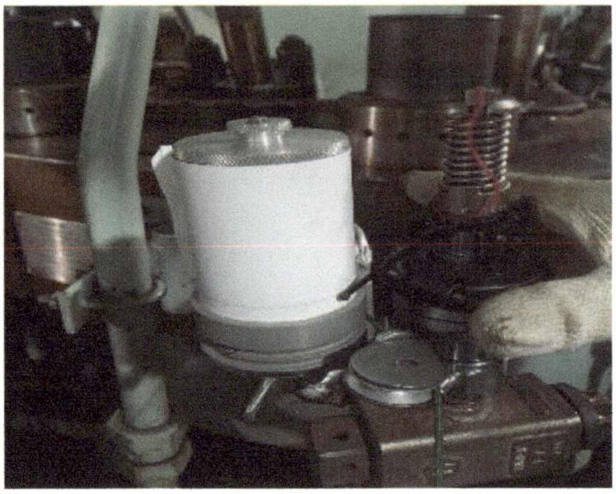

Base Line

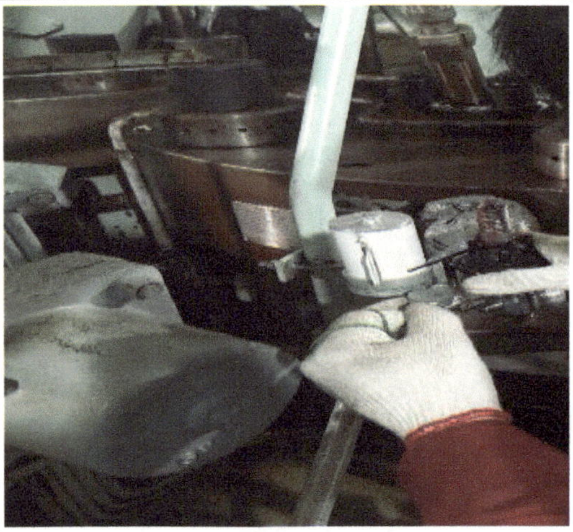

수인선도 계측

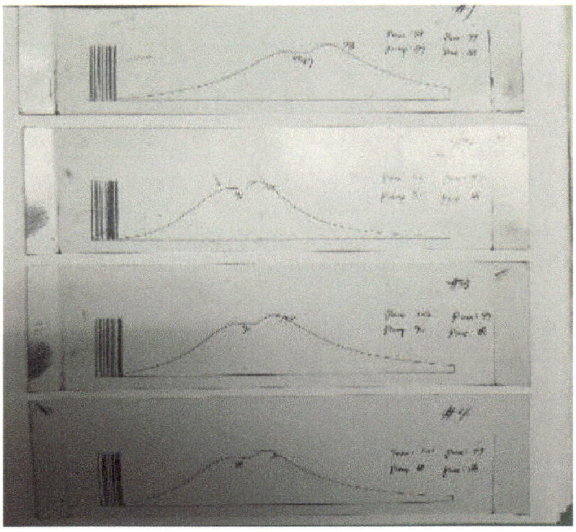

수인선도(P-max & comp)

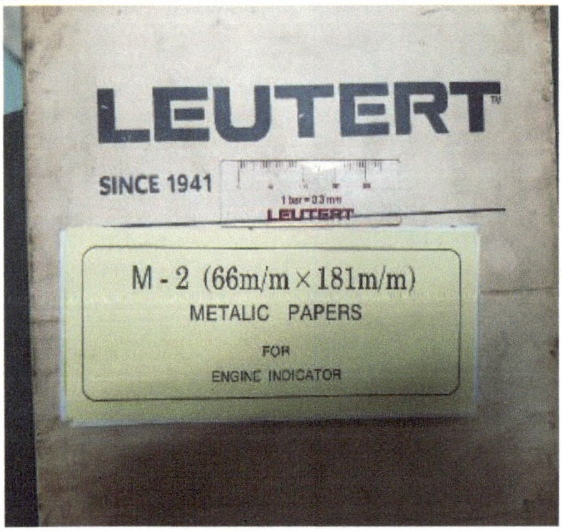

수인선도 계측 자 & 용지

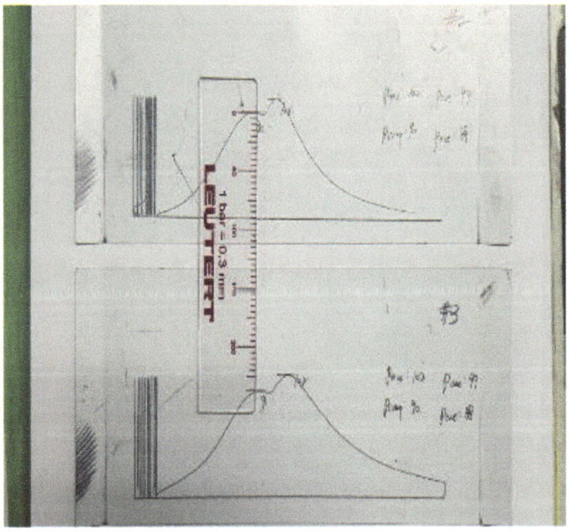

계측 값

(2) 인디케이터 테스터로 P-V선도 계측 방법

Test Cock 조금 Open하여 이물질 Blow

Indicator Tester 설치

Base Line

P-V선도 계측

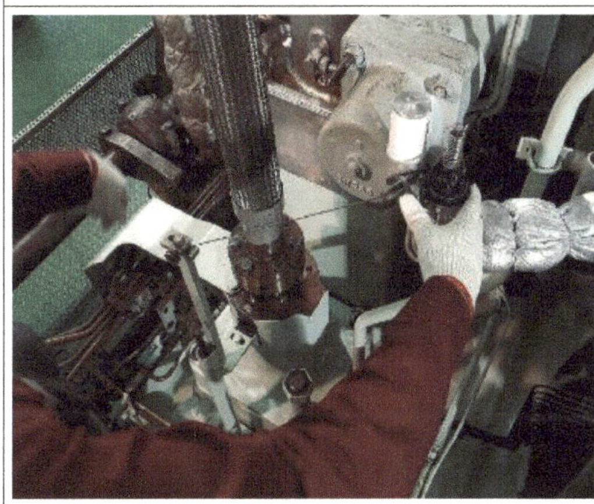

P-V선도 계측

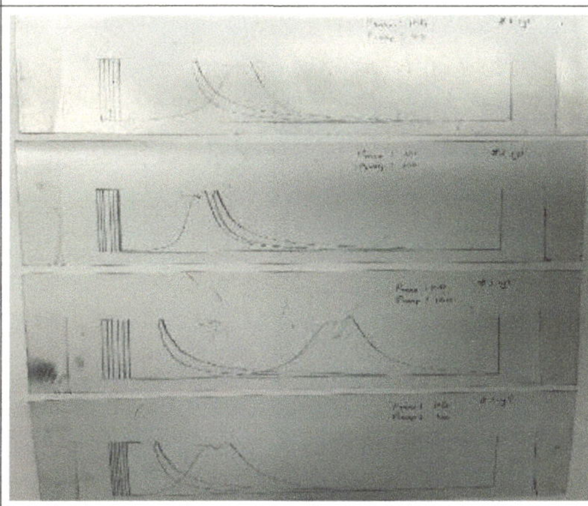

계측 값

21) 메인 엔진 퍼포먼스

M/E Performance 계측은 2달에 한번 정도로 계측을 하지만 일반적인 경우에 한 항차에 한번씩 계측을 하여 엔진 컨디션을 확인한다.

MIP 작동 방법은 쉽다.

MIP CD에 측정방법 및 Data 전송 방법이 자세하게 나와 있다.

① M/E Performance 측정 전날 미리 MIP 충전한다.
② 당일 날 측정하고자 하는 RPM까지 올린다.
③ 아래 그림처럼 Local에 있는 TDC/ Teeth 신호 Cable를 MIP에 연결한다.

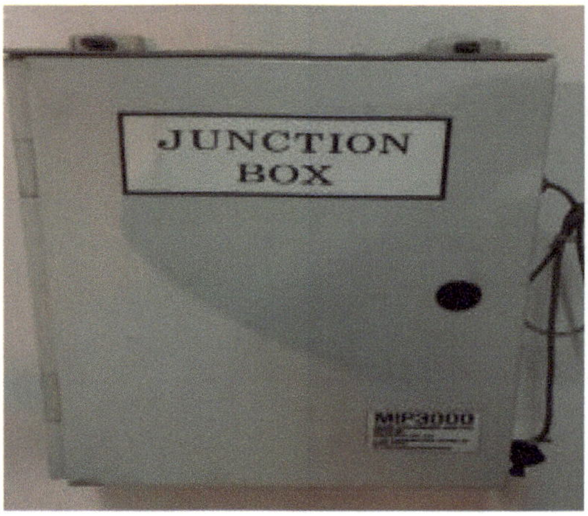

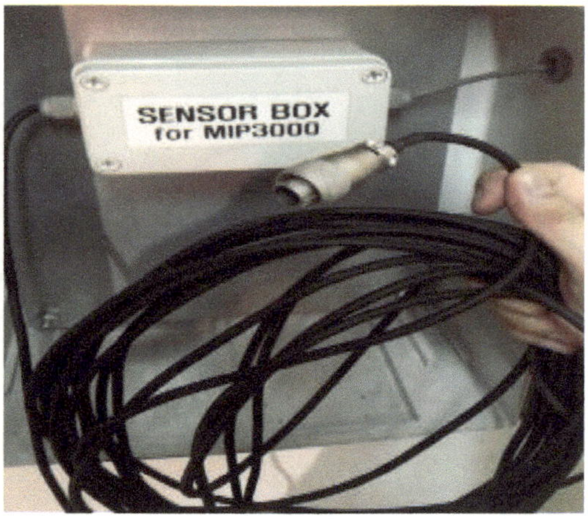

< TDC/Teeth 신호 Cable >

④ MIP CD를 넣고 미리 설치한다.
⑤ F.O Auto Filter Off 하고 각 실린더 측정 전에 Test Cock을 살짝 Open하여 Test Cock에 있는 이물질을 Blow 한다.
⑥ Test Cock 닫고 MIP와 연결한 다음 측정하면 된다.
측정 방법은 Manual을 확인한다.
측정 당시 F.O Auto Filter Off했으면 M/E Flowmeter 적어놓고 2시간 뒤에 다시 M/E Flowmeter 적는다. 총 2시간 동안 M/E Performance를 계측 하는 것이다.
전 실린더 측정 후 MIP를 ECR Computer에 연결, MIP 프로그램 작동시켜 Data를 전송한다.

측정 후 작성하는 중요 Point는 아래와 같다.

또한 Bridge 당직 Officer에게 연락하여 아래 Data들을 확인한다.

→ **Ship Speed, Displacement, Wind & Wave, Barometric Press, Ambient Temp**

ECR에서 아래의 KYMA Data를 참조한다.

→ **Total Energy Counter(SHPH), Shaft Power(SHP), RPM, Shaft Torque**

M/E Performance는 대체로 **2달** 간격으로 계측한다.

보다 자세한 설명은 Instruction Manual 다음 장의 첨부사진을 참고한다.

각 선박 마다 MIP Model이 다를 수 있으니 본선에서 사용하는 MIP Manual을 미리 확인한다.

2.8 계측 방법

**** 중요: MIP3000으로 엔진 성능 데이터를 계측 시, 사용자는 MIP3000 압력센서를 냉각하여 대기 온도와 비슷한 상태에서 사용한다.**

2.8.1 계측 "NEW"

1. MIP3000 상단의 스위치를 이용하여 전원 ON 한다.
2. MIP3000에 TDC / TEETH 신호 케이블을 연결한다.
3. MIP3000에 압력센서 케이블을 연결한다.
4. 메인 메뉴에서 "ACQUSITION" 아이콘을 터치 한다. <Fig. 2-8-1>
5. 현재 ENGINE의 추정 부하 버튼을 터치 한다. <Fig. 2-8-2>

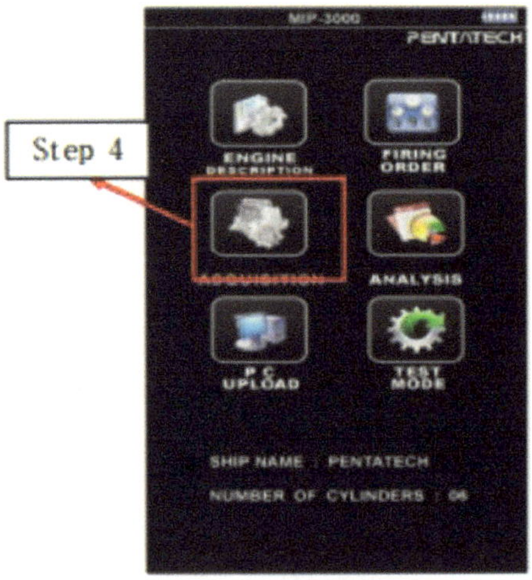

<Fig. 2-8-1>

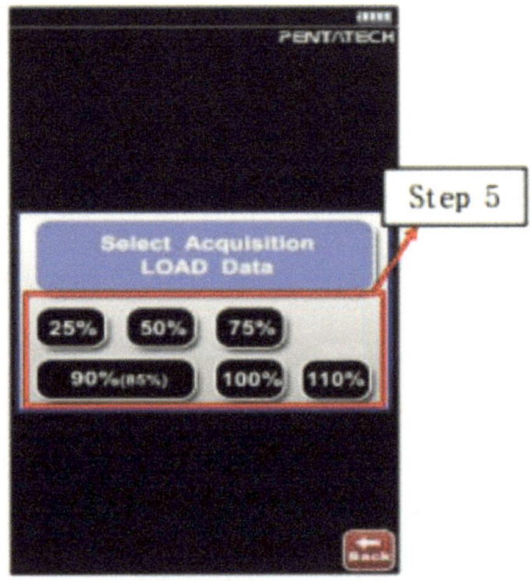

<Fig. 2-8-2>

6. "NEW"  버튼을 터치한다. <Fig. 2-8-3>
7. 계측 하고자 하는 Cylinder의 아래 흰색 부분을 터치하여 체크 표시 한다. "NEW" 계측에서는 자동으로 모든 Cylinder에 체크 표시 되어있다.  버튼을 터치하면 전체 선택, 전체 선택 해제 된다. <Fig. 2-8-4>
8. "NEXT"  버튼을 터치하여 다음 계측 화면으로 넘어간다.
9. 압력센서를 No. 1 Cylinder의 Indicator Cock에 결합한다.

10. Indicator Cock Valve를 연다.

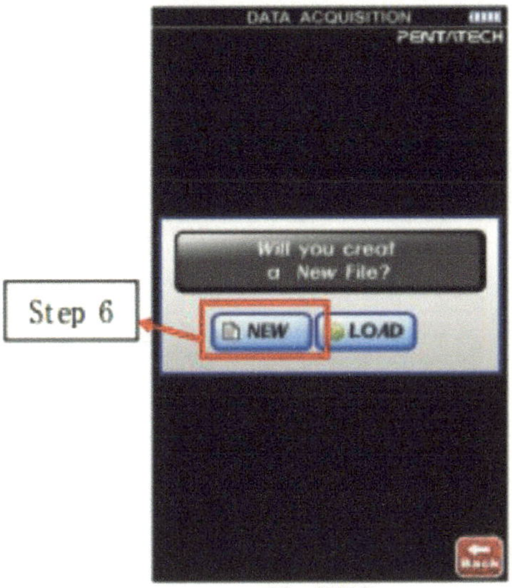

<Fig. 2-8-3>

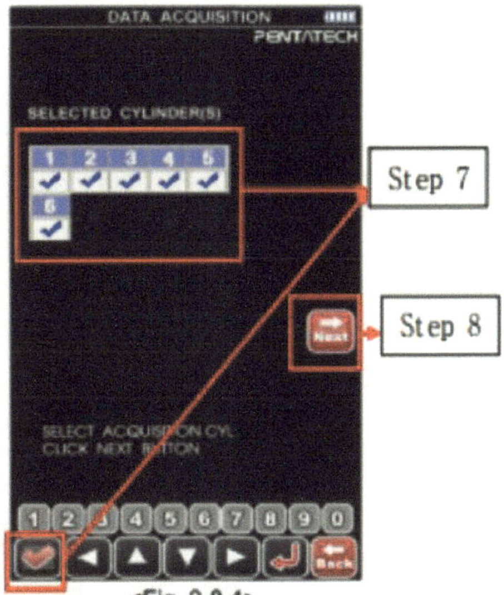

<Fig. 2-8-4>

11. 엔진 성능 데이터 계측을 위해 "Next" 버튼을 터치 한다. <Fig. 2-8-5>

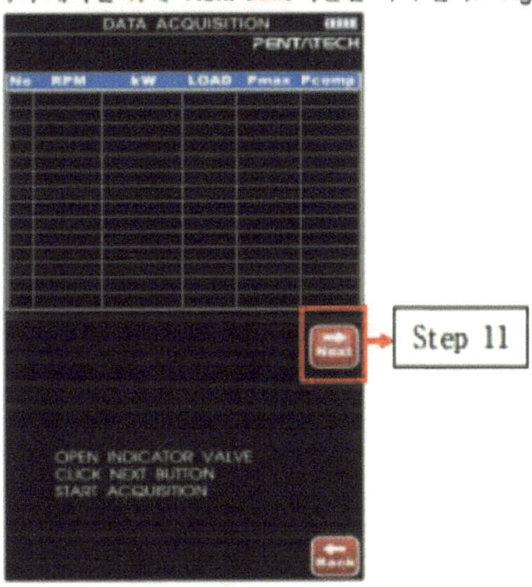

<Fig. 2-8-5>

12. No. 1 Cylinder 의 계산된 RPM과 TEETH COUNT를 확인한다. <Fig. 2-8-6>

13. 10 Cycle의 계측이 완료되고 ** **CLOSE INDICATOR VALVE** ** 메시지가 나타나면 Indicator Valve를 닫고 압력센서를 분리한다. <Fig. 2-8-7>

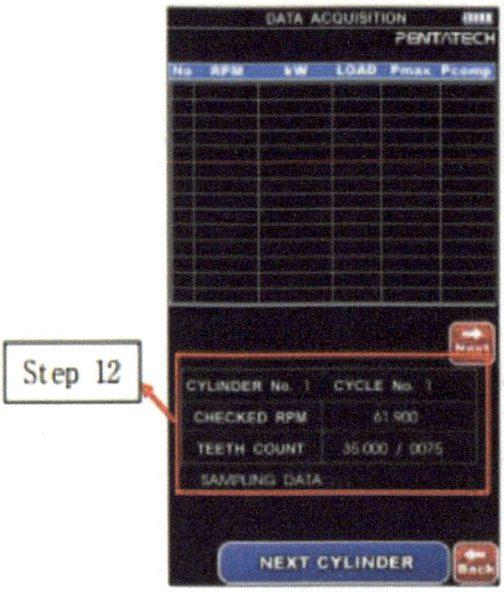

<Fig. 2-8-6>

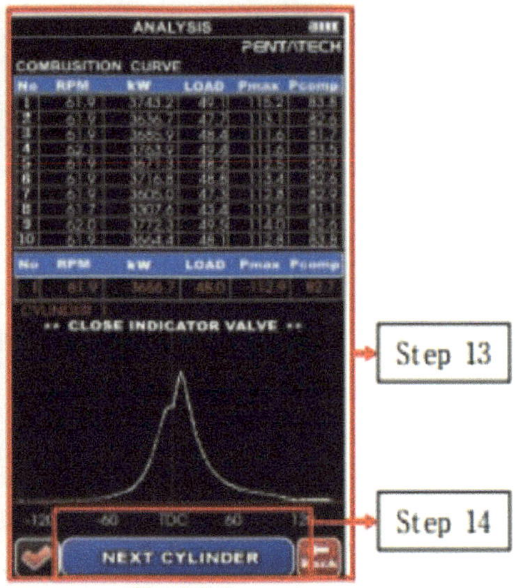

<Fig. 2-8-7>

14."NEXT CYLINDER" 버튼을 터치하면 Step11로 돌아간다.

15. 나머지 Cylinder의 계측 방법은 Step 11 에서 Step 14 반복한다.

16. 모든 Cylinder의 계측이 완료되면 "Save" 버튼을 누르면 SD CARD에 화면에 나타나는 파일 이름으로 저장된다. <Fig. 2-8-8 & 2-8-9>

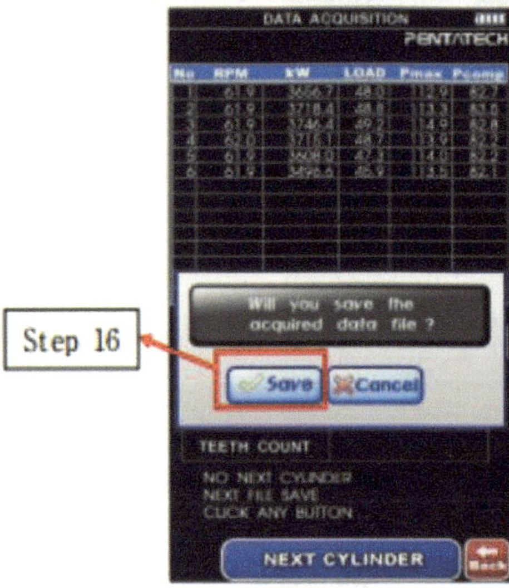

<Fig. 2-8-8>

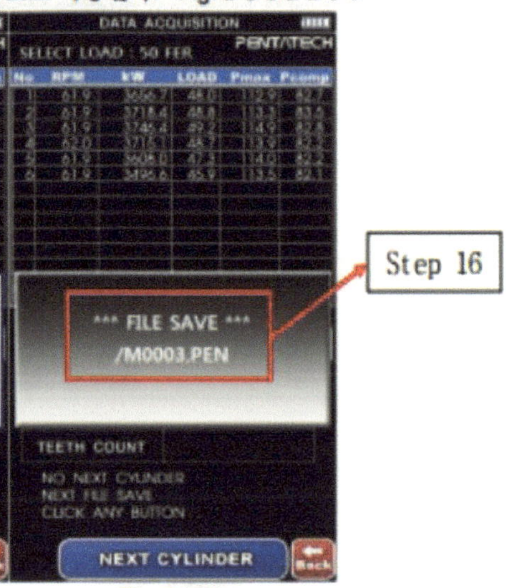

<Fig. 2-8-9>

2.8.2 재 계측 " LOAD "

1. MIP3000 상단의 스위치를 이용하여 전원 ON 한다.
2. MIP3000에 TDC / TEETH 신호 케이블을 연결 한다.
3. MIP3000에 압력센서를 케이블을 연결한다.
4. 메인 메뉴에서 "ACQUSITION" 아이콘을 터치 한다. <Fig. 2-8-10>
5. 현재 ENGINE의 추정 부하 버튼을 터치 한다. <Fig. 2-8-11>

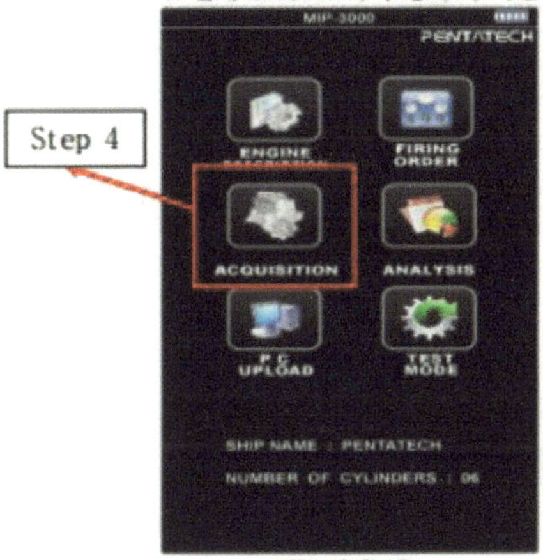

<Fig. 2-8-10>

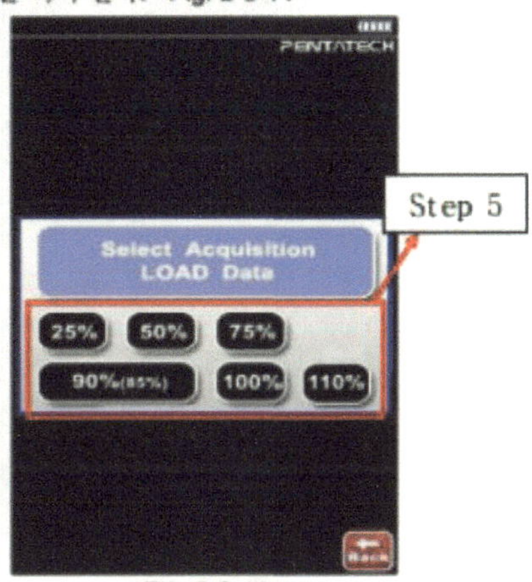

<Fig. 2-8-11>

6. "LOAD" 버튼을 터치 한다. <Fig. 2-8-12>
7. 재 계측을 원하는 Cylinder의 흰색 부분을 터치하여 체크 설정한다. <Fig. 2-8-13>

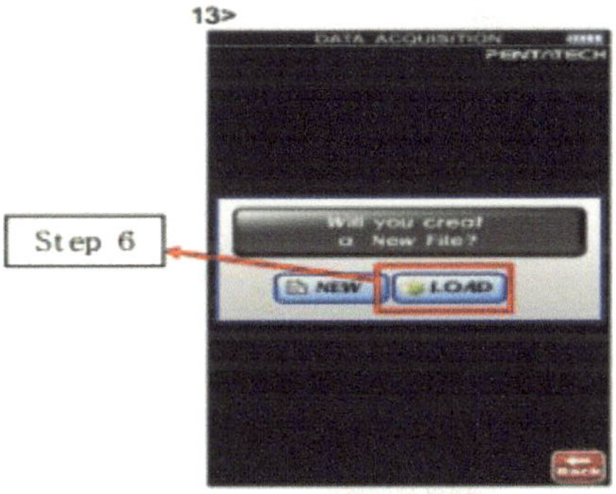

<Fig. 2-8-12>

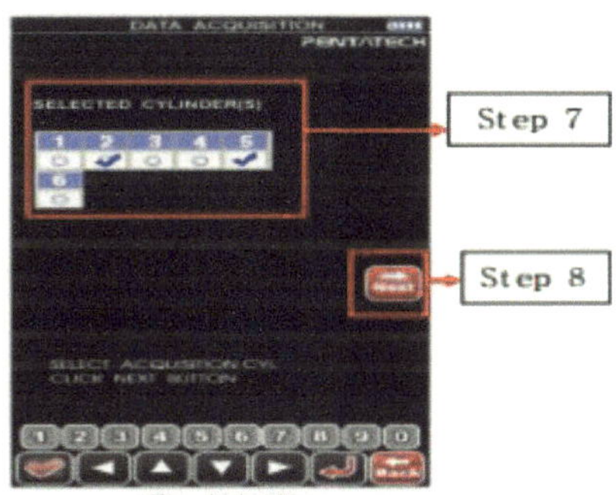

<Fig. 2-8-13>

8. "Next" 버튼을 터치한다. <Fig. 2-8-13>
9. 압력센서를 선택한 Cylinder의 Indicator Cock에 결합한다.
10. Indicator Cock Valve를 연다.
11. 엔진 성능 데이터 계측을 위해 "Next" 버튼을 터치 한다. <Fig .2-8-14>

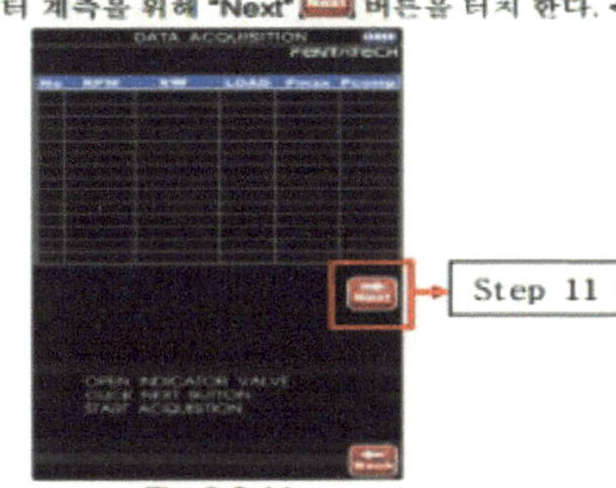

<Fig .2-8-14>

12. 선택한 실린더의 RPM과 TEETH COUNT 확인한다. <Fig .2-8-15>

13. 10 Cycle의 계측이 완료되고 *** CLOSE INDICATOR VALVE *** 메시지가 나타나면 Indicator Valve를 닫고 압력센서를 분리한다. <Fig .2-8-16>

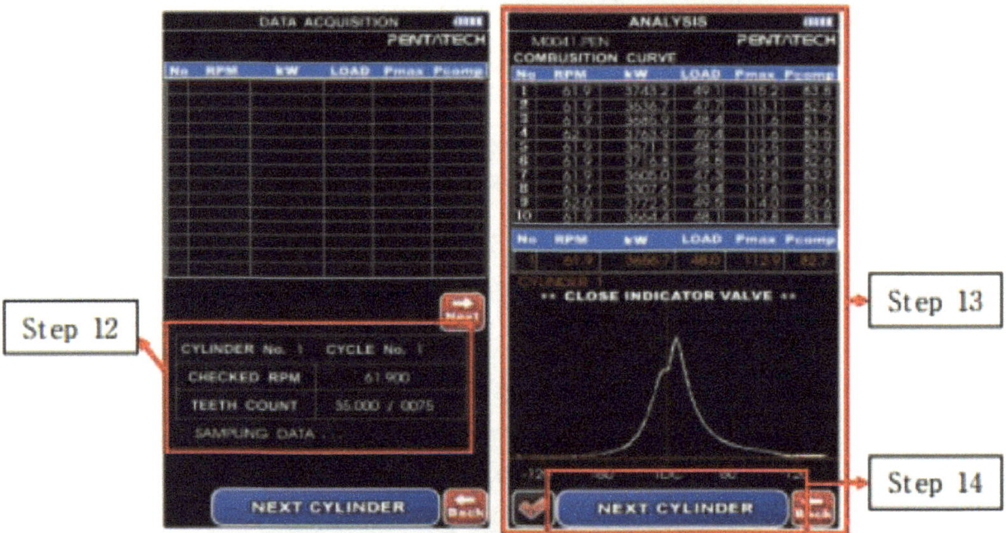

<Fig .2-8-15> <Fig .2-8-16>

14. "NEXT CYLINDER" NEXT CYLINDER 버튼을 터치하면 Step11로 돌아간다.

15. 나머지 Cylinder의 계측 방법은 Step11 에서 Step 14 반복한다.

16. "NEXT CYLINDER" NEXT CYLINDER 버튼을 터치하여 파일을 저장 할 수 있는 화면으로 넘어간다.

17. 모든 Cylinder의 계측이 완료되면 "Save" Save 버튼을 누르면 SD CARD에 화면에 나타나는 파일 이름으로 저장된다. <Fig .2-8-18>

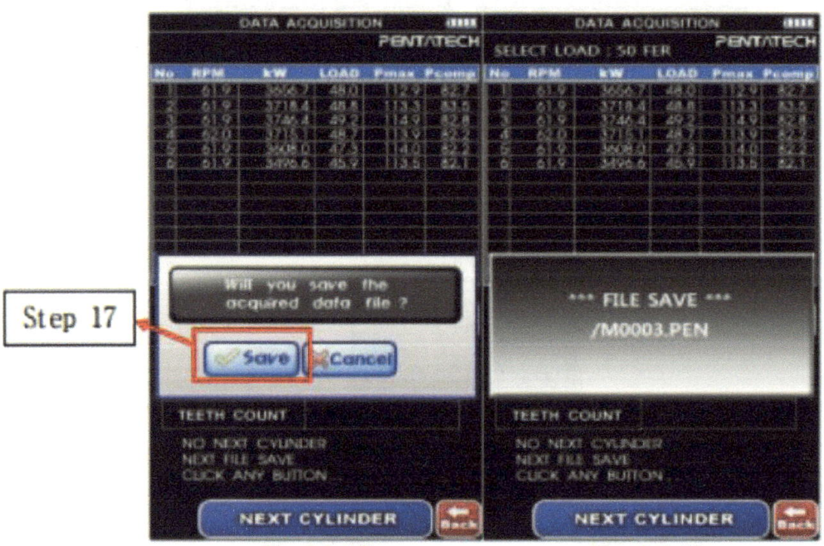

<Fig .2-8-17> <Fig .2-8-18>

3 MIP3000 설정 방법

3.1 M/E 정보 설정

MIP3000 실행 및 출력 계산에 필요한 ENGINE 데이터가 입력 되어야 한다. 최초 1회만 입력하면 되고, 추후 변경 사항이 있으면 수정 가능하다.

3.1.1 M/E Data 설정

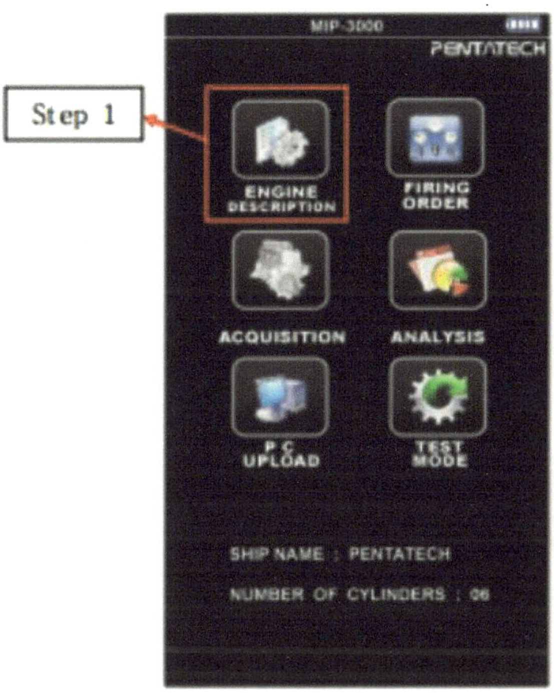

< Fig. 3-1-1 >

< Fig. 3-1-2 >

1. 메인 화면에서 "ENGINE DESCRIPTION"  버튼을 터치한다. < Fig. 3-1-1 >.
2. 입력 커서는 SHIP NAME에 나타난다.
3. 화면 하단의 알파벳과 숫자 키보드를 터치하여 SHIP NAME을 입력한다. SHIP NAME을 입력하면 "Enter"  버튼 또는  버튼을 터치하면 다음 항목인 NUMBER OF CYLINDERS 를 입력할 수 있는 위치로 커서가 이동된다.
4. NUMBER OF CYLINDERS, CYLINDER DIAMETER(cm), PISTON STORKE(cm), CONNECTING ROD LENGTH(cm) 데이터를 입력 한다.

3.2 참고

- MIP3000 전원이 OFF 상태라도 이전에 저장된 계측 파일은 모두 SD MEMORY CARD에 저장되어 있고 파일 이름은 순번으로 지정된다.
- MIP3000에서 저장된 계측 파일은 PC에 전송하여 상세 분석을 할 수 있다.
- 만일 계측 도중 압력센서 또는 TDC & TEETH 센서 문제, 계측 순서 오류일 경우 계측을 중지 하고, "LOAD" 기능으로 마지막으로 계측 한 파일을 이용하여 재 계측 할 수 있다.
- MIP3000은 32bit Micro Processor, 10 bit A/D converter, 32M x 8 bit NAND Flash Memory, SD Memory Card 1GB로 구성되어 있다.
- SD Memory Card에 저장되어 있는 계측 파일은 Serial Cable 또는 SD Card Reader를 이용하여 PC로 전송 할 수 있다.
- MIP3000 배터리가 완전 충전된 상태에서 계측을 하고, 자주 배터리를 충전하면 배터리 수명에 좋지 않다.
- MIP3000 상단에 있는 전원 스위치< Fig. 3-2-1>를 이용하여 전원 ON 하면 LCD 모니터와 PWR 램프가 ON 된다.

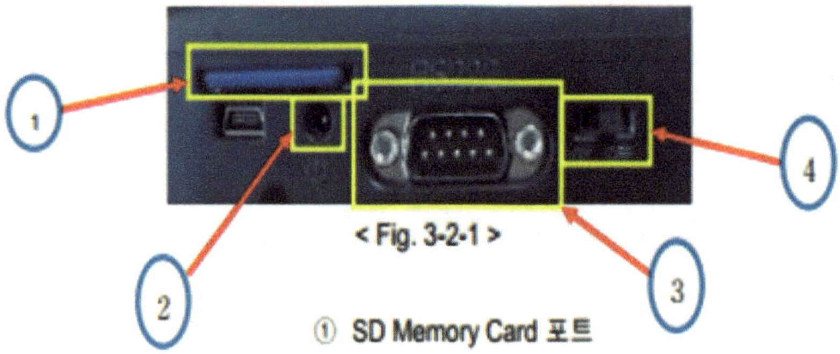

< Fig. 3-2-1 >

① SD Memory Card 포트
② 배터리 충전 연결 포트
③ RS232C 케이블 포트
④ 파워 스위치

만약에 MIP를 조작하다가 실수로 Data를 초기화 하거나, 어떠한 이유로 MIP 초기화가 되었다면 기본적인 Engine Data를 다시 Re-setting 해야 한다.

입력 Data중에 Convert Rate같은 것들은 초기 Maker에서 설정해주는 Data이다.

이전에 측정한 MIP파일 열어보면 아래와 같이 Edit Remark에 설정 Data값 확인할 수 있다.

보기 편하게 아래의 메모장에 옮겨 적었다.

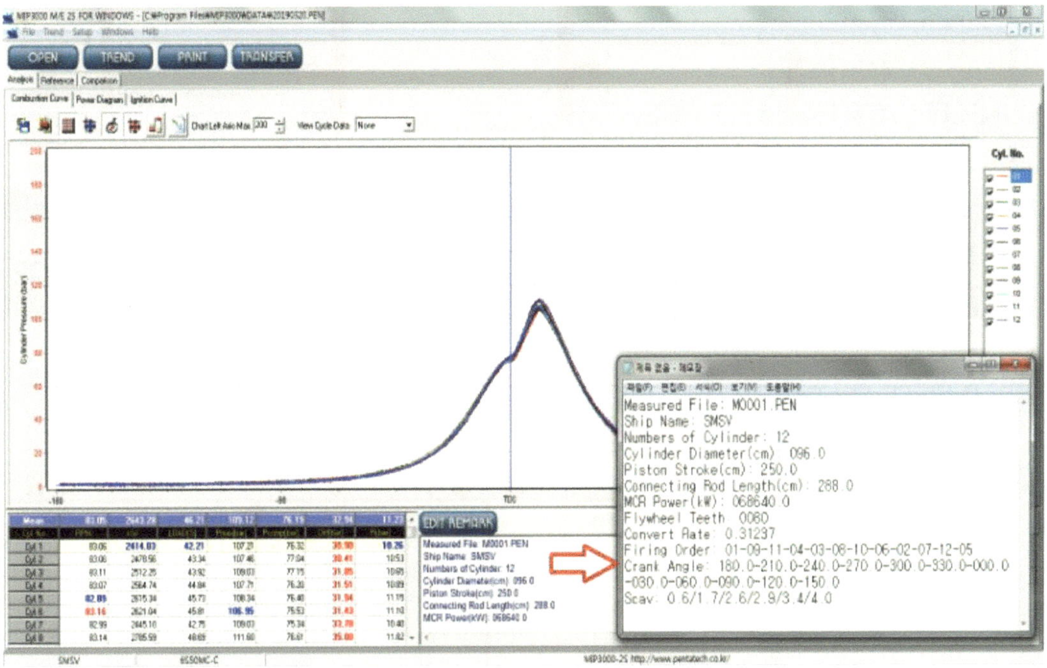

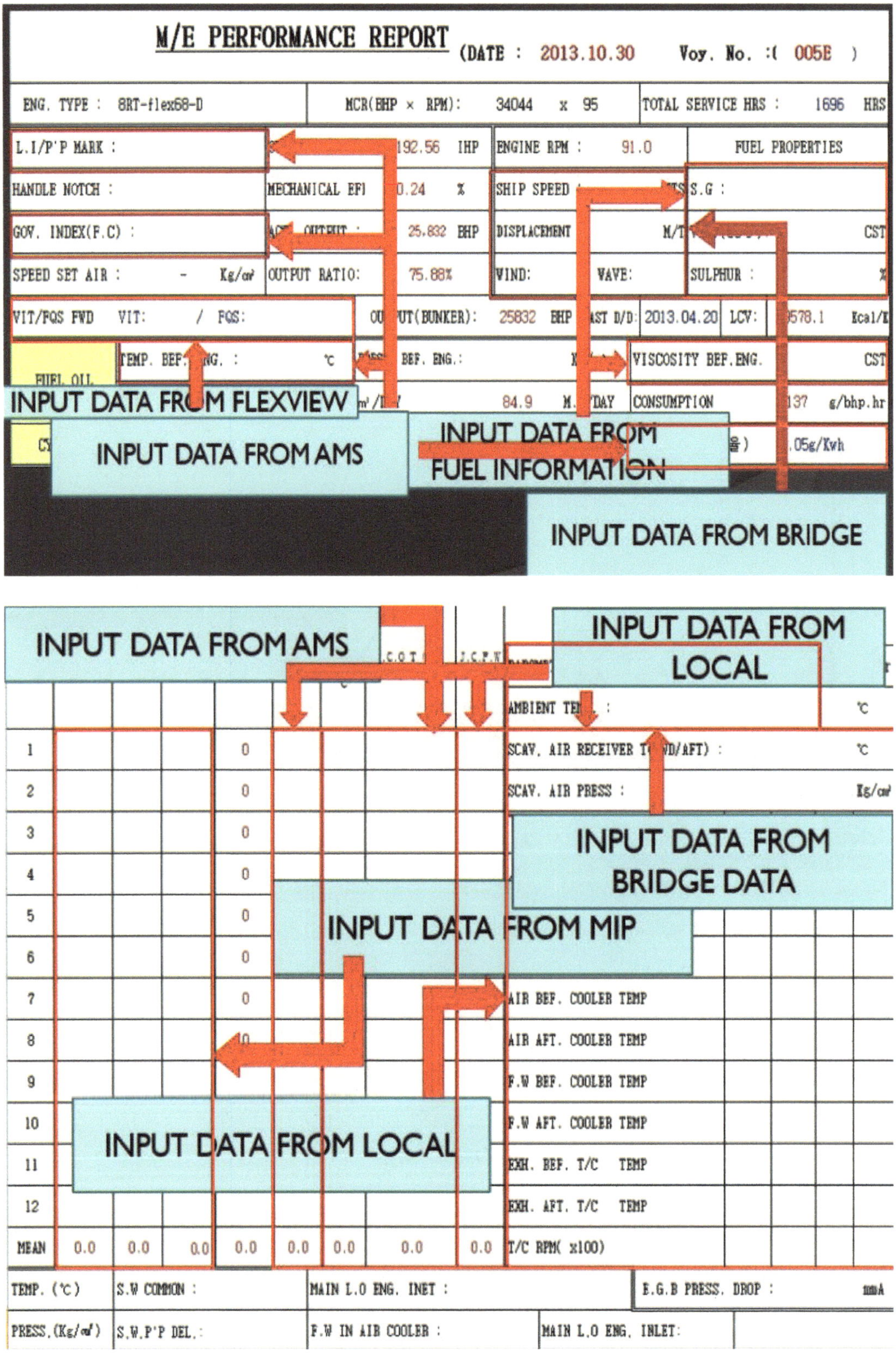
M/E PERFORMANCE REPORT
(DATE : 2013.10.30 Voy. No. :(005E)
ENG. TYPE : 8RT-flex68-D
MCR(BHP × RPM): 34044 x 95
TOTAL SERVICE HRS : 1696 HRS
L.I/P'P MARK :
192.56 IHP
ENGINE RPM : 91.0
FUEL PROPERTIES
HANDLE NOTCH :
MECHANICAL EFI
0.24 %
SHIP SPEED
S.G :
GOV. INDEX(F.C) :
25,832 BHP
DISPLACEMENT
M/T
CST
SPEED SET AIR : - Kg/㎠
OUTPUT RATIO:
75.88%
WIND:
WAVE:
SULPHUR :
%
VIT/FQS FWD VIT: / FQS:
25832 BHP
2013.04.20
LCV:
VISCOSITY BEF.ENG.
CST
84.9
CONSUMPTION
g/bhp.hr
INPUT DATA FROM FLEXVIEW
INPUT DATA FROM AMS
INPUT DATA FROM FUEL INFORMATION
INPUT DATA FROM BRIDGE
INPUT DATA FROM AMS
INPUT DATA FROM LOCAL
AMBIENT TEMP. :
℃
SCAV. AIR RECEIVER T(FWD/AFT) :
℃
SCAV. AIR PRESS :
Kg/㎠
INPUT DATA FROM BRIDGE DATA
INPUT DATA FROM MIP
AIR BEF. COOLER TEMP
AIR AFT. COOLER TEMP
F.W BEF. COOLER TEMP
F.W AFT. COOLER TEMP
EXH. BEF. T/C TEMP
EXH. AFT. T/C TEMP
T/C RPM(x100)
INPUT DATA FROM LOCAL
1
2
3
4
5
6
7
8
9
10
11
12
MEAN
0.0
TEMP. (℃)
S.W COMMON :
MAIN L.O ENG. INET :
E.G.B PRESS. DROP :
mmA
PRESS.(Kg/㎠)
S.W.P'P DEL. :
F.W IN AIR COOLER :
MAIN L.O ENG. INLET:

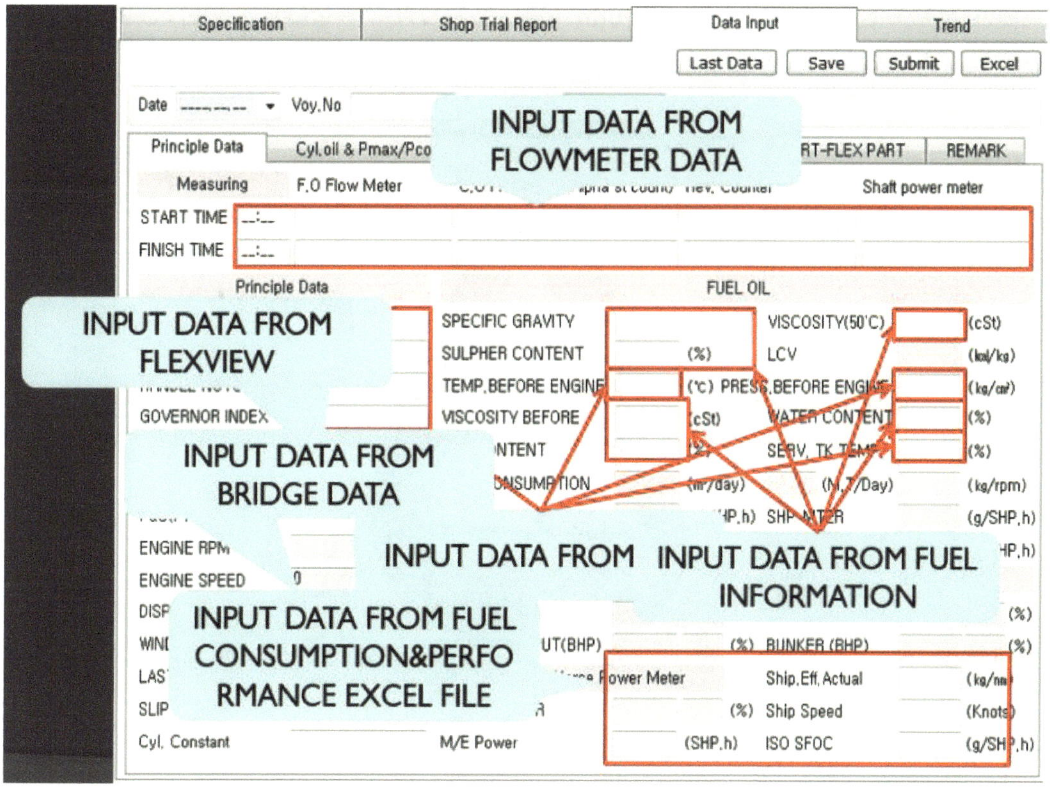
Specification
Shop Trial Report
Data Input
Trend
Last Data
Save
Submit
Excel
INPUT DATA FROM FLOWMETER DATA
INPUT DATA FROM FLEXVIEW
INPUT DATA FROM BRIDGE DATA
INPUT DATA FROM
INPUT DATA FROM FUEL INFORMATION
INPUT DATA FROM FUEL CONSUMPTION&PERFORMANCE EXCEL FILE

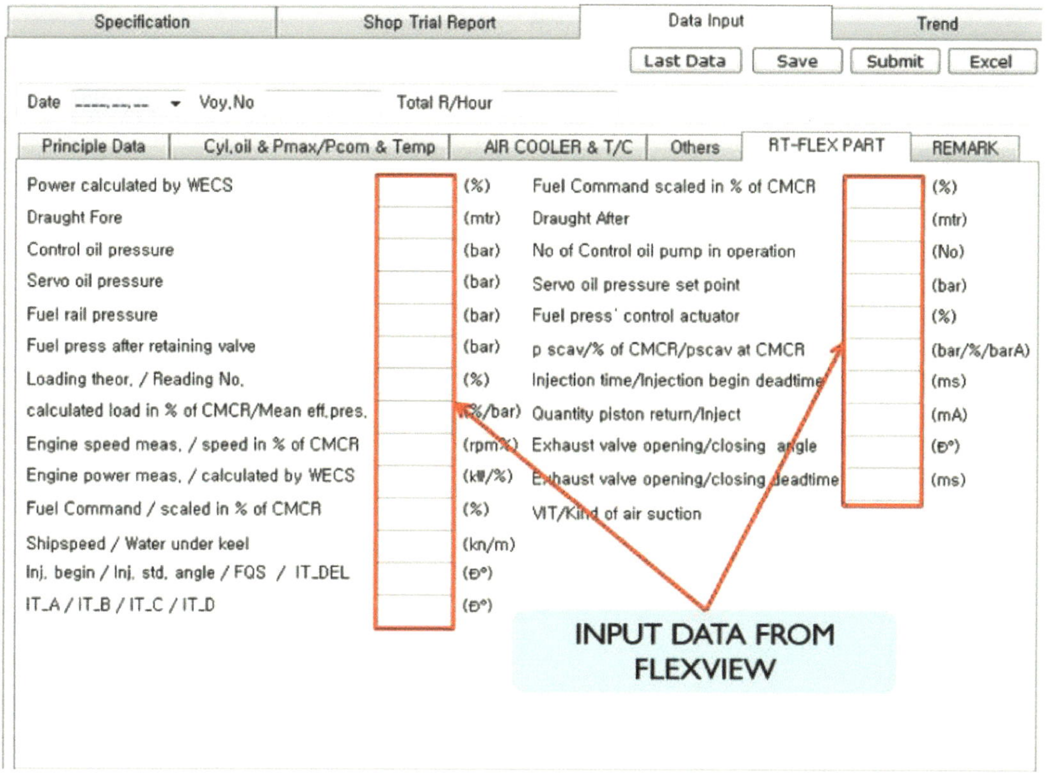
Specification
Shop Trial Report
Data Input
Trend
Last Data
Save
Submit
Excel
Principle Data
Cyl.oil & Pmax/Pcom & Temp
AIR COOLER & T/C
Others
RT-FLEX PART
REMARK
INPUT DATA FROM FLEXVIEW

상기 설명은 MIP3000에 관한 설명이며 Report 작성법은 동일하지만 만약, 아래 Type의 MIP(HLV-2005MK) 보유 중이라면 참조한다.

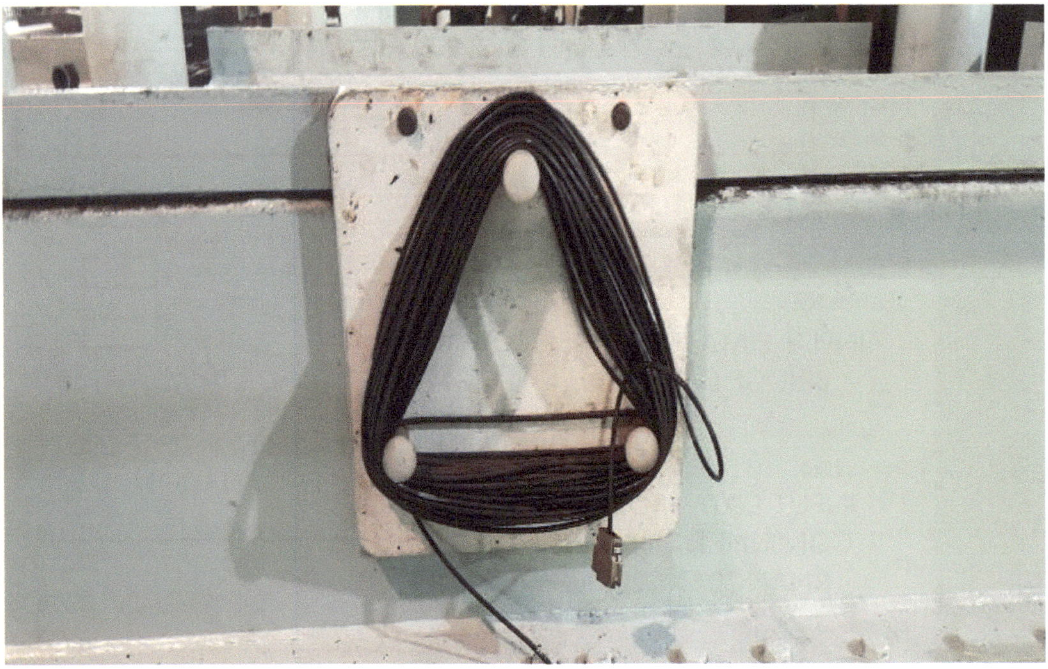

< TDC/Teeth 신호 Cable >

PC : MIP로 측정한 Data를 PC로 전송할 때 사용한다.

Mode : Single Cylinder / Complete Engine 선택할 수 있다.

E.Init : Setting된 Engine들을 확인할 수 있습니다. (EX : Main Engine / Generator Engine)

E.data : 측정할 Engine Data를 입력한다.

Options : 시간 및 세부사항들을 Setting할 수 있다.

Cal. : 측정값을 확인할 수 있다.

Measure : 측정 시에 사용한다.

Graph : 측정값을 Graph로 볼 수 있다.

MIP 충전

TDC/Teeth 신호 Cable

Test Cock 조금 Open하여 이물질 Blow

MIP Sensor Test Cock에 장착

MIP에 TDC & Teeth 신호 Cable과 Sensor장착

Test Cock Open

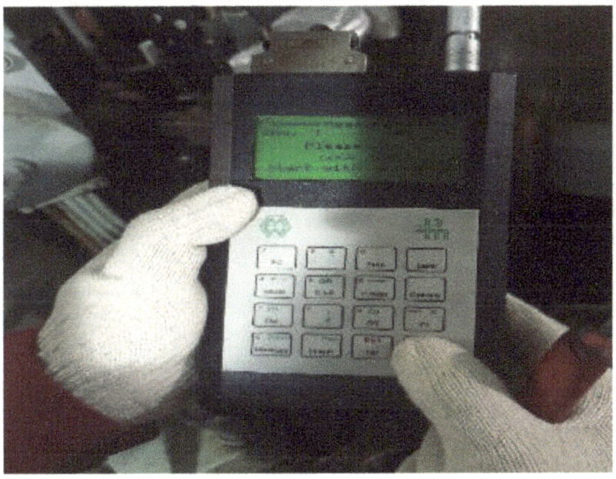

Test Cock Open

Measure in progress

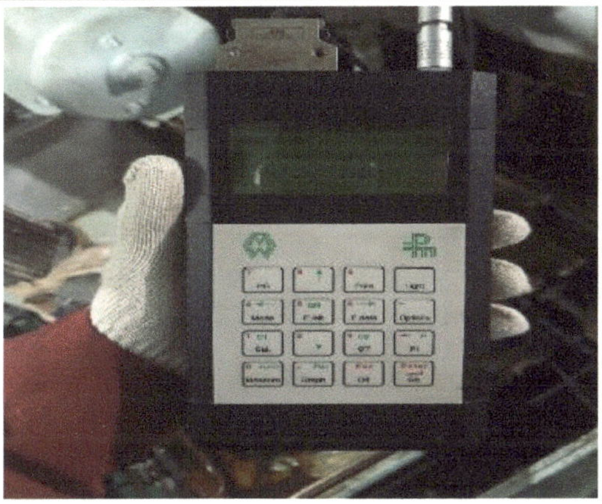

Test Cock Close

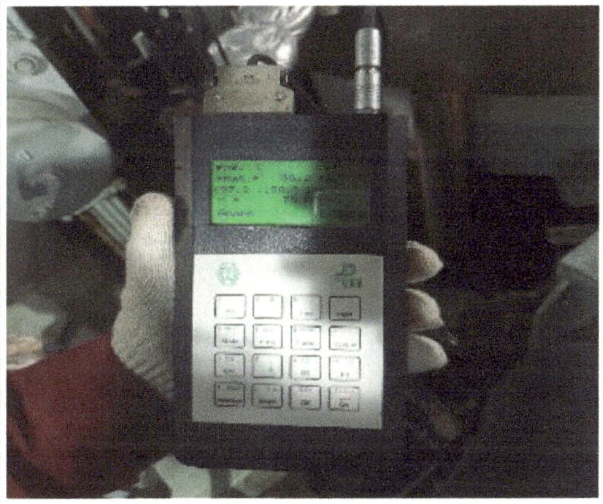

계측 Data

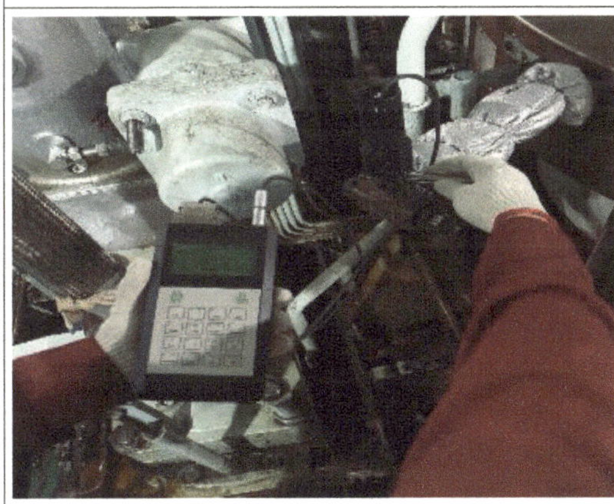

Sensor 분해

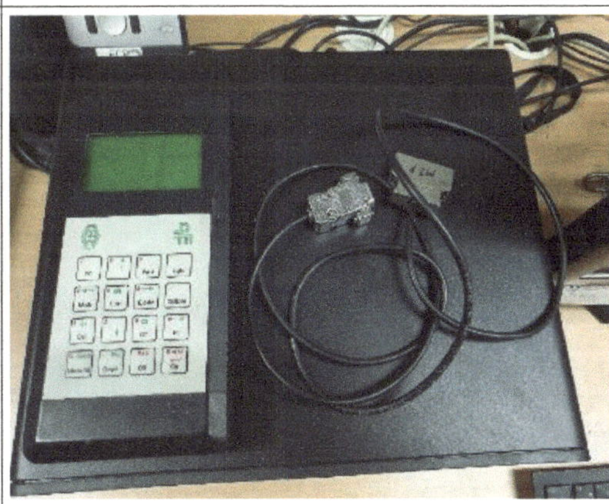

Data전송을 위한 연결 잭

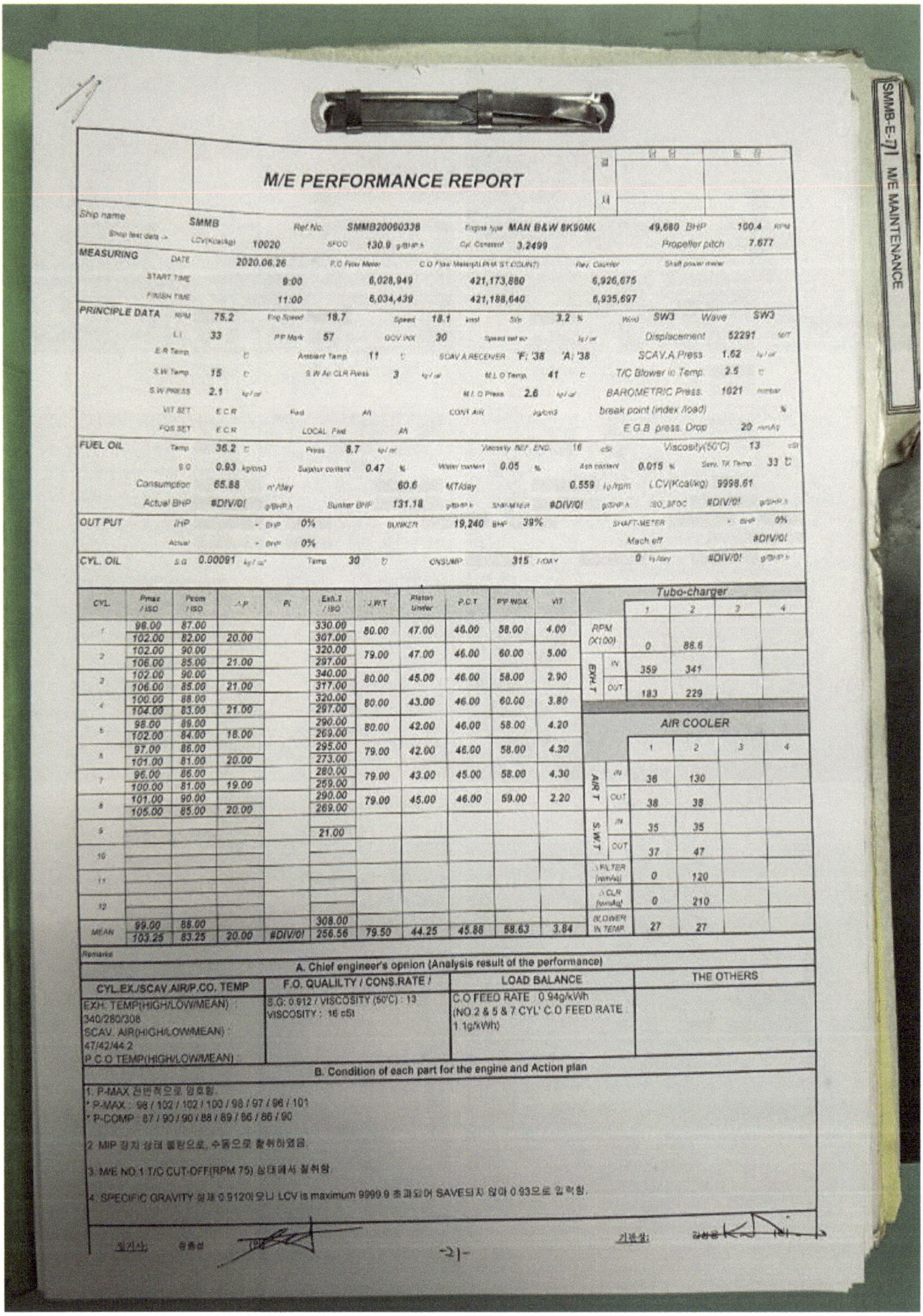

SMMB-E-71 M/E MAINTENANCE

M/E PERFORMANCE REPORT

Ship name: SMMB | Ref.No. SMMB20060338 | Engine type MAN B&W 8K90MC | 49,680 BHP | 100.4 RPM
Shop test data → LCV(Kcal/kg) 10020 | SFOC 130.9 g/BHP.h | Cyl. Constant 3.2499 | Propeller pitch 7.677

MEASURING: DATE 2020.06.26 | F.O Flow Meter | C.O Flow Meter(ALPHA ST.COUNT) | Rev. Counter | Shaft power meter
START TIME 9:00 | 6,028,949 | 421,173,880 | 6,926,675
FINISH TIME 11:00 | 6,034,439 | 421,188,640 | 6,935,697

PRINCIPLE DATA: RPM 75.2 | Eng Speed 18.7 | Speed 18.1 knot | Slip 3.2 % | Wind SW3 | Wave SW3
L.I 33 | P.P Mark 57 | GOV INX 30 | Displacement 52291 M/T
E.R Temp. ℃ | Ambient Temp. 11 ℃ | SCAV.A RECEIVER 'F: '38 'A: '38 | SCAV.A Press 1.62 kg/㎠
S.W Temp. 15 ℃ | S.W Air CLR Press 3 kg/㎠ | M.L.O Temp. 41 ℃ | T/C Blower in Temp. 2.5 ℃
S.W PRESS 2.1 kg/㎠ | M.L.O Press 2.6 kg/㎠ | BAROMETRIC Press. 1021 mbar
VIT SET E.C.R Fwd Aft | CONT AIR kg/cm2 | break point (index /load) %
FQS SET E.C.R LOCAL Fwd Aft | E.G.B press. Drop 20 mmAq

FUEL OIL: Temp. 36.2 ℃ | Press 8.7 kg/㎠ | Viscosity REF. END. 16 cSt | Viscosity(50℃) 13 cSt
S.G 0.93 kg/cm3 | Sulphur content 0.47 % | Water content 0.05 % | Ash content 0.015 % | Serv. TK Temp. 33 ℃
Consumption 65.88 m³/day | 60.6 MT/day | 0.559 kg/rpm | LCV(Kcal/kg) 9998.61
Actual BHP #DIV/0! g/BHP.h | Bunker BHP 131.18 g/BHP.h | SHAFT METER #DIV/0! g/SHP.h | ISO_SFOC #DIV/0! g/SHP.h

OUT PUT: IHP - BHP 0% | BUNKER 19,240 BHP 39% | SHAFT-METER - BHP 0%
Actual - BHP 0% | Mech eff #DIV/0!

CYL. OIL: S.G 0.00091 kg/㎠ | Temp. 30 ℃ | CONSUMP. 315 l/DAY | 0 kg/day | #DIV/0! g/BHP.h

CYL.	Pmax / ISO	Pcom / ISO	ΔP	Pi	Exh.T / ISO	J.W.T	Piston Under	P.C.T	P.P INDX	VIT
1	96.00 / 102.00	87.00 / 82.00	20.00		330.00 / 307.00	80.00	47.00	46.00	58.00	4.00
2	102.00 / 106.00	90.00 / 85.00	21.00		320.00 / 297.00	79.00	47.00	46.00	60.00	5.00
3	102.00 / 106.00	90.00 / 85.00	21.00		340.00 / 317.00	80.00	45.00	46.00	58.00	2.90
4	100.00 / 104.00	88.00 / 83.00	21.00		320.00 / 297.00	80.00	43.00	46.00	60.00	3.80
5	98.00 / 102.00	89.00 / 84.00	18.00		290.00 / 269.00	80.00	42.00	46.00	58.00	4.20
6	97.00 / 101.00	86.00 / 81.00	20.00		295.00 / 273.00	79.00	42.00	46.00	58.00	4.30
7	96.00 / 100.00	86.00 / 81.00	19.00		280.00 / 259.00	79.00	43.00	45.00	58.00	4.30
8	101.00 / 105.00	90.00 / 85.00	20.00		290.00 / 269.00	79.00	45.00	46.00	59.00	2.20
9					21.00					
10										
11										
12										
MEAN	99.00 / 103.25	88.00 / 83.25	20.00	#DIV/0!	308.00 / 256.56	79.50	44.25	45.88	58.63	3.84

Tubo-charger		1	2	3	4
RPM (X100)		0	88.6		
EXH.T	IN	359	341		
EXH.T	OUT	183	229		

AIR COOLER		1	2	3	4
AIR T	IN	36	130		
AIR T	OUT	38	38		
S.W.T	IN	35	35		
S.W.T	OUT	37	47		
ΔFILTER (mmAq)		0	120		
ΔCLR (mmAq)		0	210		
BLOWER IN TEMP.		27	27		

Remarks

A. Chief engineer's opnion (Analysis result of the performance)

CYL.EX./SCAV.AIR/P.CO. TEMP	F.O. QUALILTY / CONS.RATE /	LOAD BALANCE	THE OTHERS
EXH. TEMP(HIGH/LOW/MEAN) : 340/280/308 SCAV. AIR(HIGH/LOW/MEAN) : 47/42/44.2 P.C.O TEMP(HIGH/LOW/MEAN) :	S.G: 0.912 / VISCOSITY (50℃) : 13 VISCOSITY : 16 cSt	C.O FEED RATE : 0.94g/kWh (NO.2 & 5 & 7 CYL' C.O FEED RATE : 1.1g/kWh)	

B. Condition of each part for the engine and Action plan

1. P-MAX 전반적으로 양호함.
* P-MAX : 98 / 102 / 102 / 100 / 98 / 97 / 96 / 101
* P-COMP : 87 / 90 / 90 / 88 / 89 / 86 / 86 / 90

2. MIP 장치 상태 불량으로, 수동으로 채취하였음.

3. M/E NO.1 T/C CUT-OFF(RPM 75) 상태에서 채취함.

4. SPECIFIC GRAVITY 실제 0.912이오나 LCV is maximum 9999.9 초과되어 SAVE되지 않아 0.93으로 입력함.

1기사: 기관장:

-21-

< M/E Performance Report >

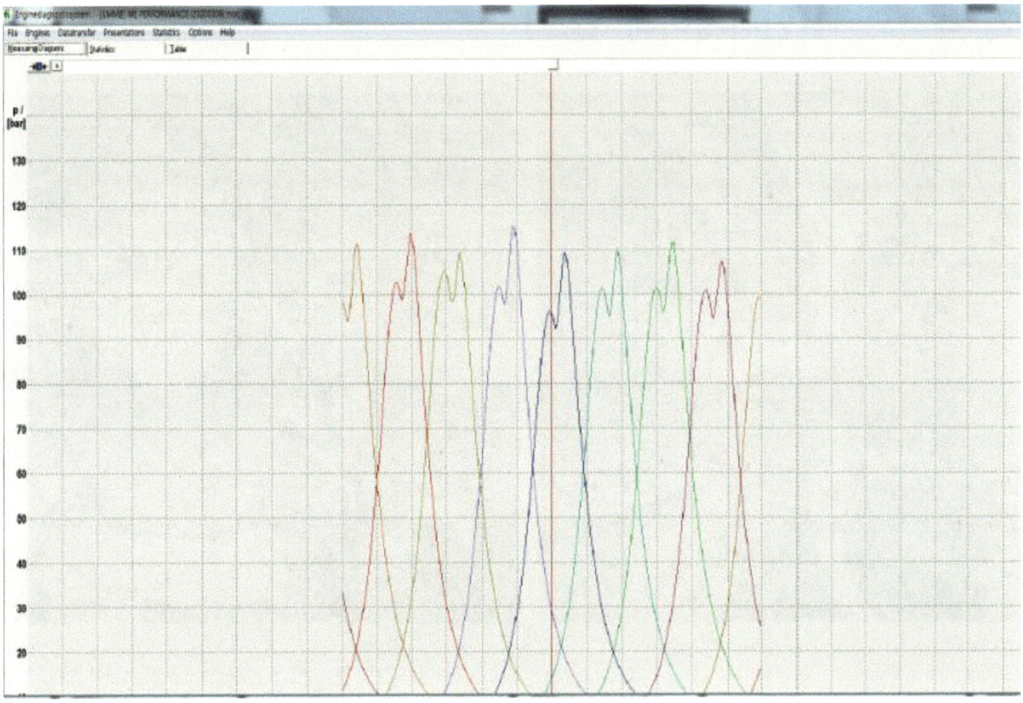

Engine type : 8K90MCC

Cylinder count : 8

Length of con. rod : 3500 mm

Half stroke : 2300 mm

Bore : 900 mm

p. after cooler : 2.0 bar

T before cooler : 130.0 캜

T after cooler : 39.0 캜

Stroke : 2

Swept volume : 2926394 cm?

Compression room : cm?

Blower revs : rpm

T after blower : 230 캜

KISTLER
measure. analyze. innovate.

Cyl.	Date/Time	n [rpm]		pmax [bar]		pcomp [bar]		pi [bar]		Power [kW]		Rack Set [mm]	T.exh. [캜]
1	9.3.20 21.48	72.3	-3.5%	109.2	-1.6%	95.9	+414.7%	9.9	-8900.0%	3501.5	-5445.8%	64.0	340
2	9.3.20 21.51	74.0	-1.2%	113.6	+2.4%	0.0	-99.9%	-29.8	+26388.9%	-10771.3	+16344.7%	63.0	330
3	9.3.20 21.53	75.7	+1.0%	111.7	+0.7%	2.0	-89.0%	46.9	-41768.9%	17320.0	-26542.7%	69.0	350
4	9.3.20 21.56	74.9	+0.0%	107.1	-3.4%	0.5	-97.4%	19.3	-17255.6%	7036.8	-10843.2%	69.0	310
5	9.3.20 21.58	76.6	+2.3%	109.2	-1.5%	6.5	-65.2%	-52.6	+46655.6%	-19640.4	+29885.3%	67.0	320
6	9.3.20 22.00	75.9	+1.3%	115.3	+3.9%	29.6	+58.8%	-40.0	+35455.6%	-14792.3	+22483.7%	66.0	300
7	9.3.20 22.03	74.1	-1.0%	111.4	+0.4%	0.6	-96.7%	-4.8	+4166.7%	-1746.1	+2565.8%	66.0	290
8	9.3.20 22.05	75.8	+1.2%	109.9	-0.9%	13.9	-25.3%	50.2	-44722.2%	18567.8	-28447.8%	63.0	295
		Ø 74.9		Ø 110.9		Ø 18.6		Ø -0.1		Ø -65.5 / Σ -524		Ø 65.9	Ø 317

Data 전송하여 PC에서 상기와 같이 표시되며, 상기 사진은 본선 MIP기기 불량으로 Data 값이 비정상인 경우를 나타낸다. 이런 경우 MIP Software 업데이트 또는 수리를 맡긴다.

22) 테스트 콕(인디케이터 밸브)

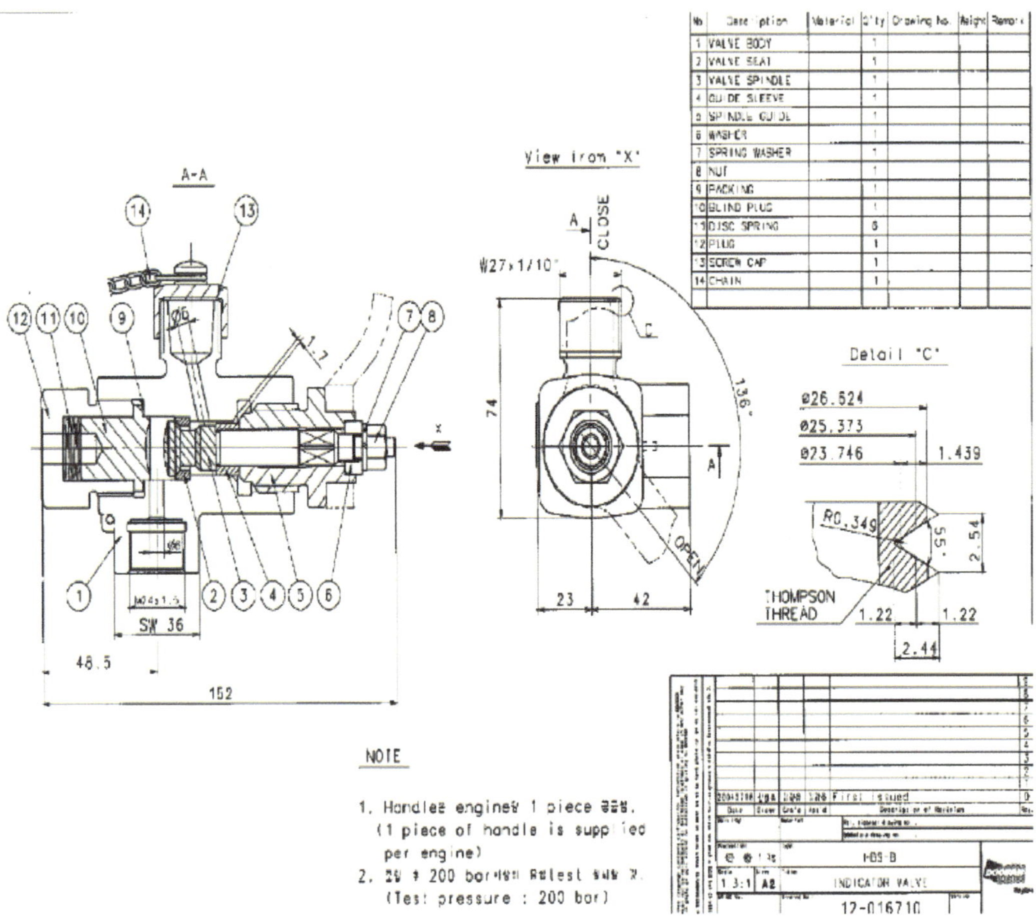

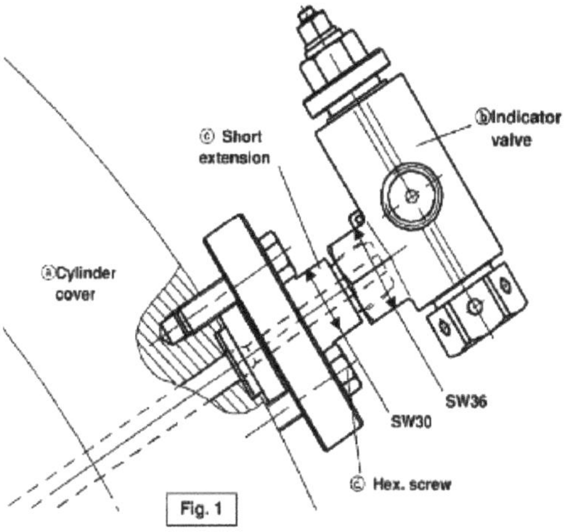

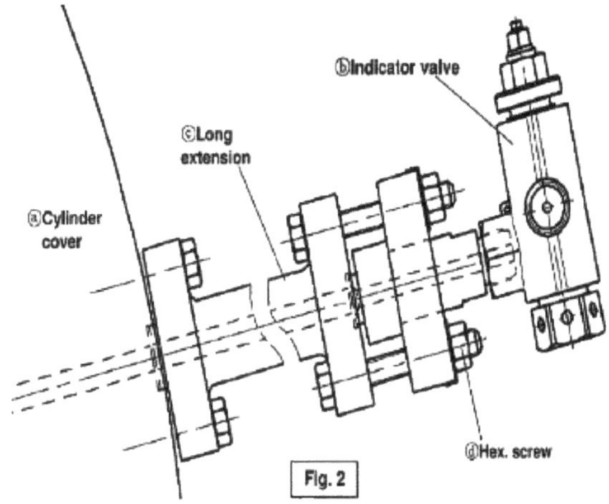

< Short Type / Long Type >

Test Cock(Indicator Valve)은 Short Type과 Long Type이 있다.

운항 중에 Test Cock이 심하게 누설하면 즉시 Engine Stop하고 예비품으로 교환하며, 미량 누설하는 경우엔 가까운 항구에 접안 후 교환한다.

취외한 Test Cock은 관통 소제 및 필요 시 Lapping Tool 이용하여 내부 Lapping 후 보관한다.

비상상황에 대비하여 예비품 1~2개 정도 보유하는 것이 좋다.

교환 및 내부 소제는 따로 설명할 것 없으며 예비품이 없는 비상 상황에서는 아래를 참조한다.

7. Emergency

In case of happening of gas leakage continuously, the below actions as temporary countermeasure are followed. However in order to avoid the damage of the back-seat, it is recommended that the leaky valve is replaced with spare.

1. Put packing(tool, no. 5) into Screw cap(⑮)

2. Stop up the thread installed indicator with Screw cap(⑮)

3. Open the valve and close the back-seat.

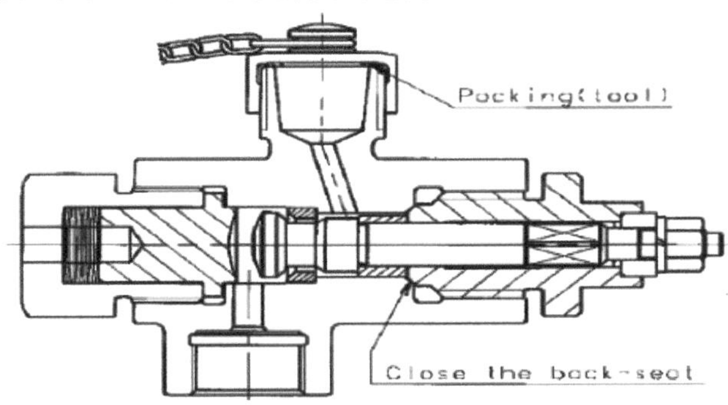

23) 센서

M/E Fuel Oil Leaky Sensor, Cylinder Jacket Outlet Temp Sensor, Exhaust Gas Outlet Temp Sensor Piston Cooling Oil Non-flow Sensor 순서로 설명하겠다.

먼저 **M/E Fuel Oil Leaky Sensor**

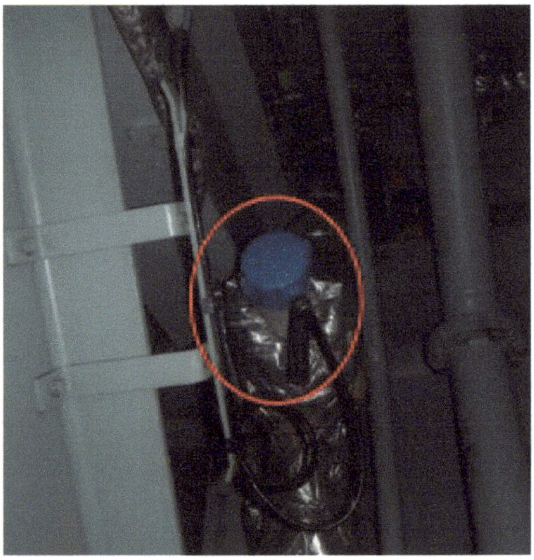

< Fuel Oil Leaky Sensor >

Sensor에는 **두 가지 Type**이 있다.

첫 번째 사진과 같이 **내부 Float가 위로 움직이면서 Alarm을 발생시키는 Type**

두 번째 사진과 같이 **내부 PCB에 의해서 Alarm이 울리는 Type**

첫 번째 Float Type은 Push Button 아래방향으로 누르면 내부의 Float가 위로 올라가면서 Alarm Test 할 수 있다.

두 번째 PCB Type은 Sensor Cover Open하여 내부 Button을 누르거나 내부 Switch를 변경하여 (On & Off) Alarm Test할 수 있다.

만약에 전체 Leaky Sensor Test 및 Alarm이 울리지 않을 경우 공급 전원을 점검한다.

하나 또는 일부의 Sensor가 Alarm Test 되지 않더라도 공급 전원을 우선 점검하고 전원이 공급되나 Test가 안되면 Sensor 불량이므로 교환한다.

실제로 Leaky가 일어나는 경우엔 Leaky Part를 찾아내고, Leaky 발생하지 않다면 내부 라인에서 막혀서 Drain 되지 않는 것이므로 Sensor 취외하고 내부 Air Line으로 관통한다.

다음은 Cylinder Jacket Outlet Temp Sensor 과 Exhaust Gas Outlet Temp Sensor

Cylinder Jacket Outlet Temp & Exhaust Gas Outlet Temp는 Engine Slow-down 항목

AMS(Alarm Monitoring System)에서 비정상적인 온도 Hunting, Sensor Fail 또는 타 기통에 비하여 지나치게 온도가 높거나 낮을 경우, 먼저 Local 온도 Gauge 확인한다.

Cylinder Jacket Temp Sensor의 경우 Local에서는 괜찮다면 Sensor Power Cable Re-tight 하거나 Sensor 교환한다.

Sensor 교환해도 동일한 현상이 발생한다면 AMS상에 문제이므로 AMS Alarm Card 확인한다.

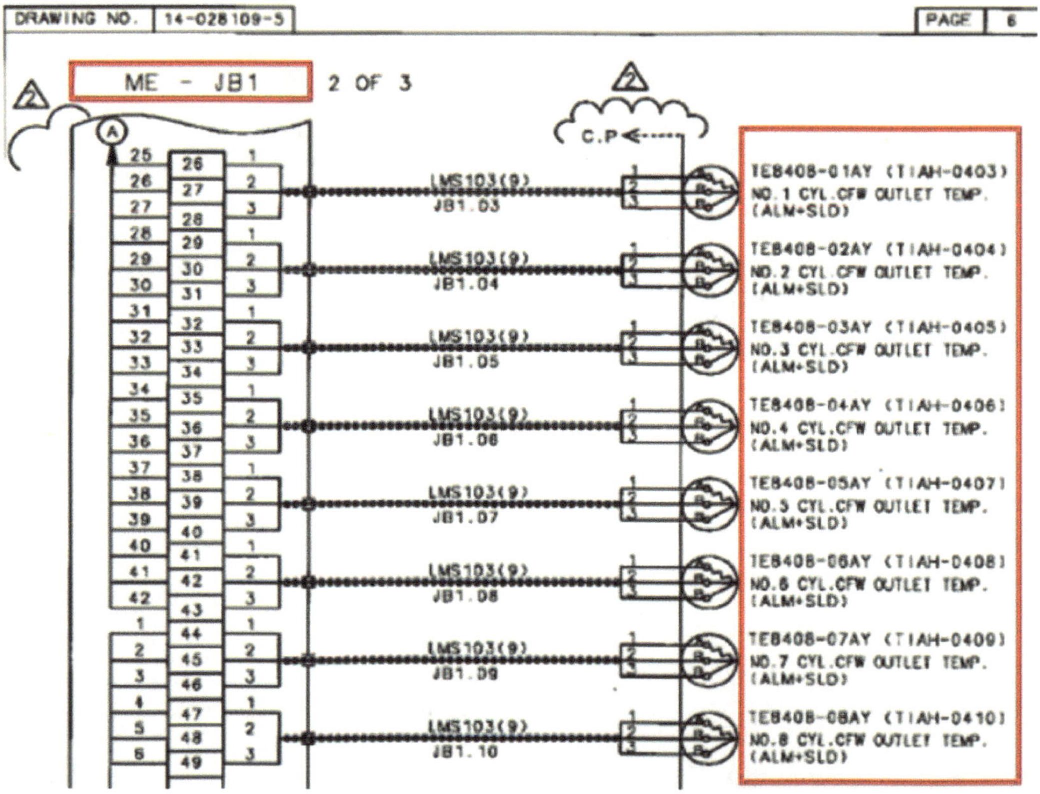

〈 Cylinder Jacket Outlet Temp 도면 〉

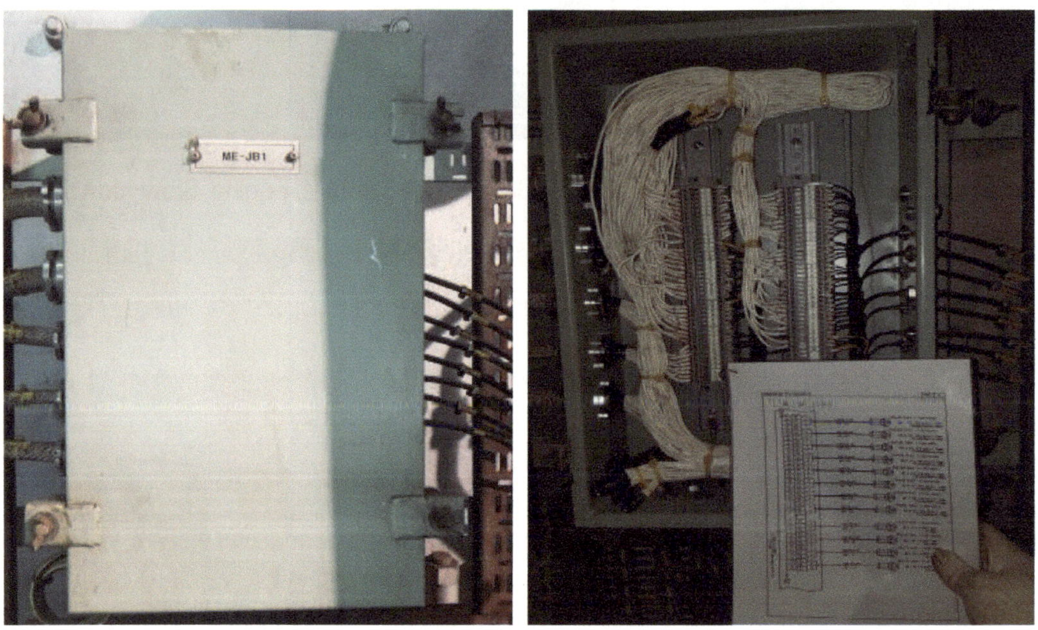

〈 JB1 – Junction Box1 〉

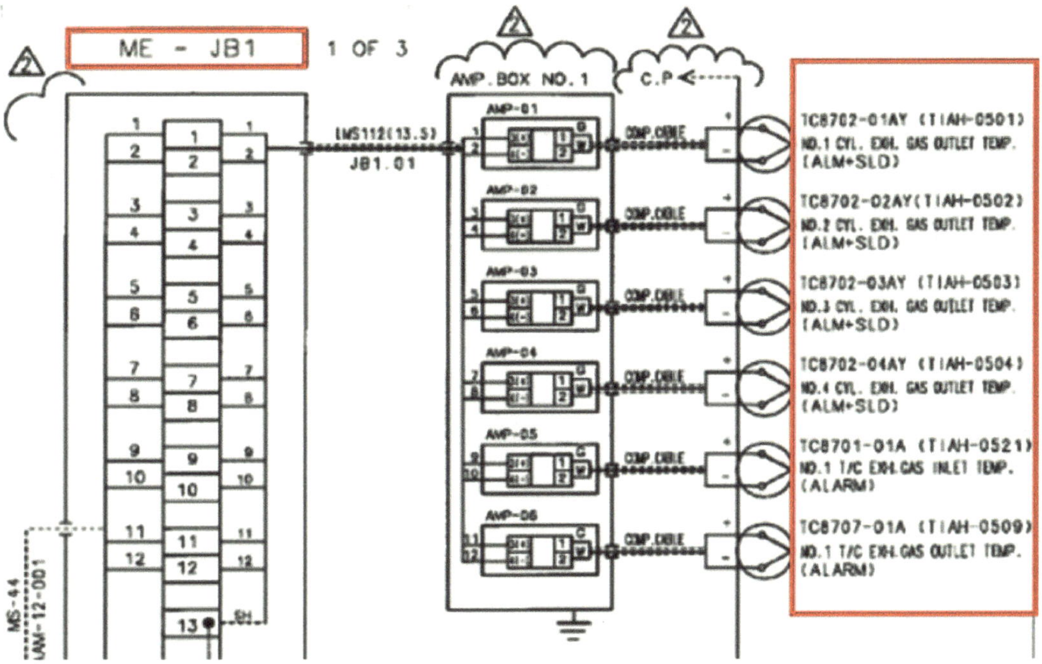

< Exhaust Gas Outlet Temp 도면 >

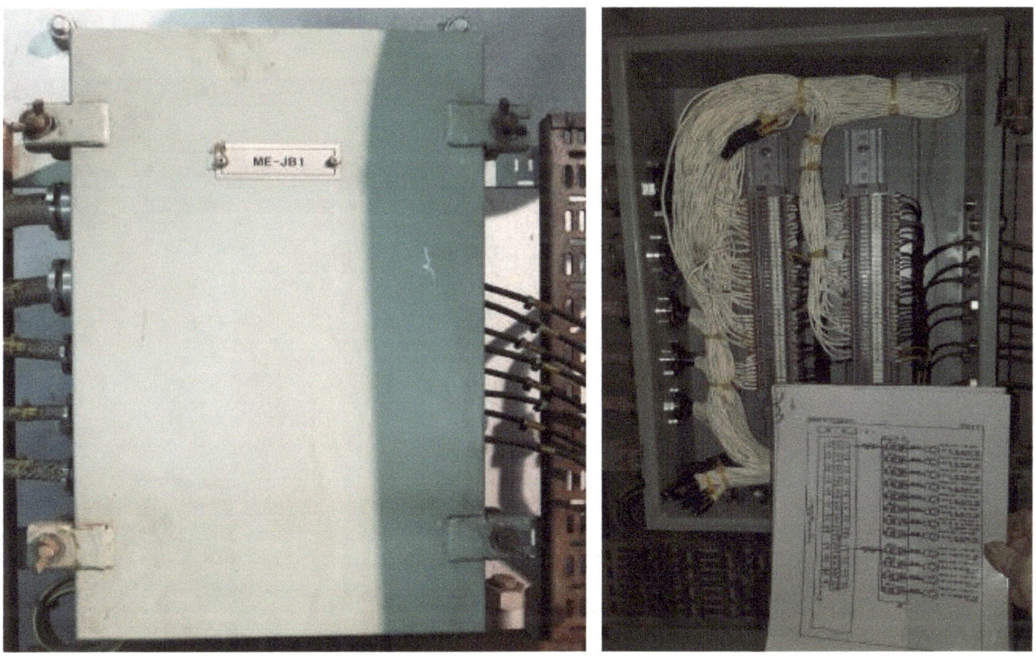

< JB1 - Junction Box1 >

Exhaust Gas Temp Sensor의 Power Cable 또한 JB1(Junction Box1) 내부에 있다.

Power Cable Re-tight 먼저 해보고 상태 동일하면 Amplifier문제이거나 Sensor 문제로 접근한다.

< Exhaust Gas Outlet Temp Amplifier Box >

< Exhaust Gas Outlet Temp Amplifier >

Amplifier Box는 8기통 Engine 기준 2개 있다.

No.1 Box : No.1~4 Cylinder Exhaust Gas Outlet Temp & No.1 T/C Inlet / Outlet Temp

No.2 Box : No.5~8 Cylinder Exhaust Gas Outlet Temp & No.2 T/C Inlet / Outlet Temp

예를 들어 2번 Exhaust Gas Temp 값이 비정상적일 경우, 2번과 3번 Exhaust Gas Temp Alarm을 AMS에서 Alarm Block 하고, NO.1 Exhaust Gas Temp Amplifier Box Cover Open 한다.

3번 Amplifier의 Power Cable과 2번 Amplifier Cable을 맞바꿔서 연결한다.

AMS에서 2번 Alarm Clear되고 3번 Alarm 발생하면 Amplifier 불량한 것이고, 상태 동일하다면 Sensor가 불량한 것이다.

또는 문제가 있는 2번 Amplifier Power Cable을 취외한 상태에서 3번 Amplifier Power Cable을 2번에 연결했을 때 Alarm Clear되면 Amplifier 불량, 상태 동일하다면 Sensor 불량이다.

결론적으로 Amplifier 불량인지 먼저 확인하고, 그 다음에 Sensor 불량인지 확인한다.

어떤 작업이든 기본적으로 쉽게 확인할 수 있는 것부터 진행한다.

마지막으로 **Piston Cooling Oil Non-flow Sensor**

< 위쪽 Piston Cooling Oil Temp Sensor /
아래쪽 Piston Cooling Oil Non-flow Switch >

P.C.O(Piston Cooling Oil) Temp & Non-flow는 Engine Slow-down 항목

Non-flow는 Sensor Fail Alarm이 발생하며 Non-flow & Temp Sensor 경우도 앞서 설명한 Sensor 문제발생 시 점검 사항과 동일하다.

< 주로 Spring 불량 또는 Flow 불량이며 문제발생 시 신품으로 교환한다. >

* 메인 엔진 피스톤 쿨링 오일 논-플로우 알람이 발생했을 때 조치 방법

예를 들어 NO.1 CYL의 P.C.O Non-flow Alarm이 발생했다고 가정해보자.

① Alarm 발생 후 M/E Slow-down 발생한다.

② AMS에서 문제 발생한 CYL P.C.O Outlet Temp'를 확인하여 타 기통과 온도 차이 확인한다.

③ 입, 출항 S/BY와 같이 Engine Stop 할 수 없을 경우에는 Shut-down Cancel 한다.

④ 온도차이가 급속도로 발생하면 실제 L.O가 Non-flow가 된다는 뜻이며, Non-flow 되는 원인을 찾아서 수리를 해야 한다.

⑤ 실제로 Non-flow되는 것은 아니지만 AMS상 문제이거나 Sensor 자체 문제일 경우 Alarm Block하고 가까운 항구에서 접안 후 점검 및 교환한다.

Alarm Block하기 애매한 상황이면 Junction Box내 Cable Jumping하여 Alarm Clear 한다.

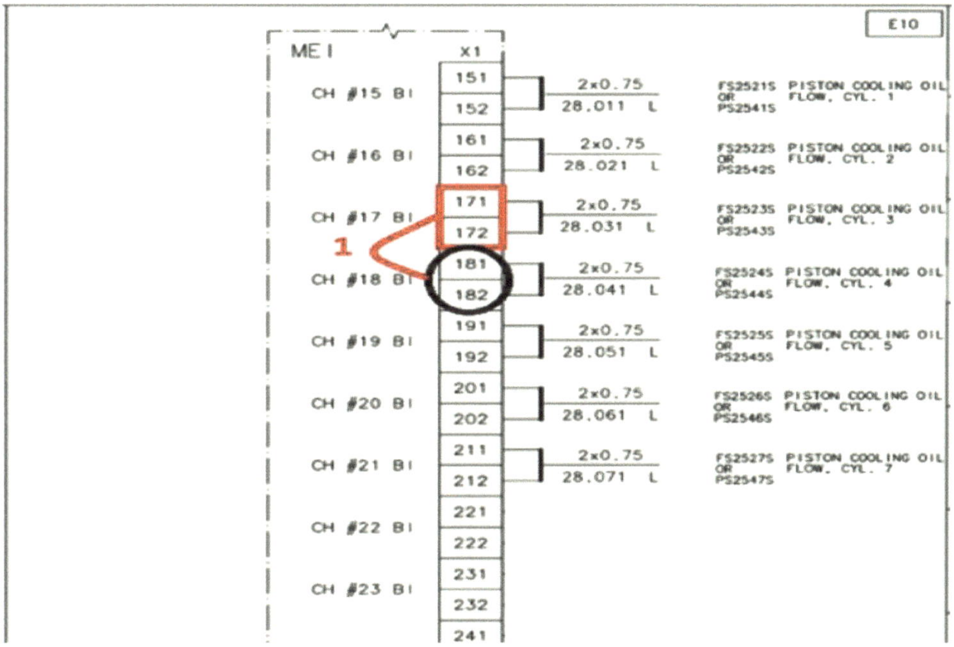

상기와 같이 Jump Cable로 181,171을 Jumping 하고 172,182를 Jumping하면 Alarm Clear 된다.

모든 Sensor Cover에는 Power Cable의 결선에 대한 설명 또는 기호가 있으니 참조한다.

24) 연료 교환

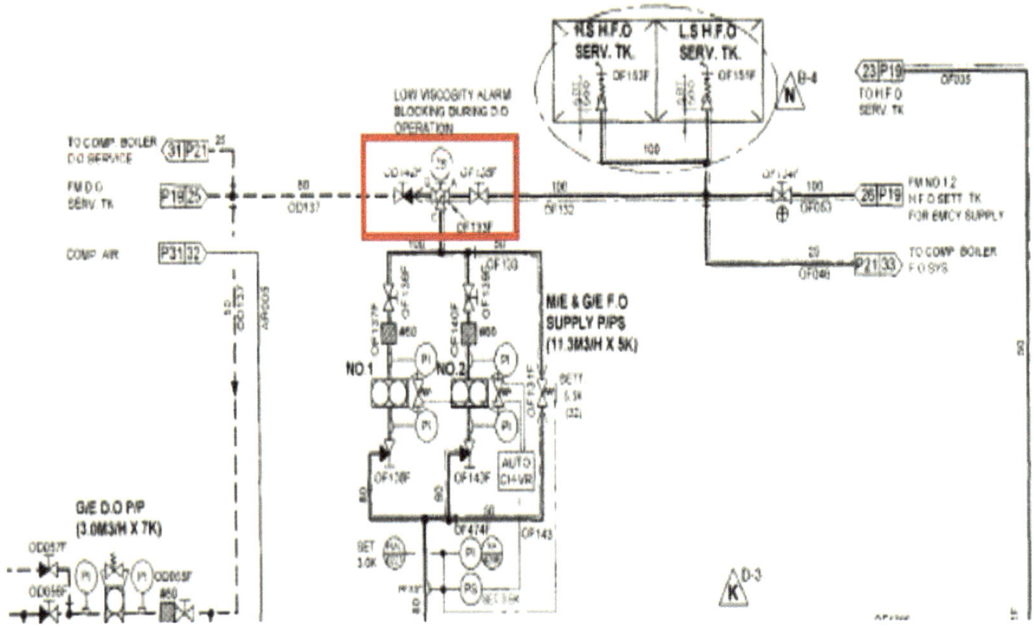

2020년 저유황유 규정으로 인하여 HFO는 더 이상 사용하지 않는다.

최근 0.5% 미만인 VLSFO를 사용하고 ECA구역에서는 LSGO를 사용한다.

예전과 같이 Viscometer로 Steam Valve개도 조절하여 온도를 낮추거나 올릴 필요가 없다.

바로 Valve Change하여 Bunker Change 한다.

Change Program으로 계산하여 M/E & G/E & Boiler 예상 시간을 확인하고 ECA(Emission Control Area) 진입 시간 고려해 Bunker Change 한다.

VLSFO, LSGO 비중 차이가 크진 않다.

단, Bunker Change 할 때 Valve Change 하고 바로 올라가는 기관사들이 있는데, 반드시 Bunker Change하고 기관실 순찰을 돌아 누설 유무를 확인한다.

25) 점도계

(1) 점도 센서 소제 방법

① By-pass Valve를 열고 In & Out Valve를 닫아 Sensor측 F.O 흐름을 차단한다.

② Interface Box 전원을 Off 한다.

③ 1번 Nut(41mm Spanner)를 Housing에서 풀어낸다.
(F.O가 나올 수 있으므로 주의!)

④ 간혹 Viscosity Sensor 하부의 플라스틱 Cap을 제거하지 않은 채 설치하는 경우가 있다.
(Cap이 있다면 제거한다.)

⑤ Viscosity Sensor를 꺼낸 후 3번 Set Screw 2개를 풀어 4번 Tube를 벗겨낸다.
(Set Screw가 작으므로 분실하지 않도록 주의!)

⑥ Viscosity Sensor의 목이 휘어지지 않도록 조심해서 깨끗한 헝겊으로 기름때를 닦아 낸다.

⑦ 재 조립 시 Tube 측면의 Hole이 F.O Inlet 방향을 향해야 한다.
(5번 Pin과 같은 방향)

⑧ Tube가 움직이지 않도록 양쪽 Set Screw 2개를 확실히 잠근다.

Tube가 까딱까딱 움직이면 측정치에 영향을 미친다. **(값이 상승)**

⑨ 2번 O-ring이 정상인지 확인하고, 경화 및 절손 되었다면 교체한다.

⑩ 조립 시 1번 Nut에 Seal Tape를 감으면 안된다. **(값이 상승)**

⑪ Viscosity Sensor 고정용 Pin 5번을 정확하게 Housing의 Hole에 맞추어 조립한다.

⑫ 조립이 끝나면 In & Outlet Valve Open하여 Viscosity Sensor 조립 부에 F.O 누설 확인한다.

Interface Box에 전원을 넣은 후 By-pass Valve Close 한다.

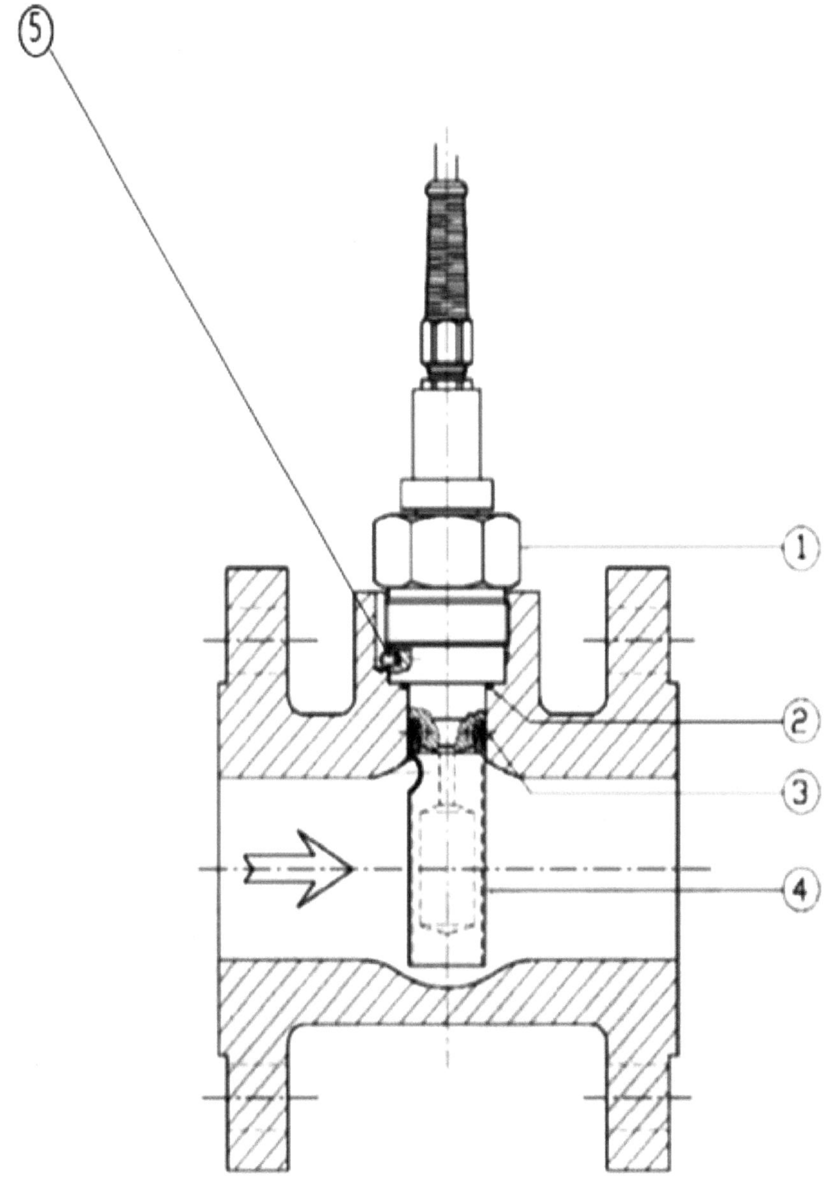
⑤
①
②
③
④

(2) 점도계 제로 셋팅

Zero Setting 관련해서는 아래 먼저 참조.

8.1 INITIAL START-UP

1. Close valves A and B and open equalizing valve C at the dP-transmitter manifold (see Figure 9).
2. Allow diesel oil to enter fuel system.
3. Start Viscotherm® motor.
4. Vent system by unscrewing pipe coupling in ' - ' (minus) sensing line between sensor and dP- transmitter. (See Figure 9 and label on sensor housing).

⚠ Diesel oil will flow out. Make sure that system pressure is less than 1 bar and protect eyes before unscrewing coupling.

5. Retighten coupling when the air bubbles stop.Open manifold valves A and B (Figure 9).
6. Check if viscosity reading is at 0 mPa·s.If reading is not 0 mPa·s, adjust zero point of dP-transmitter following instructions in section 9.2.
7. Close equalizing valve C (Figure 9).
8. When further Viscotherm® system components have been supplied by VAF Instruments, refer to Technical Manual 773 for additional start-up and operating instructions.

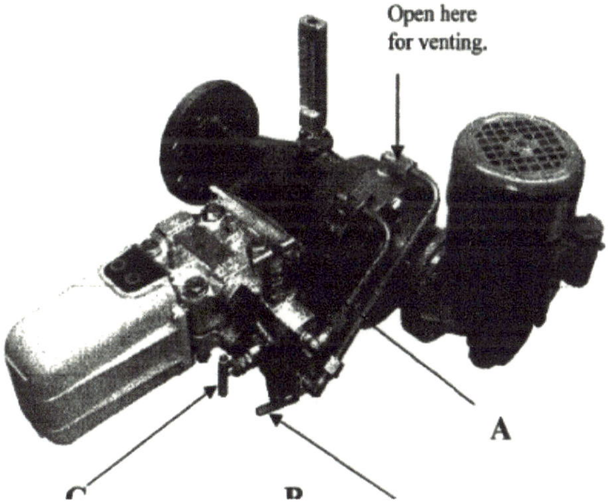

9.1 GENERAL

Under normal conditions the Viscotherm sensor/dP-transmitter assembly requires a minimum of maintenance. "Normal conditions" means:

a. A clean operating environment.

b. Viscotherm components installed in accordance with the installation instructions given.

c. Operation of the viscosity control system in accordance with this manual and other related publications issued by VAF Instruments B.V.

d. Air supply for pneumatic system parts is free from oil, moisture and dirt.

e. Uninterrupted electric power supply at the normal specified values.

☞ Keep record of all maintenance operations, e.g. using the sheet of section 11.

9.2 MONTHLY ZERO CHECK OF DP-TRANSMITTER

1. Open equalizing valve on transmitter manifold (C, Figure 9).
2. Switch off electric motor of Viscotherm® sensor.
3. Check that viscosity reading is 0 (zero) mPa·s. If it is not, adjust zero point at transmitter as follows:
   a) Move slide on side of dP-transmitter upward and turn zero adjustment screw (Fig. 10) until reading at viscosity control station is 0 mPa·s.
   b) Close slide on dP-transmitter.
4. Close equalizing valve on dP-transmitter manifold (C, Figure 9).
5. Start electric motor of Viscotherm® sensor.

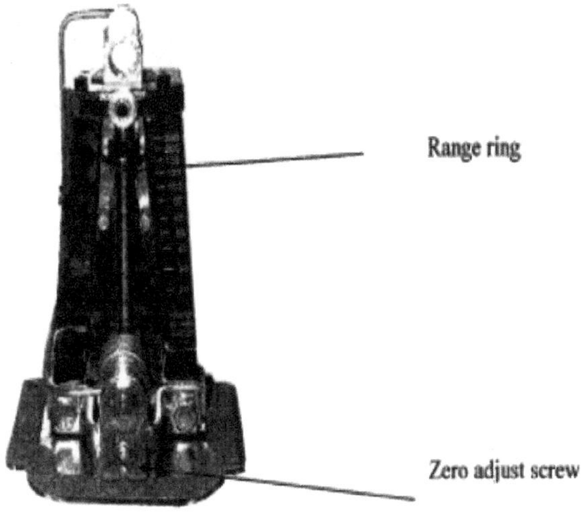

Figure 10, Location of range and zero adjustment points of dP-transmitter

Except where noted all dimensions are in millimetres.

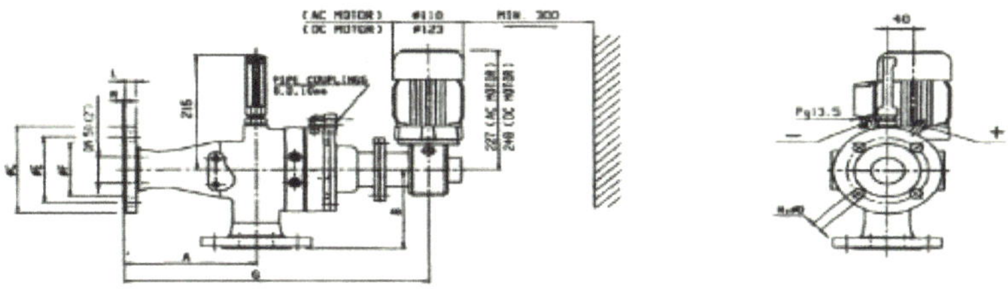

For dimensions see table in figure 7

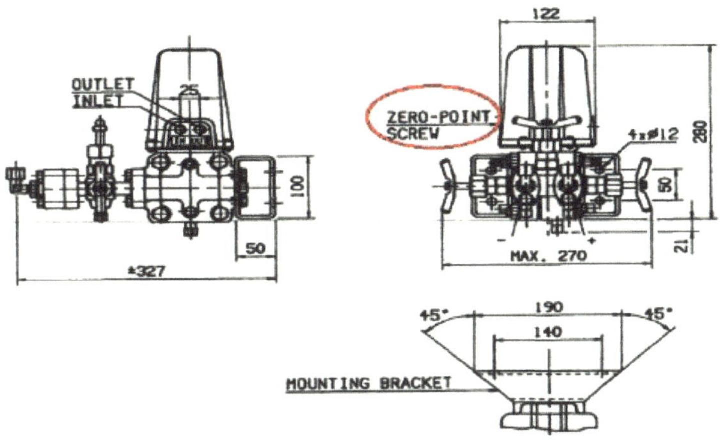

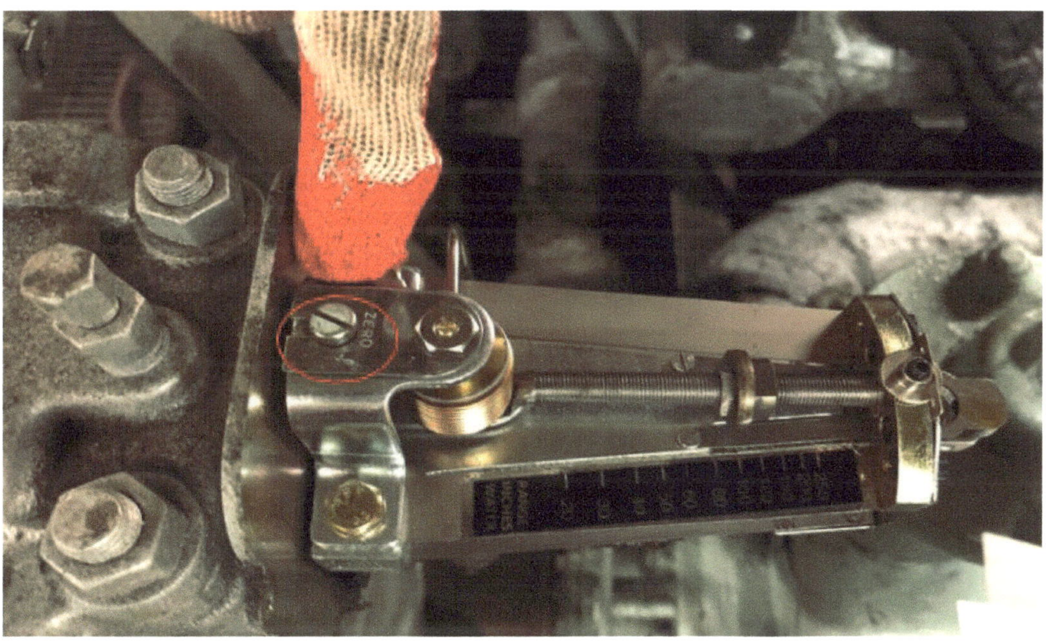

→ 영점 조정은 상기 사진처럼 조절하면 된다.

본인이 승선했던 선박에서는 갑작스럽게 M/E Viscometer 조절이 되지 않아 영점 조정했다.

① Viscometer Motor가 작동 중이라고 가정한다.
(영점 조정 때 Motor 끄면 점도는 0이 된다.)

② 트랜시버로 기관실, Local에서 대기하고 ECR에서는 Viscometer Controller 자동에서 수동으로 바꿔 놓는다.

③ Local에서 영점 조절 스크류로 **조금씩 조절**하여 기관실과 통신하며 원하는 Viscometer 값에 맞춘다.

④ 기관실에서 Viscometer Controller로 PV값이 변하는 것을 확인한다.

⑤ Controller에서 원하는 PV값에 고정되면 다시 수동에서 자동으로 바꿔준다.

원래는 영점 조정하는 것이 맞지만 간편하게 아래와 같이 조절하는 방법도 있다.

자동으로 조절 안되니 Local에서 영점 조절 스크류로 값을 조금 변경해주고 Controller 자동으로 다시 바꿔서 값이 일정하게 고정되면 작업 종료하면 된다.

26) 오일 미스트 디텍터

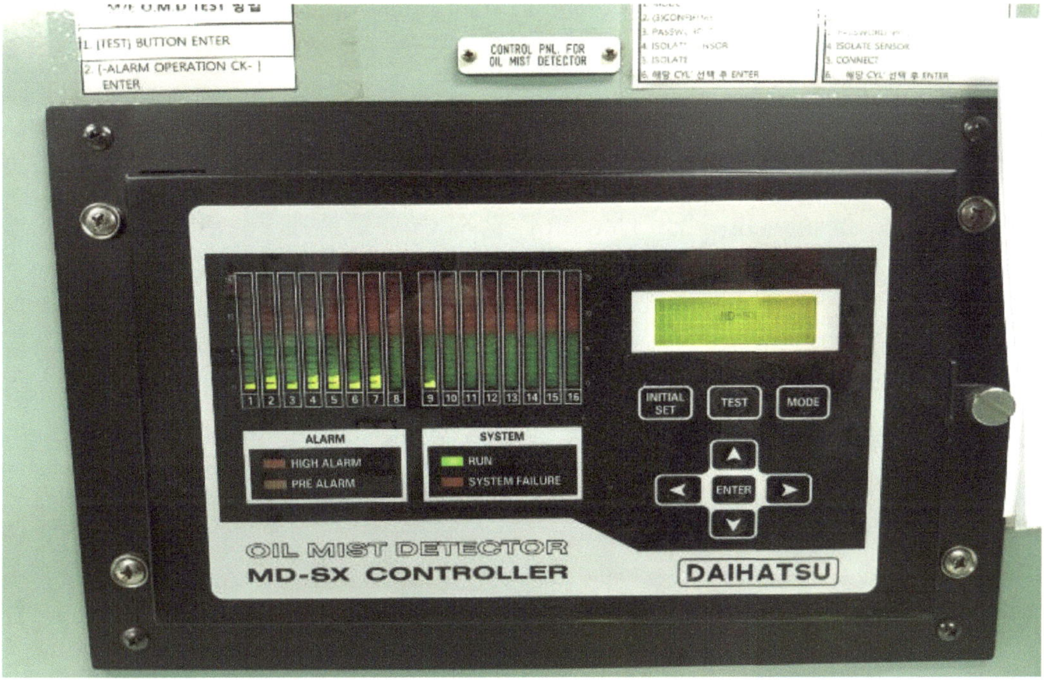

*** ISOLATE 방법**

① MODE
② (3)CONFIGURE
③ PASSWORD '9999' (선박 마다 다름)
④ ISOLATE SENSOR
⑤ ISOLATE
⑥ 해당 CYL'선택 후 ENTER

*** ISOLATE 해제**

① MODE
② (3)CONFIGURE
③ PASSWORD '9999' (선박 마다 다름)
④ ISOLATE SENSOR
⑤ CONNECT
⑥ 해당 CYL'선택 후 ENTER

*** M/E O.M.D TEST**

: [TEST] Enter → [ALARM OPERATION CK] Enter

(1) 교환 방법

① M/E 운전 중에 Oil Mist Detector Alarm & Sensor Fail 발생할 경우 상기와 같이 해당 Oil Mist Detector Isolate 한다.
(Local Oil Mist Detector에도 Alarm 또는 Failure에 점등되어 있다.)

② Local에서 Power Plug 분해하고 Oil Mist Detector 고정 Nut 분해하여 취외한다.

③ 정비 또는 소제한 Oil Mist Detector 앞쪽 Cover 분해하여 취외품과 동일하게 Master Number 설정한 다음 취외한 구멍으로 집어넣고 고정 Nut Tight한 다음 Power Plug 연결한다.

④ 운항 중에도 교환 정비를 할 수 있으며 정비품 없는 경우 Oil Mist Detector 취외한 구멍으로 System Oil 밖으로 나오지 않게 천으로 구멍을 막는다.
(단, 구멍 안으로 천이 들어가지 않도록 주의!)

*** Oil Mist Detector에도 2가지 Type이 있다.**

Oil Mist Detector **내부 Fan 장착되고 PCB 교환할 수 있는 Type**
Oil Mist Detector **내부 Fan 없고 PCB 교환할 수 없는 Type**

첫 번째, 두 번째 경우 Alarm발생하면 Sensor 부분부터 WD-40과 부드러운 천으로 소제한다.

첫 번째 경우 Fan을 Air로 불어서 작동상태 확인하고 정상이면 Sensor 소제 후 조립하여 Alarm 발생 여부 확인한다.

소제 후 Alarm 발생하면 PCB교환하고 상태 동일하면 해당 Oil Mist Detector 수리 신청한다.

두 번째 경우에는 Sensor 소제 후 Alarm 발생하면 해당 Oil Mist Detector수리 신청한다.

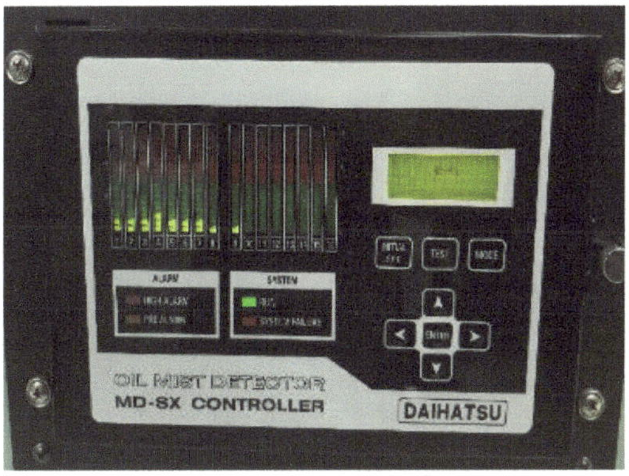

NO.8 OMD Isolate 전

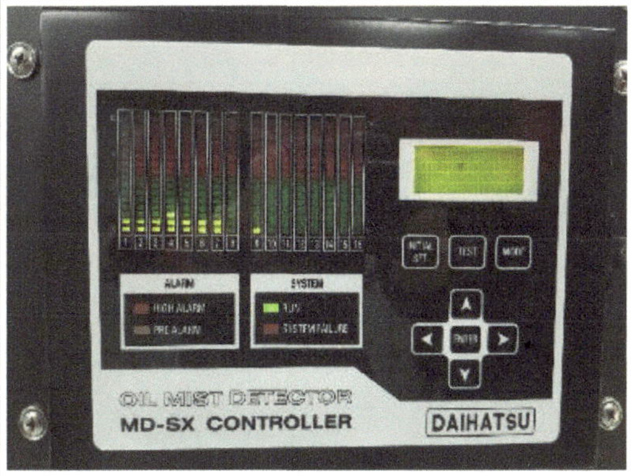

NO.8 OMD Isolate 후

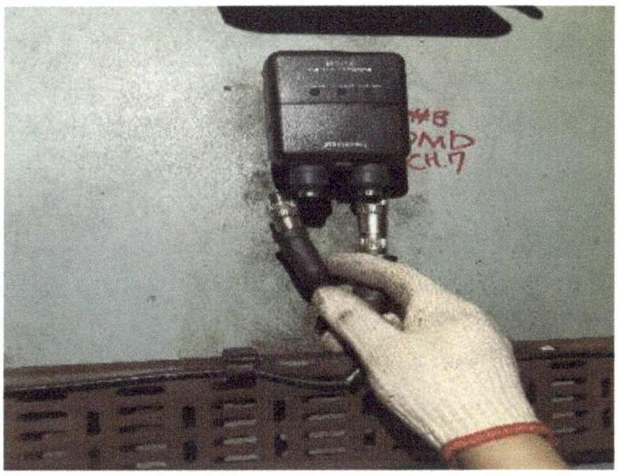

Power Cable 분해

Oil Mist Detector 고정 Nut 분해

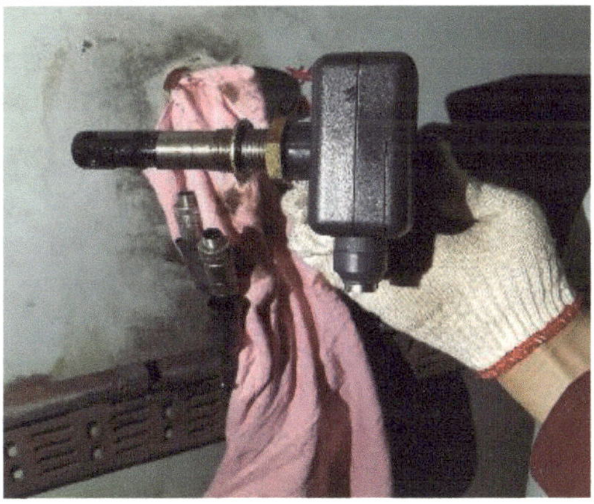

Oil Mist Detector 취외

Master Number 설정 후 조립

(2) 소제 방법

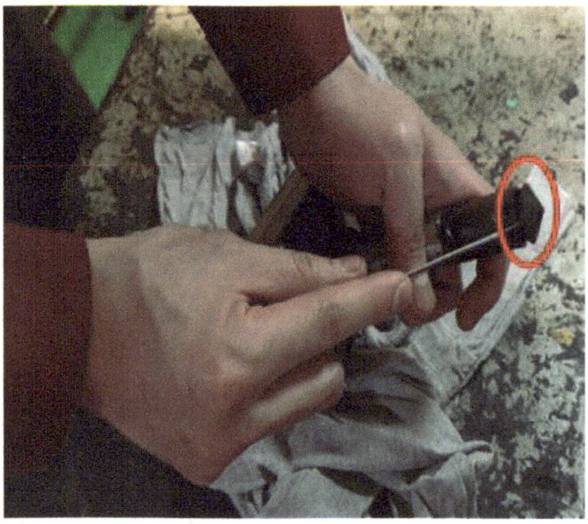

Oil Mist Detector Sensor Cap 취외

Oil Mist Detector Sensor Cap 취외

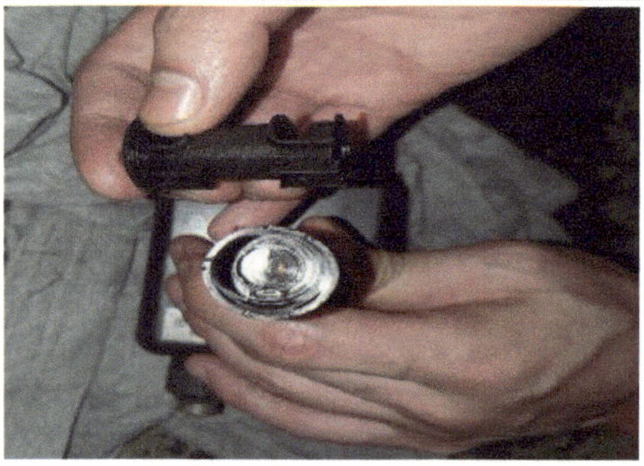

Oil Mist Detector Sensor

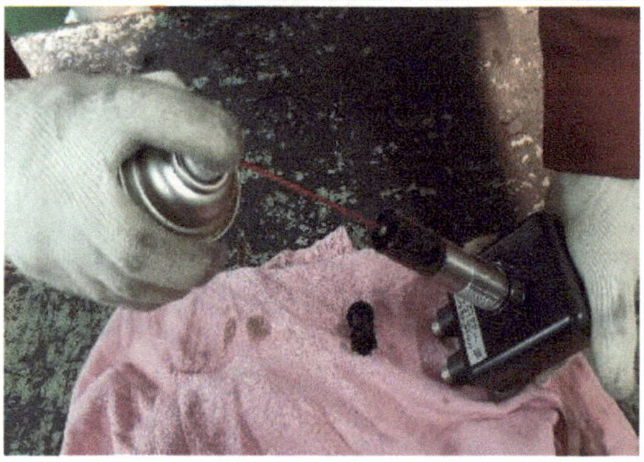

WD-40 이용하여 Sensor 소제

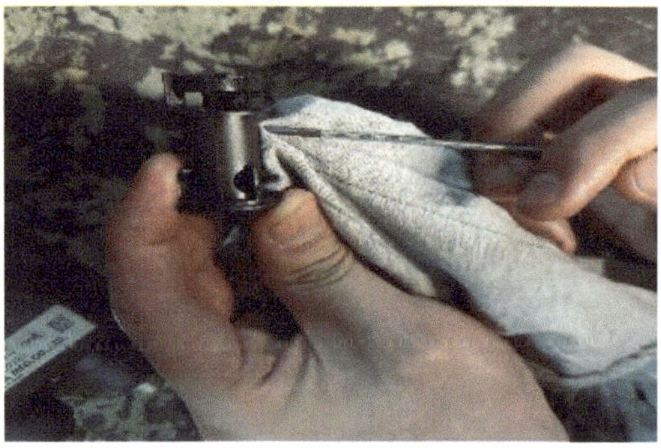

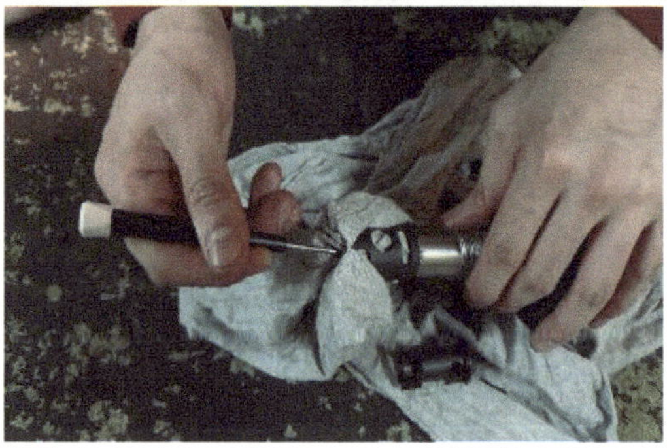

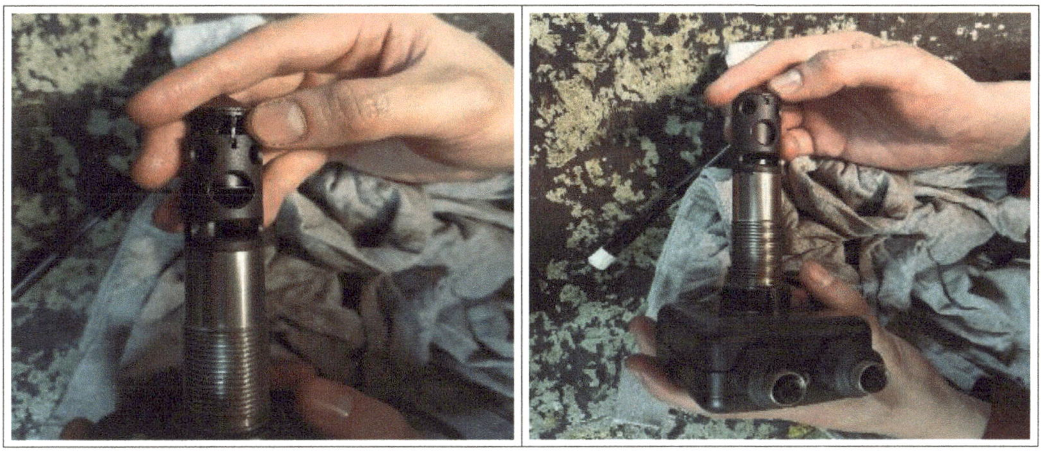

내부 Fan & PCB 교환하는 Oil Mist Detector Type은 사진 자료가 없어서 추가하지 않았지만 몇 개의 볼트만 풀어내면 쉽게 분해된다.

소제하여 조립하여도 불량하거나 Alarm이 다시 빈번하게 발생하는 Oil Mist Detector들은 따로 모아 놓고서 Detector 수리 신청한다.

[3] MAIN ENGINE 주기적인 점검사항

1) 피스톤 링 & 라이너

① Instruction Book에는 Piston Ring & Liner, Scavenge Air Chamber & Air Tank는 Monthly 점검,

② Crankcase는 Weekly ~ Monthly점검,

③ Top Bracing, Crankshaft Deflection, Camshaft Chamber는 8000H점검,

④ Chain Drive는 4000H마다 점검

⑤ Engine Stop 후 Scavenge Chamber Door를 Open하여 1-2시간 Potable Fan을 설치한다.

⑥ 내부 Cooling후 M/E Piston Ring & Liner를 점검한다.
(부득이하게 내부 Cooling할 시간이 짧을 경우 Scavenge Chamber Door Open후 AUX. Blower Manual 기동하여 내부 Hot Gas를 불어내고 점검한다.

참고로 L.O Pump Stop후 Spring Air 차단하여 EXH′ Valve Spindle Open한 상태에서 점검하면 보다 시원한 상태에서 점검할 수 있다.)

⑦ Turning Gear Remote Controller 들고 Scavenge Air Chamber로 들어가서 Firing Order에 따라 Piston Ring & Liner를 점검한다.

⑧ Turning Gear를 이용하여 Piston을 내려서 Liner 점검하고,

Piston을 다시 올리면 Piston Ring들이 중력으로 인해 아래로 내려온다.

⑨ 부드러운 천으로 Piston Ring을 닦아서 상태 점검 후 Ring Groove를 계측한다.
(Cylinder Liner & Piston Ring 판단 기준 참조)

⑩ Piston Ring 상부 틈새를 Peeler Gauge 로 상,하,좌,우 4부분과 Top Ring의 CL-Groove를 계측한다.

⑪ Black & White Mark는 주유량을 늘리면 제거된다.

⑫ 주유량을 무작정 계속 늘리는 것이 좋은 것이 아니고 경험 상 최대 1.3g/kwh까지만 올리며, 1.4g/kwh까지 올리면 Cylinder Oil Deposit이 생겨 Liner 및 Piston Ring 마모를 유발한다. 본선 T/C Cut-in & Cut-off Condition 및 Piston Ring & Liner를 고

려하여 적절히 조절한다.

계측 치 및 계측 방법은 아래 도면 참조

Piston

SAFETY PRECAUTIONS *For detailed sketch, see 900-2*

X	Stopped engine
X	Shut off starting air supply – *At starting air receiver*
X	Block the main starting valve
X	Shut off starting air distributor/distributing system supply
X	Shut off safety air supply – *Not ME engines*
X	Shut off control air supply
X	Shut off air supply to exhaust valve – *Only with stopped lubricating oil pumps*
X	Engage turning gear
X	Shut off cooling water
X	Shut off fuel oil
X	Stop lubricating oil supply
X	Lock the turbocharger rotors

Data

Ref.	Description	Value	Unit
D01-05	Cylinder cover stud, check distance	184+0.5	mm
D02-01	Test pressure	7	bar
D02-04	Piston rod/crown stud, screwing-in torque	300	Nm
D02-05	Piston skirt, tightening torque	380	Nm
D02-06	Cooling oil pipe, tightening torque	190	Nm
D02-08	Piston ring new, radial width	28	mm
D02-09	Piston ring worn, min. radial width	24	mm
D02-10	Groove No.1, max. vertical height	20.3	mm
D02-11	Groove Nos. 2, 3 and 4, max. vertical height	15.8	mm
D02-13	Piston ring new, height of ring No. 1	19.4	mm
D02-14	Piston rings new, height of ring Nos. 2, 3 and 4	14,9	mm
D02-15	Minimum free ring gap (before dismantling)	50	mm
D02-16	Minimum ring gap, ring No. 1 (new ring in new liner)	6.3	mm
D02-17	Minimum ring gap, ring Nos. 2, 3 and 4 (new ring in new liner)	4.5	mm
D02-18	Vertical clearance, new parts	0.47	mm
D02-19	Vertical clearance, worn parts, max.	0.97	mm
D02-20	Piston complete	3900	kg
D02-21	Piston crown	1300	kg
D02-22	Piston rod	2000	kg
D02-23	Piston skirt	280	kg
D02-24	Piston cooling pipe	65	kg
D02-25	Lifting tool, tightening torque	150	Nm
D02-46	CPR ring CL groove, min. depth	2.0	mm
D02-47	Piston top centre, max. permissible burn-away	12	mm
D02-48	Piston top centre diameter	420	mm
D02-49	Piston top outer ring area, max. permissible burn-away	8	mm
D02-59	Cooling Insert, tightening torque+angle	10/75	Nm/°
D02-60	Cooling Insert	14	kg
D13-01	Hydraulic pressure, mounting	900	bar
D13-02	Hydraulic pressure, dismantling	800 -990	bar

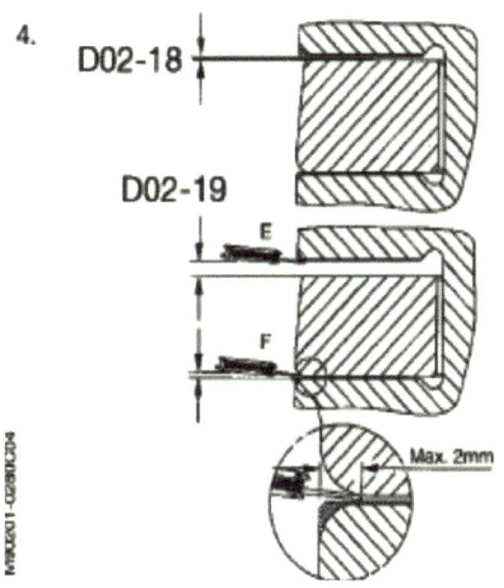

Piston Ring 간극 (Max : 0.97mm)

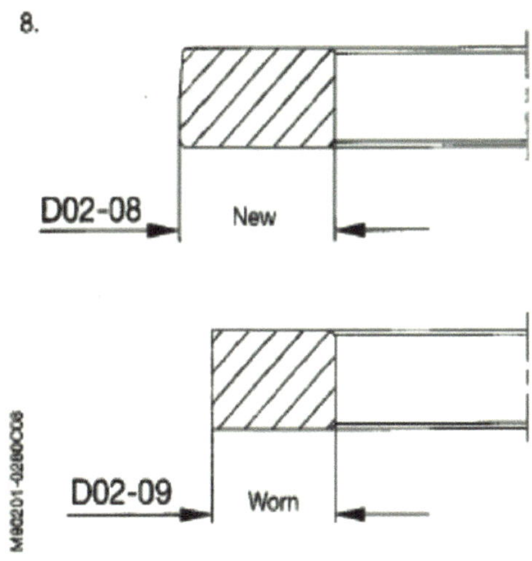

Piston Ring 두께

Piston Ring 점검 시 Ring Edge 부분이 각져있는지 확인한다.
마모가 진행될수록 Edge 부분이 사라지고 Ring 표면이 평면으로 매끄럽게 된다.

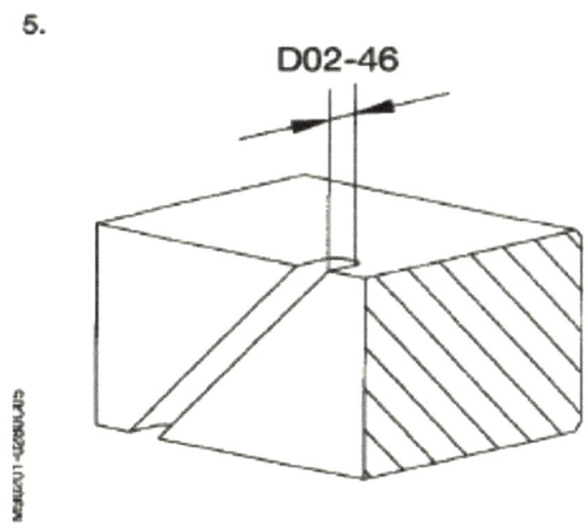

< CL - Groove점검 >

Digital Caplier 이용하여 CL-Groove 간극 점검하고 Limit(2.0mm) 초과하는지 점검한다.

Ring Clearance를 계측 하였을 때 보통 Top Ring 기준 **0.95-1.0mm이상**이 되면 Piston OV'HL을 하여 Piston Ring을 교환한다.

Peel off가 발견된 기통은 주유량 증가하여 진전되는 것을 막을 수 있다.

관련 기통은 점검 시에 Peel off 진행 여부 주의 관찰한다.

Scavenge Air Chamber

Lantern Space

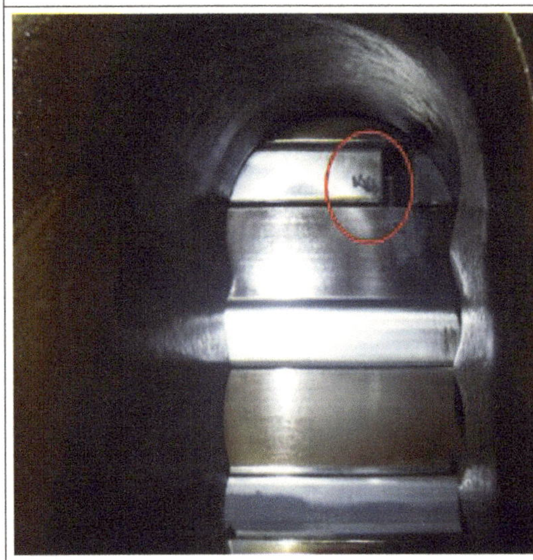

Peel Off

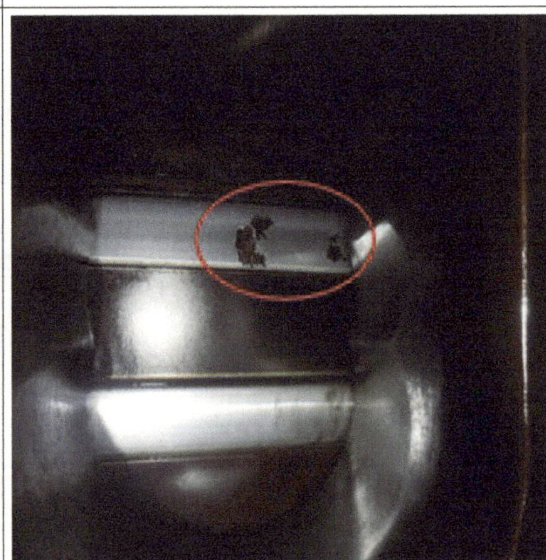

Peel Off

DATE : 09TH-APR-2020 / CAVAN CYL'NO. : 2

-> PISTON CROWN TOP SIDE & SCAV' AIR PORT

-> LINER WALL

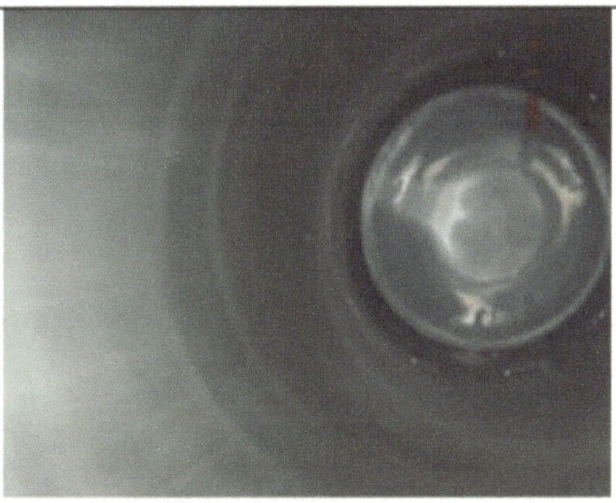

-> LINER OVERALL

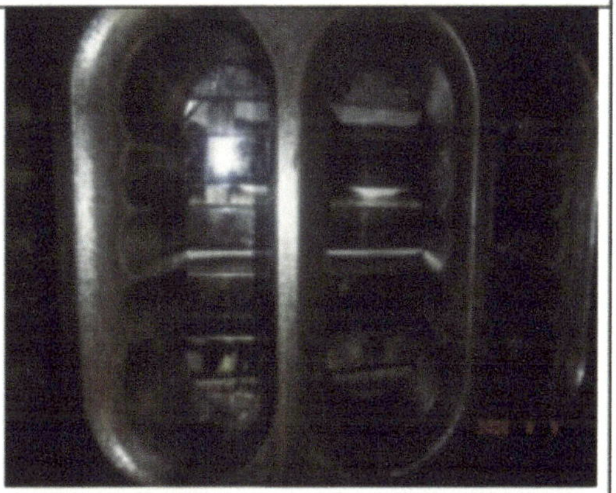

-> TOP, 2ND, 3RD & 4TH RING(BEFORE CLEANING)

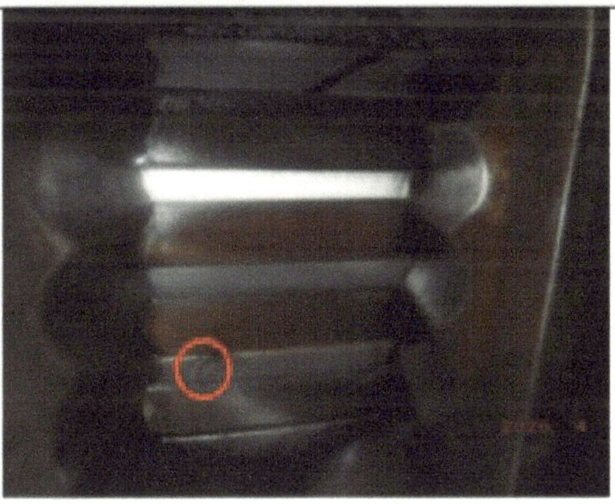

-> 4TH RING PEEL-OFF (AFTER CLEANING)

-> 4TH RING PEEL-OFF (AFTER CLEANING)

※C.O FEED RATE : 100(1.10g/kWh)

* 실린더 라이너 판단 기준

Normal condition

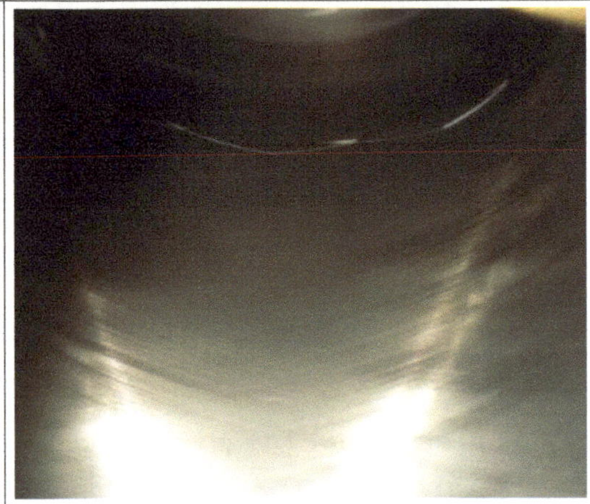

Normal condition

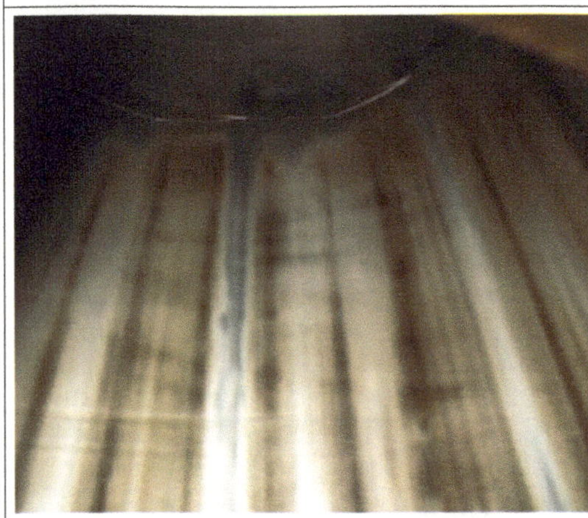

Strong corrosion and lacquer formation

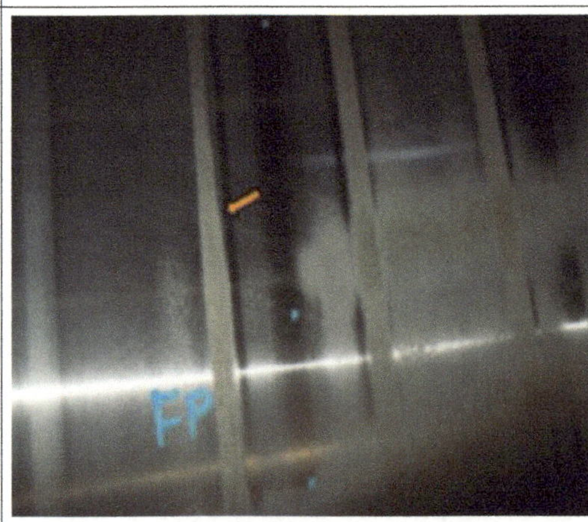

Corrosion stripes

Polishing marks by deposit on piston head

Minor scratch

	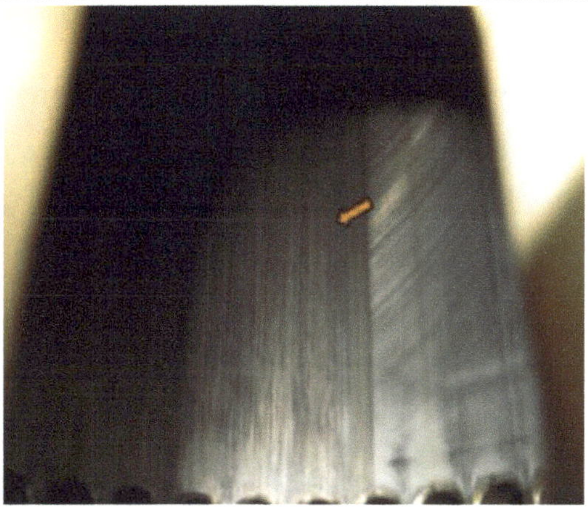
Scoring mark	Scuffing

Scuffing	Excessive scratching marks

Hard contact marks	Hard contact marks

* 피스톤 링 판단 기준

Abrasion
: Caused by cat fines in fuel oil or foreign hard Particles (sand in suction air)

Recovered scuffing by abrasion due to Load change, increased cylinder lubrication

Hard contact

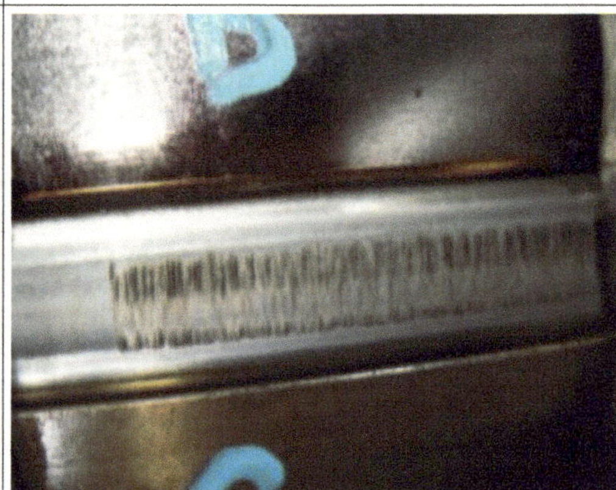

Adhesive wear

Scuffing

Scuffing

Scratch, scoring and scuffing marks

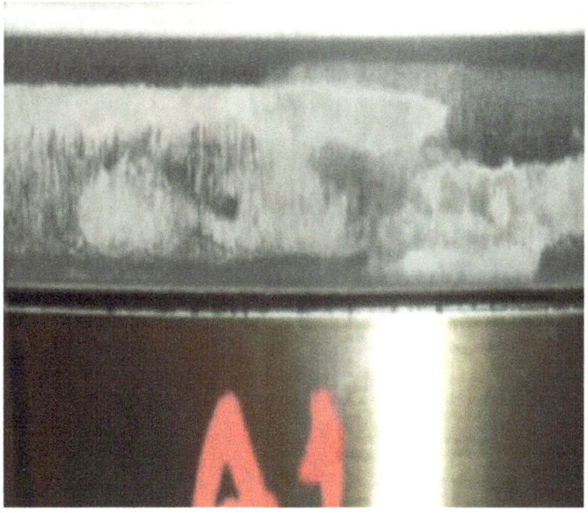

Corrosion

Micro crack

Ring broken

Excessive deposit

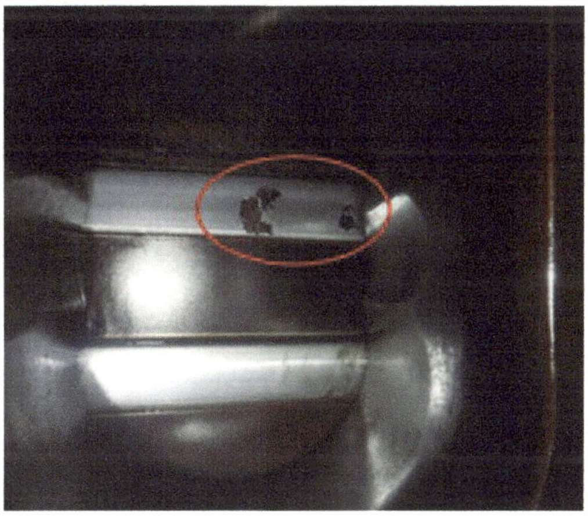

Peel off

2) 소기 챔버 & 슬러지 오일 트랩

Scavenge Air Chamber Flap의 경우 AUX. Blower 기동하면 Valve가 열려 Scavenge Air 공급되고, Engine RPM이 올라가서 AUX. Blower 정지하면 Chamber 내부 압력으로 Flap Valve가 닫힌다.

Scavenge Air Chamber Flap 상태 불량, Crack 있다면 신품으로 교환하고, Lantern Space 소제 후 전 기통 확인하여 소제 도구 및 Waste Rags 놓고 나온 것은 없는지 확인한다.

주기적으로 Scavenge Air Tank(Sludge Oil Trap) 확인하여 Sludge 과다 축적되지 않도록 한다.

Tank Outlet Valve는 Normal Close이며 Tank High Alarm Test 및 Overflow Line 상태 점검한다.

Scavenge Air Chamber

Flap Check Valve

Flap Check Valve

Lantern Space

Lantern Space

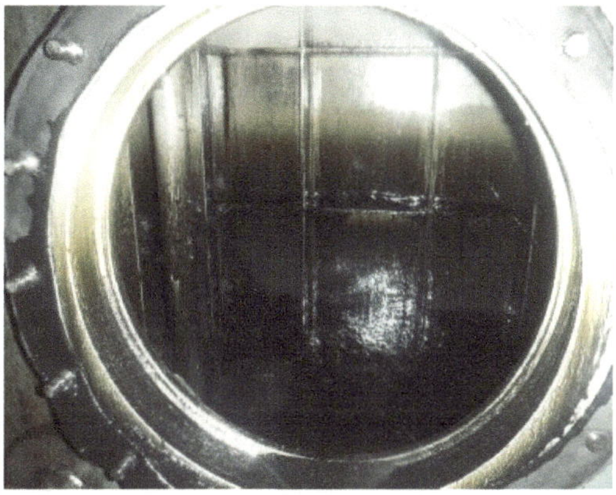

Scavenge Air Tank 내부

Scavenge Air Tank 내부

Scavenge Air Tank High Alarm Switch

3) 크랭크 샤프트 챔버

반드시 Main L.O Pump Stop후 Engine Turning하면서 전 기통 Chamber 점검을 시행한다.

Metal Piece, Sludge, 각 부품의 상태 및 수분을 Check 한다.

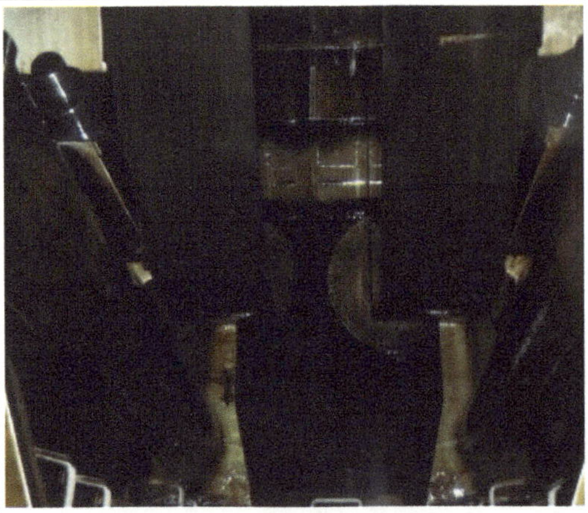

Crankshaft Chamber

Hydraulic Nut & Main Bearing

Crankshaft Chamber 내부

Crankshaft Chamber 내부

처음과 끝이 하나의 통로로 연결된 RT-FLEX Engine과 달리 MAN B&W Engine은 각 Chamber 독립적으로 되어있다.

따라서 각 Chamber 별도로 점검한다.

4) 크랭크 샤프트 디플렉션

선박의 적하 상태나 메인 베어링의 마모 등으로 인하여 크랭크 암은 회전 시 개폐 작용한다.

개폐 작용이 과도하면 Crankshaft가 절단되는 대형사고도 발생할 수 있으므로, 이를 막기 위해서 안전하게 운전 가능한 범위인가를 확인하기 위해서 Crankshaft Deflection 계측한다.

다시 말해 Crank Deflection은 Crankshaft 중심선이 바뀌었는지를 확인하기 위해 계측한다.

정기적으로 Crankshaft Deflection을 계측하여 종전의 기록과 비교하면 그 변화추이를 유추할 수 있다.

Crankshaft Deflection 계측 요령을 숙지하고, 계측된 수치가 안전한 한도 범위 내에 존재하는가 등 판단이 필요하다.

안전한 허용한도를 벗어났을 경우 담당 감독과 업무협의 및 Maker에 문의하며, 참고로 M/E Crankshaft Deflection의 계측 주기는 VMS상으로 8000H이다.

(1) 크랭크 DEFLECTION의 점검

① 정상적인 상태에서 계획 정비에 의거 일년에 1회

② 선박이 충돌, 좌초, 좌주 된 후

③ Main Bearing을 교환하였을 때와 사용 후 약 100HRS이 경과한 시점

④ Main Bearing의 이상징후를 인지한 때

⑤ Propeller또는 Rudder의 손상 후

⑥ 기타 기관장이 필요하다고 판단될 경우

(2) 측정을 위한 조건

① 인디케이터 밸브들은 Open되어야 한다.

② Heeling및 Trip 조정하여 Even Keel 상태

③ Crankshaft는 모든 Main Bearings위에 완벽히 안착되어 있어야 한다.

④ **측정에 미치는 영향 요소**

㉠ **Cold 상태인 지 Hot 상태인지**

㉡ **선박의 Loading된 상태**

㉢ **Air와 Water 온도의 차이**

㉣ **강한 햇빛(E/R 온도)**

㉤ **선체 또는 엔진의 어떤 응력**

(3) 측정 절차는 아래와 같다.

Deflection 계측하기 위해서는 Crankshaft는 엔진의 운전 방향이 있음에도 불구하고 아래와 같이 flywheel 1과 Turning Gear의 pinion 2가 화살표로 표시된 것과 같이 회전되어야 한다.

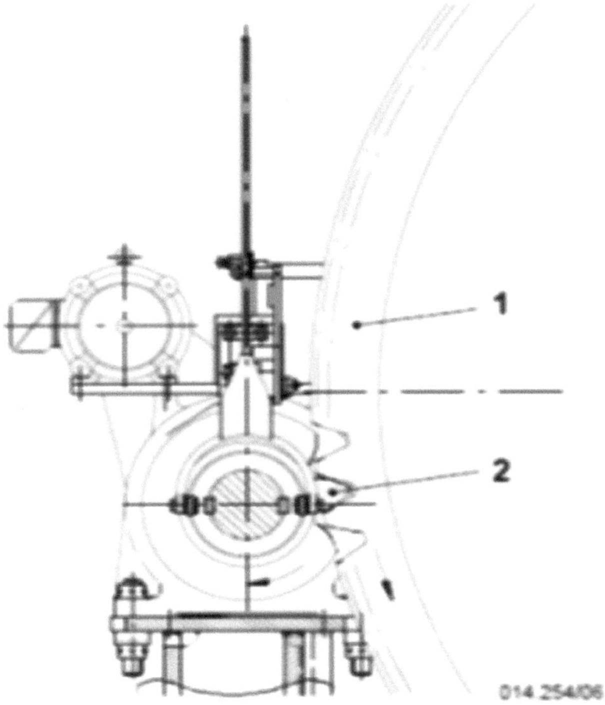

측정되는 크랭크는 B.D.C를 향해 회전되도록 한다.

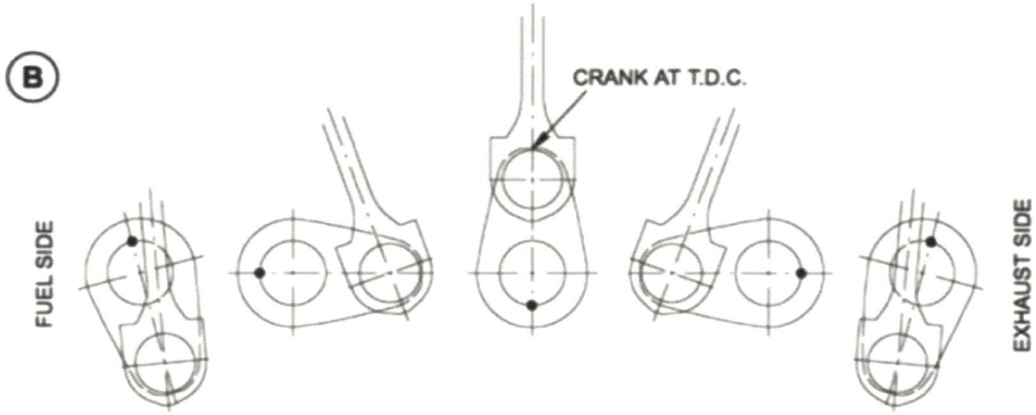

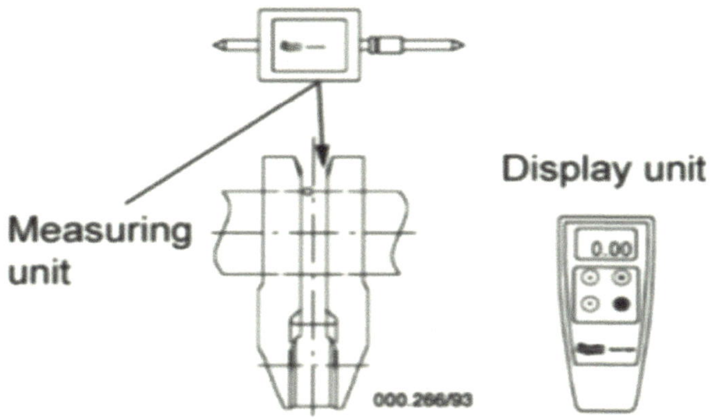

< Digital Gauge Type >

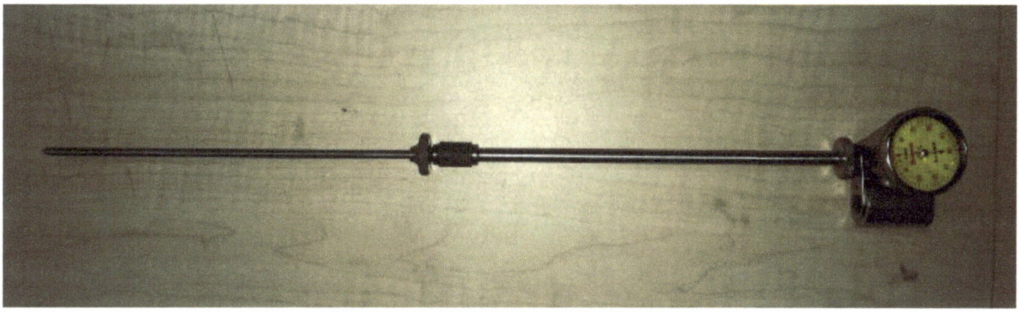

< Analogue Gauge Type >

Analogue Gauge 사용방법이다.

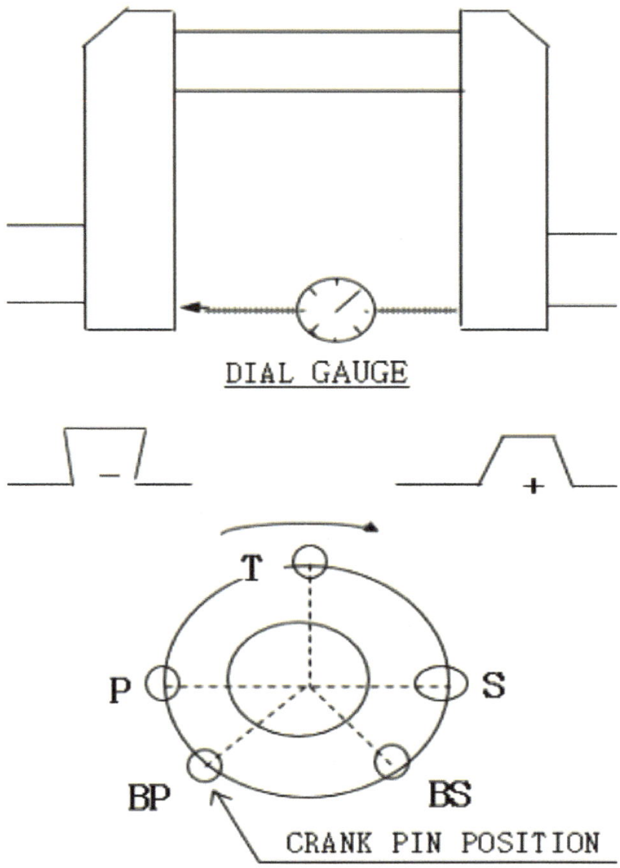

Turning Gear 이용하여 Turning을 시작하고 Crank Pin의 위치가 TDC로 갔을 때 Gauge 설치한다.

(Gauge 설치할 때 Crank arm에 보면 구멍이 있다.

그곳에 Gauge를 설치하면 되고 Gauge를 설치 시 느슨하지 않게 Tension을 주어 설치한다.)

Gauge 설치했다면 상기 그림에서 보듯이 Turning을 하여 먼저 Crank Pin을 BP(Bottom Port)에서 멈추고 Gauge **'Zero Setting'**을 한다.

따라서 계측 표에서 BP(Bottom Port)의 값은 무조건 **'0'** 이다.

BP(Bottom Port) → P(Port) → T(Top) → S(ST'BD) → BS(Bottom ST'BD)의 순서로 Engine을 Turning하여 Gauge에서 지시되는 값을 읽도록 한다.

마지막 BS(Bottom ST'BD)까지 계측했다면 반대로 Turning 하여 Gauge를 분해하고 다음 기통을 계측한다.

(4) 계측 결과의 판단

계측이 끝나면 각 기통의 V값을 구하여 전회치와 비교 검토하고 Limit 범위 내에 있는지 확인한다.

1. V = T - B
2. B = (BP+BS)/2

Engine의 사양에 따라 Limit 값이 다르므로 정확한 허용 값은 본선 Manual을 확인한다.

계측된 값들이 최대 허용 제한 치 위에 있다면 원인을 찾아야 하고 필요한 조치를 취해야 한다.

(Main Bearing 손상, 선체 손상에 의한 엔진 정지, Holding Down 볼트의 풀림, Propeller shaft bearing 손상 등등)

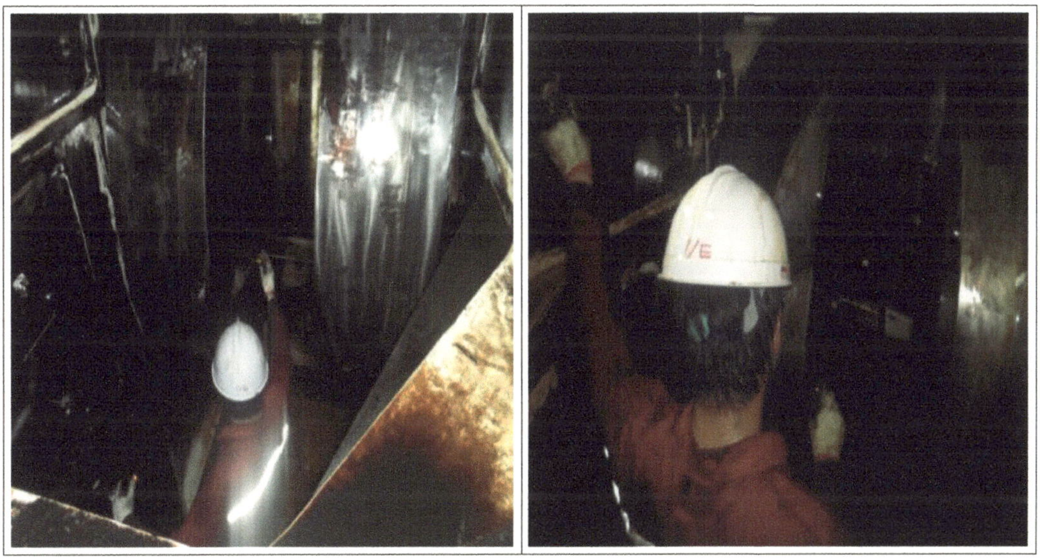

< Crankshaft Deflection 점검 >

M/E CRANK DEFLECTION REPORT

Header

Date 2019.08.04 — Draft: F 11.5 M 12 A 13.5 — ○ Empty

Engine Room Temperature 37 'C — Temperature in Crank Case 40 'C — ◉ Loaded

Crank Deflection

Max. Deflection 18 mm — UNIT: 1/100 mm

	1	2	3	4	5	6	7	8	9	10	11	12
BP	0	0	0	0	0	0	0	0	0	0	0	0
P	2	6	4	0	7	1	-2	4	-2	5	-1	2
T	18	14	8	12	18	14	8	10	6	1	2	10
S	0	4	2	4	4	13	8	12	9	6	4	4
BS	-1	6	-4	4	12	11	10	12	11	8	6	0
Absolute value	18.5	11	10	10	12	8.5	3	4	0.5	3	1	10

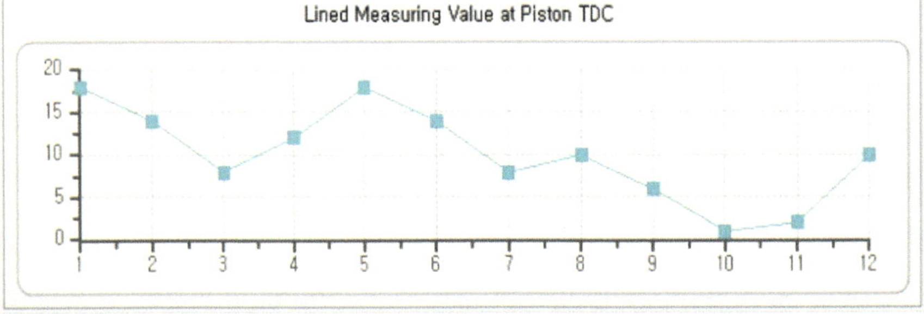

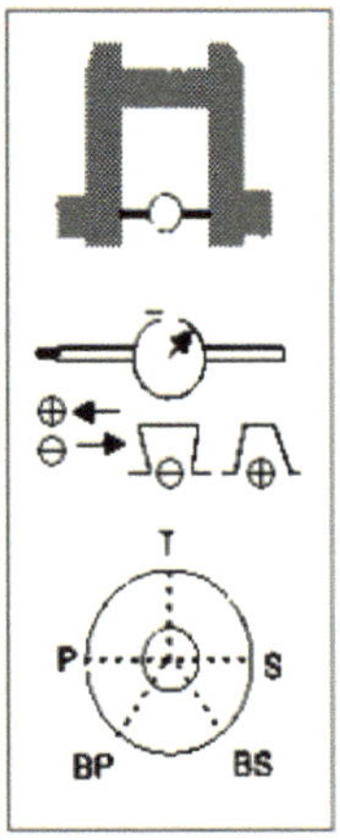

Remark

LIMIT
CYL. 1 & 12 : +17~-33 (1/100 mm)
CYL. 2 ~ 11 : +-17 (1/100 mm)
INNSA항에서 계측함.

5) 체인 드라이브

Arrangement of Chain Drive and Camshaft

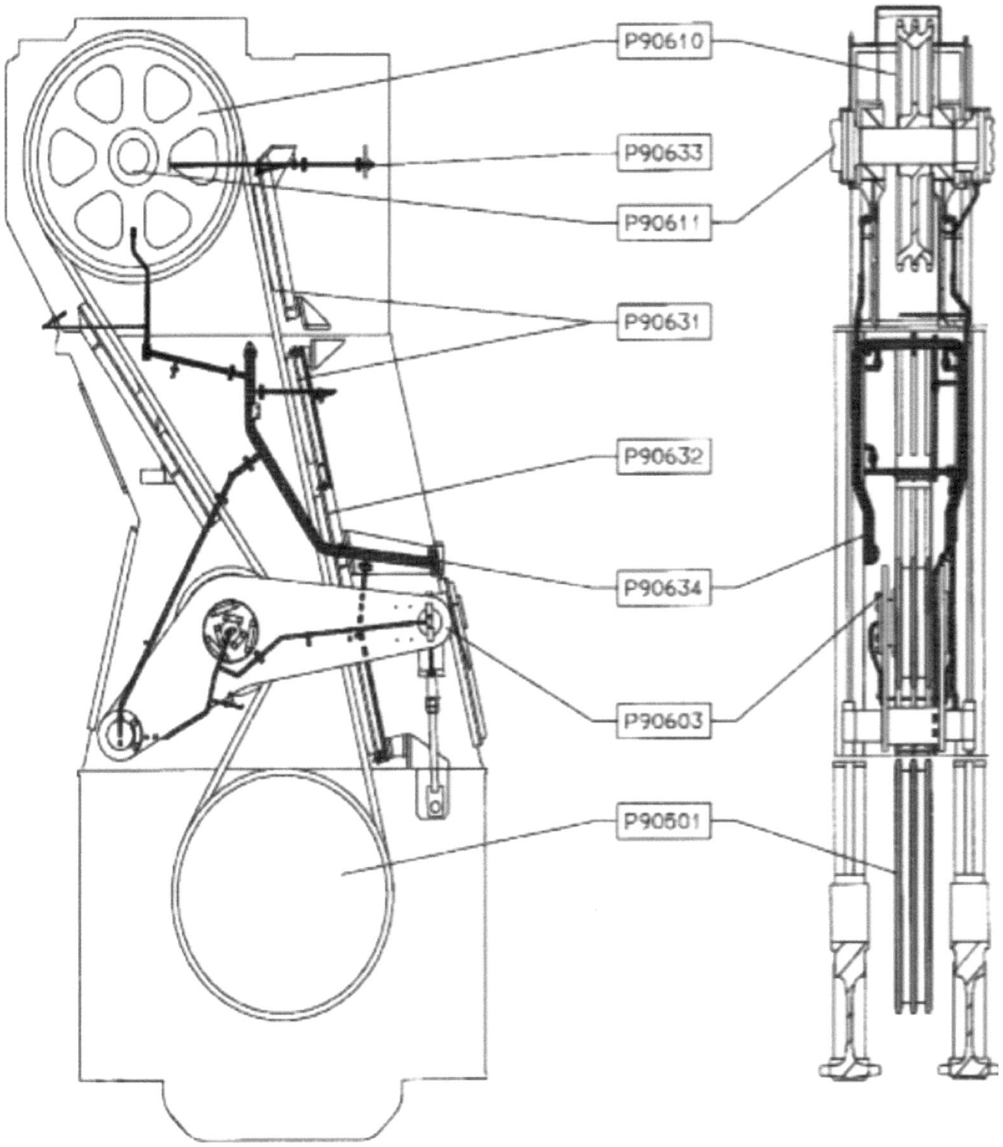

Chain Tightener

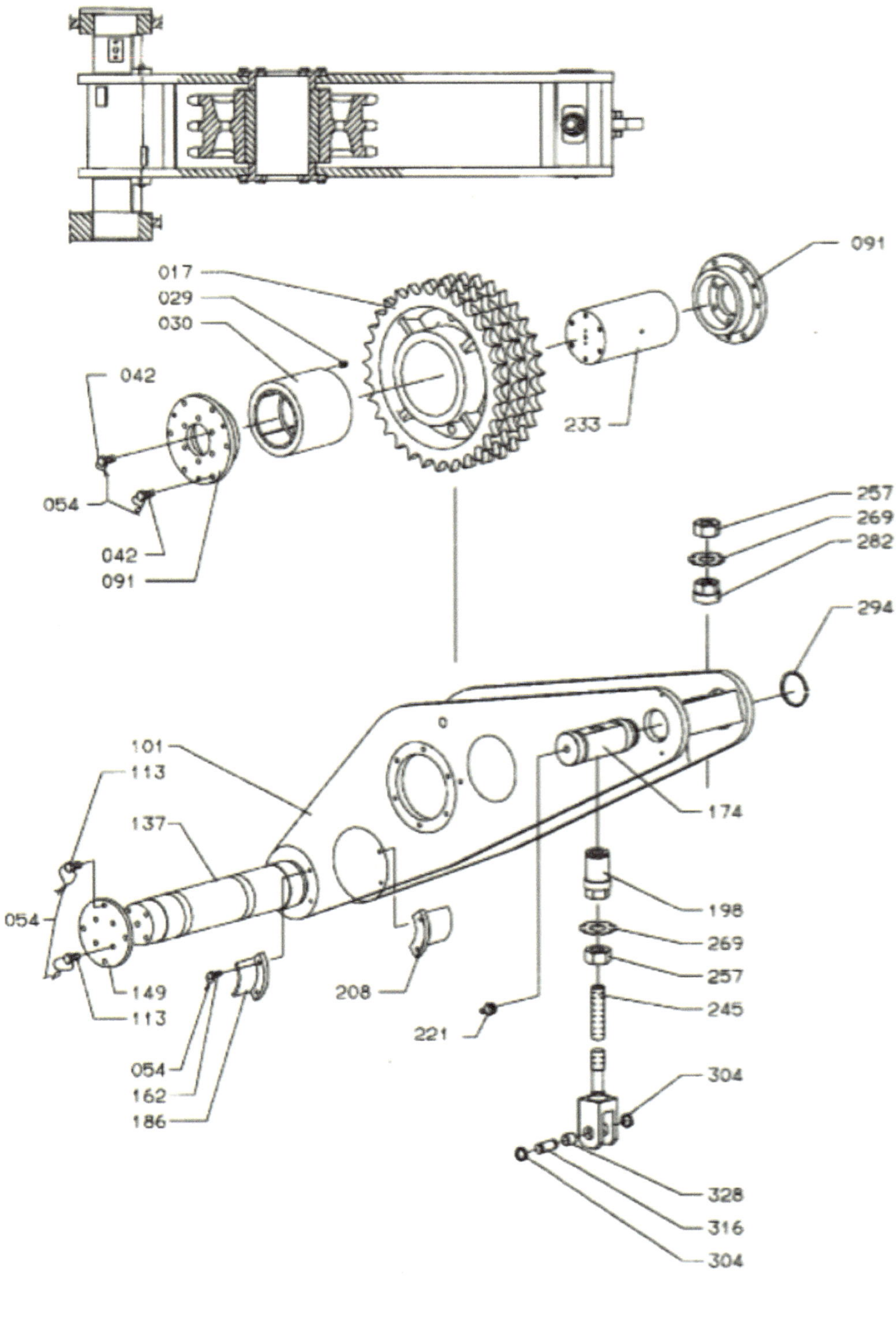

Chain

SAFETY PRECAUTIONS *For detailed sketch, see 900-2*

- [X] Stopped engine
- [X] Shut off starting air supply – *At starting air receiver*
- [X] Block the main starting valve
- [X] Shut off starting air distributor/distributing system supply
- [X] Shut off safety air supply – *Not ME-engines*
- [X] Shut off control air supply
- [X] Shut off air supply to exhaust valve – *Only with stopped lubricating oil pumps*
- [X] Engage turning gear
- [ ] Shut off cooling water
- [ ] Shut off fuel oil
- [X] Stop lubricating oil supply
- [ ] Lock the turbocharger rotors

Data

Ref.	Description	Value	Unit
D06-18	Teeth on chain wheel, max. wear	4	mm
D06-19	Original length (chain pitch x 10 links)	1143	mm
D06-20	10 links measurements + 1% of a tensioned chain = scrapping of chain	1154.4	mm

Chain Tightener

SAFETY PRECAUTIONS *For detailed sketch, see 900-2*

- [X] Stopped engine
- [X] Shut off starting air supply – *At starting air receiver*
- [X] Block the main starting valve
- [X] Shut off starting air distributor/distributing system supply
- [X] Shut off safety air supply – *Not ME engines*
- [X] Shut off control air supply
- [ ] Shut off air supply to exhaust valve – *Only with stopped lubricating oil pumps*
- [X] Engage turning gear
- [ ] Shut off cooling water
- [ ] Shut off fuel oil
- [X] Stop lubricating oil supply
- [ ] Lock the turbocharger rotors

Data

Ref.	Description	Value	Unit
D06-21	Tightening angle = 12 hexagons	720	°
D06-25	Mechanical or blocked chain tightener, clearance	0,1	mm

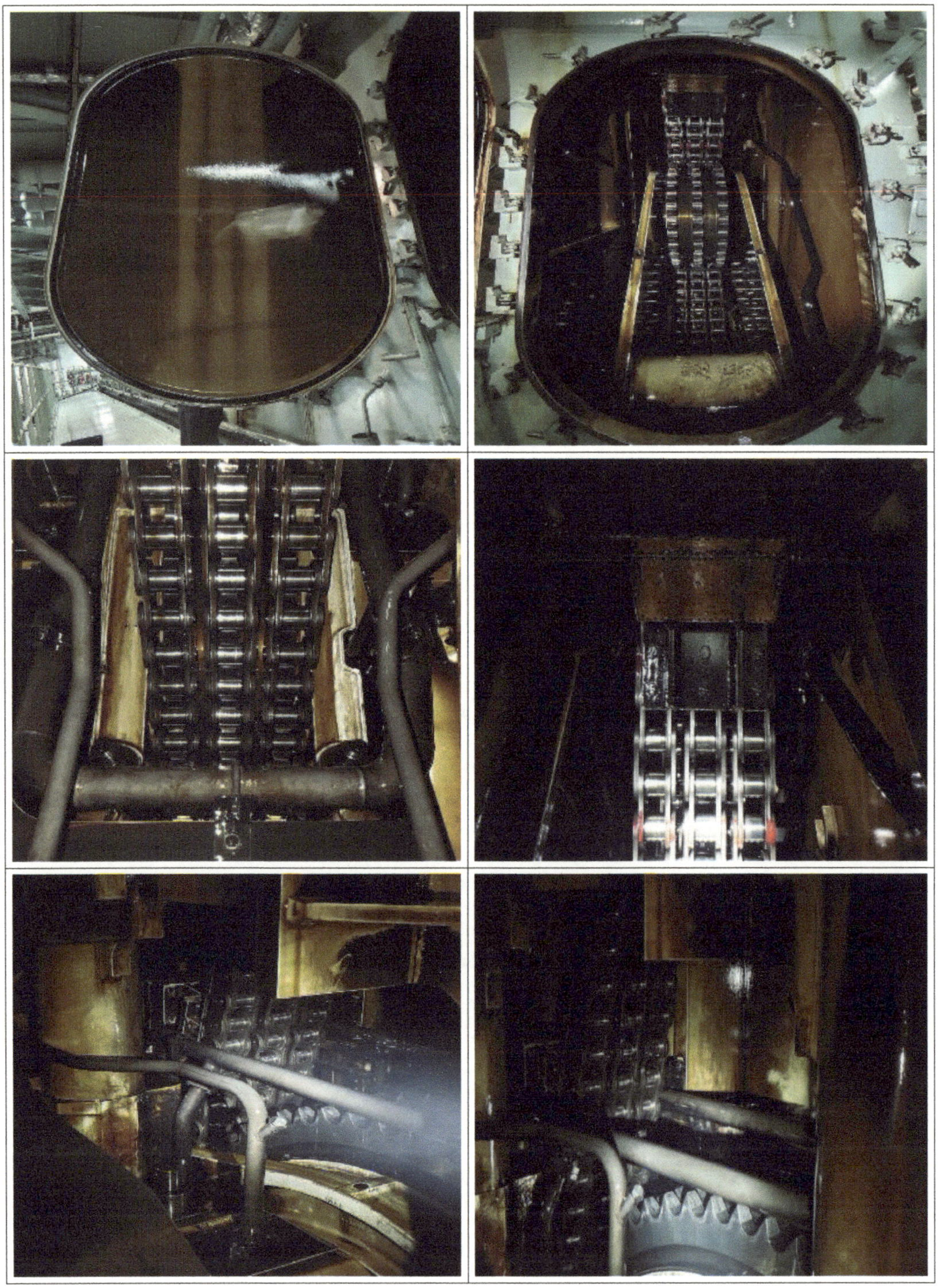

보통 Crankshaft Chamber 점검 시 같이 시행

Camshaft는 Crankshaft로부터 Chain Drive에 의하여 구동되며 Fuel Oil Pump와 EXH' Valve, 작동시켜 주는 Cam이 설치되어 있다.

4 ~ 8 Cylinder Engine : Engine의 후단에 설치

9 ~ 12 Cylinder Engine : Engine의 중앙에 설치

Chain Drive는 Crankshaft와 Camshaft에 설치된 Chain Wheel 통하여 구동되는 두 줄 Roller Chain으로 구성되어 있다.

Crankshaft와 Intermediate Wheel 사이의 Chain Casing에 설치되어 있는 Chain Tightener로 Chain을 조여준다.

Chain의 길게 늘어진 부분은 Guide Bar를 따라 움직이며, Guide Bar와 Chain Wheel에 설치된 Spray Pipe를 통해서 L.O가 공급된다.

Main L.O Pump Stop한 상태에서 Crankshaft Chamber Cover Open하고 Engine Turning하면서 전반적인 상태 점검한다.

① Bolt & Screw 풀린 것이 있는지 전반적으로 검사한다.

② L.O Pipe가 손상되지 않았는지 확인하고, Spray Nozzle이 막혀 있는지 찌그러진 곳은 없는지 검사한다.

③ Guide Bar의 균열된 곳이나 기타 다른 손상에 대하여 검사한다.

④ Chain Wheel의 톱니 상태를 점검한다.

6) 탑 브래싱

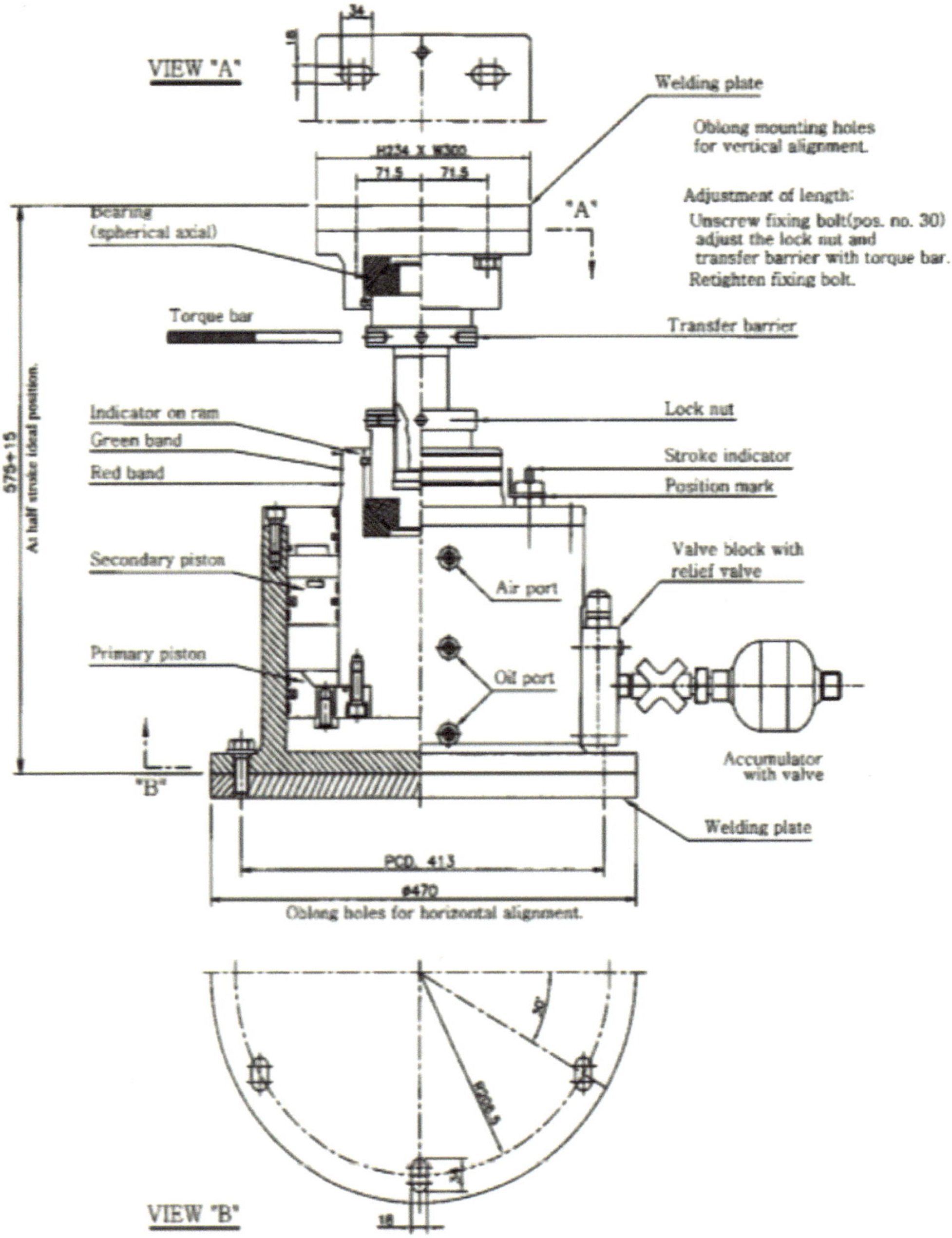

VIEW "A"
34
Welding plate
Oblong mounting holes for vertical alignment.
H234 X W300
71.5
71.5
"A"
Adjustment of length:
Unscrew fixing bolt(pos. no. 30) adjust the lock nut and transfer barrier with torque bar.
Retighten fixing bolt.
Bearing (spherical axial)
Torque bar
Transfer barrier
Lock nut
Indicator on ram
Green band
Red band
Stroke indicator
Position mark
575+15
At half stroke ideal position.
Secondary piston
Valve block with relief valve
Air port
Primary piston
Oil port
Accumulator with valve
"B"
Welding plate
PCD. 413
ø470
Oblong holes for horizontal alignment.
30
VIEW "B"
18

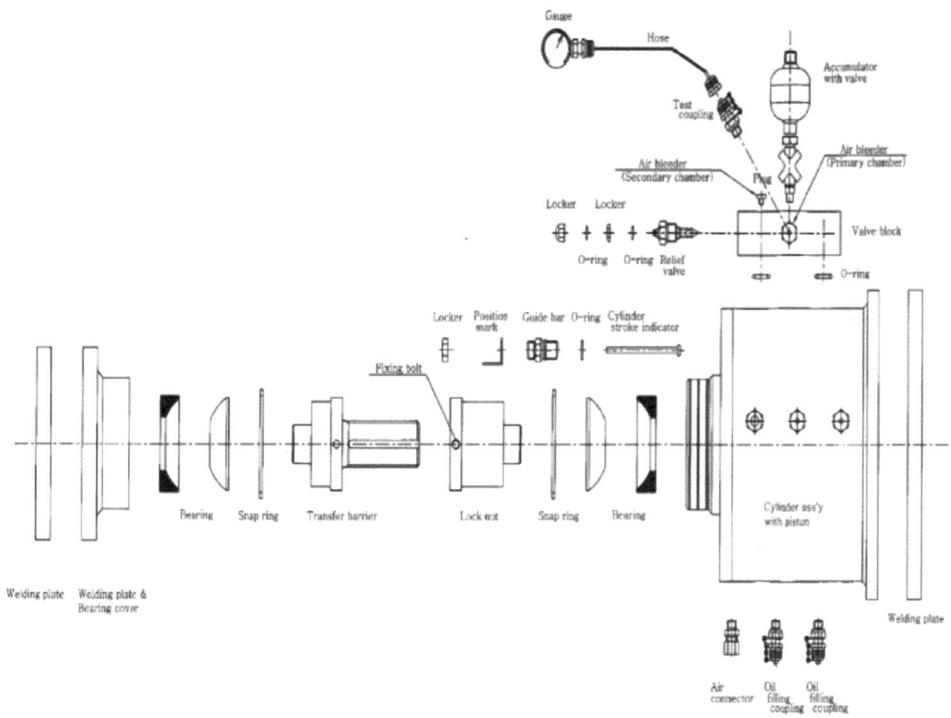

	Range of activate	Function description
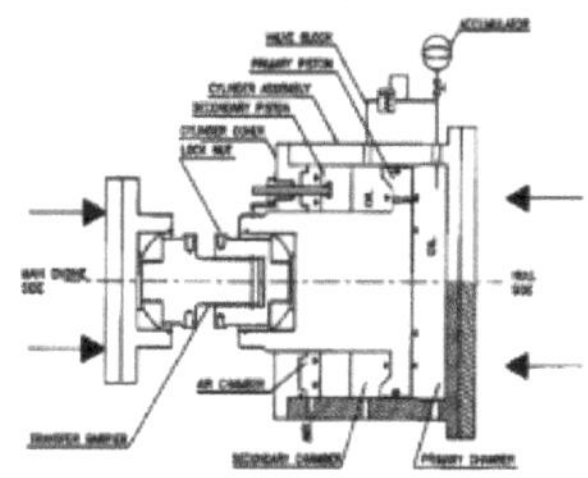	LOW LOAD 1. Static condition 2. Light load	When the force from engine and/or hull is less than 85 KN, top bracing keeps preadjusted condition. But, the force is between 85 KN and 183.8 KN, oil pressure in primary chamber will increase. In this case, accumulator absorbs the force. * Relief valve set pressure : 26 bar Accumulator set pressure : 15 bar Air regulator set pressure : 12 bar (Remark : It is changeable depend on ship's condition.)
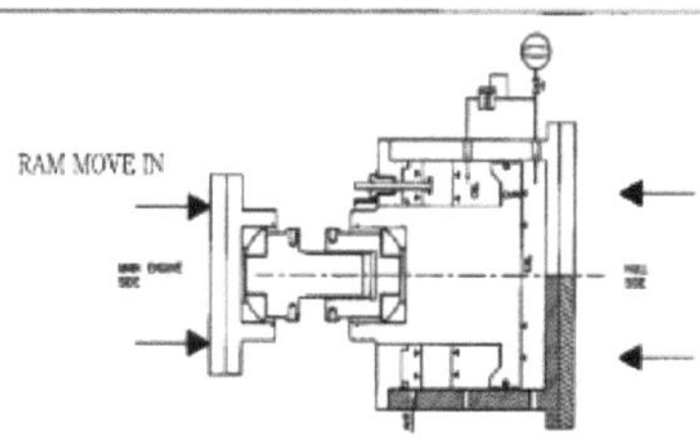	HEAVY LOAD	When the force from engine and/or hull is higher than setting pressure of relief valve(force : 183.8 KN), the relief valve is open. At the same time, oil in primary chamber flows to secondary chamber. During the oil in primary chamber flows to secondary chamber, the secondary piston move backward until the pressure of secondary and air chamber is equal. So, air chamber absorb the force.
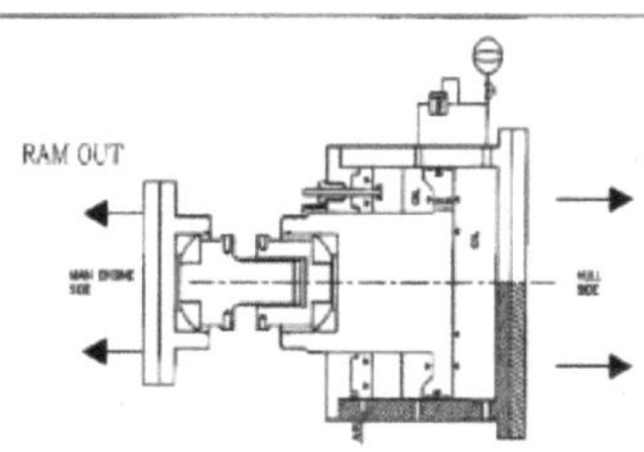	REVERSE LOAD (Negative force)	When engine and hull moves away from each other the oil pressure in primary chamber is decreased. At the same time, compressed air in air chamber push the secondary piston forward and the oil in secondary chamber flows back to primary chamber through the check valve until the pressure in primary, secondary and air chamber is equal.

REV.	DATE	DESCRIPTION	REMARK

b) Piston setting and air venting

Standard position

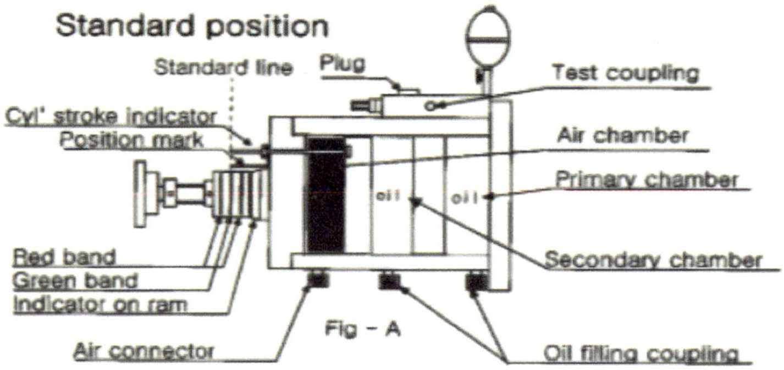

Fig - A

Ram out position

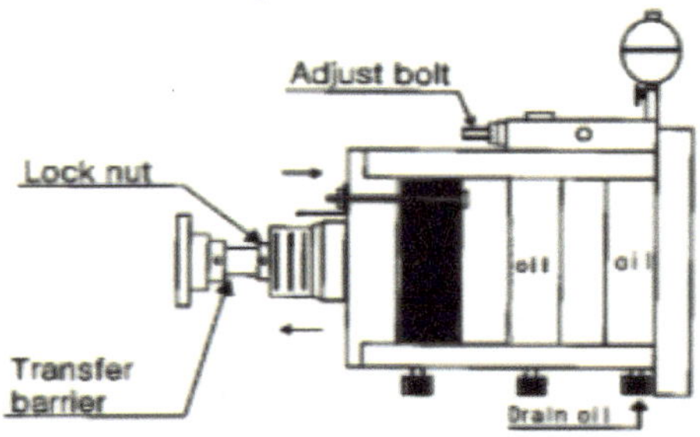

Fig - B

1) Stop air supply.
2) Connect hyd' hose to test coupling.
3) Tighten up the transfer barrier and lock nut until the green band meet position mark. During this work, air will be come out first then oil will be come out through hyd' hose. (primary chamber air venting)
4) Disconnect the hyd' hose from test coupling when 3) is done.
5) Check the position of secondary piston by cyl' stroke indicator.
6) Loosen up the plug a little on the valve block for secondary chamber air venting.
7) If cyl' stroke indicator is shorter than the end of position mark, fill oil to secondary chamber until cyl' stroke indicator meet the end of position mark.
   If cyl' stroke indicator is longer than the top of position mark, supply air to air chamber over 4bar. Do 2) to secondary oil filling coupling then oil will be come out. Drain oil until cyl' stroke indicator meet the end of position mark.

Ram in position

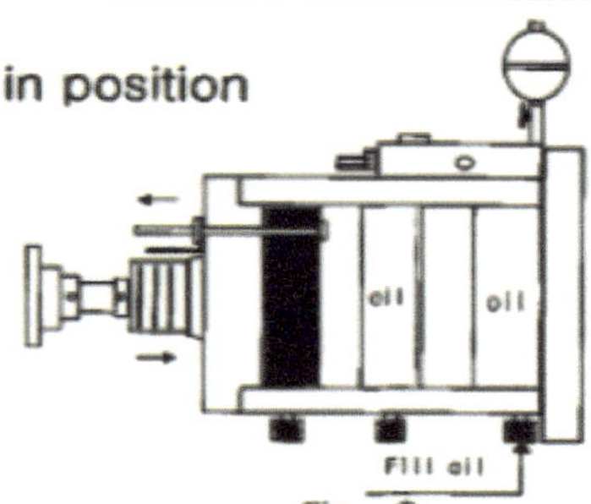

Fig - C

1) Connect hyd' hose to primary oil filling coupling and drain off until oil is not coming out.
2) Loosen up stoppper and transfer barrier enough.
3) Supply air over 4bar, until the green band meet the end of position mark. Then stop air supply.
4) Do 3) in order of Fig-B.
5) Do 4), 5) in order of Fig-B.

* 뉴메틱 시스템(PNEUMATIC SYSTEM) : 2.5 YEARS 주기

2.5년 주기로 외주 업체를 통하여 수리한다.

Engine Stop 후 L.O Pump Stop, Control Air & Spring Air Close 한다.

수리 완료 후 가능하다면 Engine을 짧게 Start하여 누설 및 이상유무를 확인하고, Engine을 시동 할 수 없는 상황이라면 Air-blow라도 시행하고 출항 후 Engine 시동할 때 주의 관찰한다.

[4] 보조 기기

1) 센트럴 쿨러

S.W Pipe Corrosion Flange 교환, Strainer 소제 & Backwashing 순서로 설명하겠다.

(1) CORROSION FLANGE 교환 방법

먼저 Central Cooler S.W Pipe 내부에 Corrosion Flange가 장착된다.

경험 상 **보통 1년마다 교환했으며,** 저 RPM으로 운항 중에 Cooler 하나만 사용해도 Engine 및 보조기기 운용에 무리가 없다면 운항 중에도 작업할 수 있지만, 가급적이면 정박 중에 작업한다.

Cooler S.W Inlet & Outlet Valve Close하여 Cooler 하나를 Stop 한다.

주변 Pipe Line들 모두 분리하고, 내부 Gasket 또는 Packing을 교환 하도록 한다.
(예비품 Packing 있으면 교환하고, 없으면 Hippa Gasket 얇은 것으로 교환한다.)

Corrosion Flange를 조립하고 나면 반드시 Cable도 잊지 말고 꼭 연결 하도록 한다.

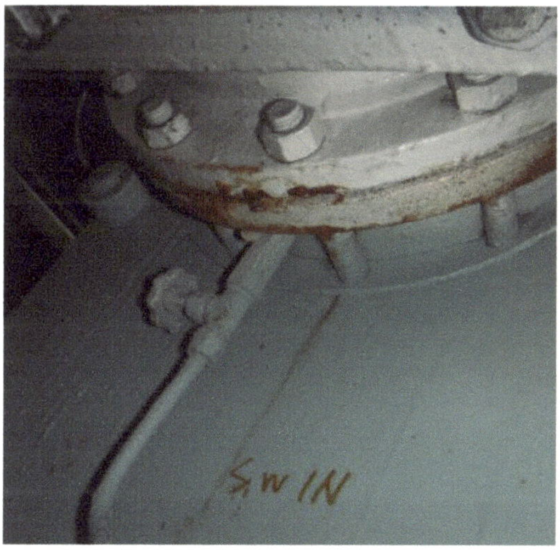

장착 중인 S.W Line에서 해수 미량 누설

신품(좌측) 및 취외품(우측) Corrosion Flange

취외한 Corrosion Flange

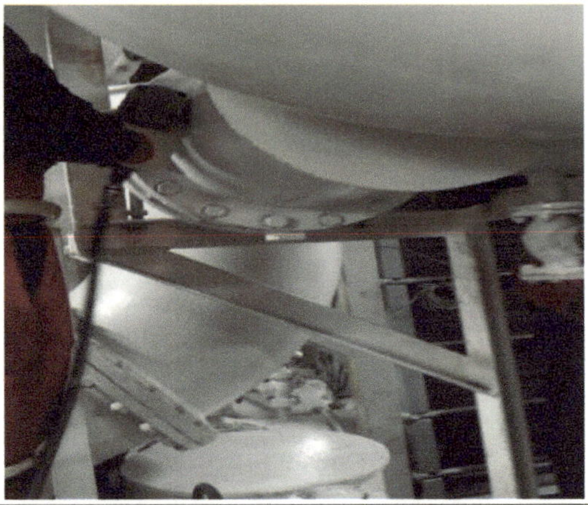

Gasket

Cable

(2) 해수라인 스트레이너 소제 & 백워싱

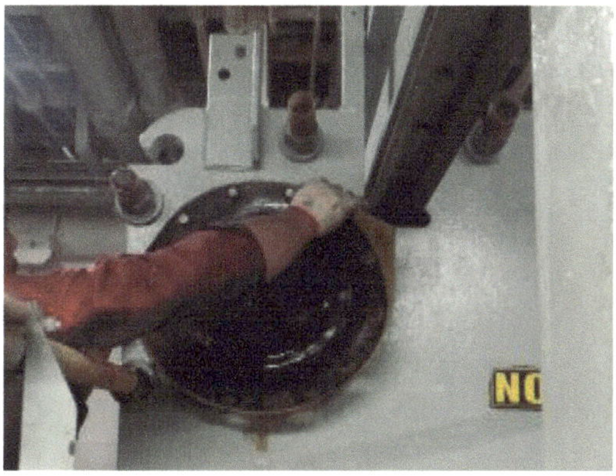

Strainer Cover Rubber Gasket

Strainer

Venting Valve

Central Cooler S.W Outlet Valve

해수 및 뻘이 튀어나오므로
Wood Plate 같은 것으로 막아준다.

① 저 RPM으로 운항 중에 Cooler 하나만 사용해도 Engine 및 보조기기 운용에 무리가 없다면 운항 중에도 작업 가능하지만 가급적이면 정박 중에 작업하다.

② Cover Bolt 풀어내고 Cover 살짝 풀어내어 S.W(Sea Water) Line의 해수를 Drain 한다.

Coaming Drain Hole에 물만 Drain 될 수 있도록 여과지나 필터 같은 것들을 집어넣어 준다.

Bilge Well High Alarm 확인하여 필요 시 이송한다.

주로 Central Cooler는 ST'BD 측에 위치하고, Cooler Coaming은 ST'BD Bilge 또는 MID Bilge에 연결된다.

③ Strainer를 취외하고 Strainer 및 S.W Line 내부 소제한다.
(예비품 Strainer가 있는 경우는 맞교환 후 취외한 Strainer 소제한다.)

④ S.W Pressure 2.0 kg/㎠ 이하일 경우, Main Cooling Sea Water Pump 한대 더 기동하여 압력을 올린 상태에서 S.W Strainer 취외한 Cooler의 S.W Outlet Valve Open하여 Backwashing한다.
(Backwashing 시 나무판자 같은 것으로 주변이 오손되지 않게 막아준다.)

⑤ Backwashing 후 소제한 Strainer 집어넣고 Cover Close 한다.

S.W Inlet Valve Open 및 Venting Valve Open하여 Air Venting 한다.

Venting이 끝나면 Valve 원상 복귀하고 주변부 소제한다.

*** Cooler 상태 확인하여 Strainer 소제 하여도 Cooler 성능이 개선되지 않는다면 Backwashing 또는 Chemical Cleaning을 한다.**

Backwashing 방법에는 두 가지가 있다.

첫 번째 방법은 상기 설명한 바와 같이 Cooler Strainer 소제 작업 시 Backwashing 하여 뻘들을 제거하는 방법이다.

가장 효과적이고 작업 시간도 단축되는 방법이지만 주변 작업공간이 오손될 수 있다.

두 번째 방법은 S.W Line Strainer 취외하는 방법까지는 똑같다.

① 소화전 라인을 NO.1 Central Cooler S.W Outlet측 Flange에 연결한다.

S.W Inlet측 Flange와 NO.2 S.W Outlet 측 Flange에 또 다른 소화전 라인을 연결한다.

그리고 Flange에 있는 Valve들 및 소화전 Valve를 Open 한다.

② G.S Pump를 기동하여 S.W Outlet → Inlet 방향으로 Backwashing을 1시간 정도 한다.

③ Backwashing 끝났으면 G.S Pump Stop 한다.

소화전 Valve, Flange Valve Close 한다.

S.W Inlet Valve 및 Venting Valve Open하여 Air Venting 한다.

④ S.W Valve 원상복귀 하고 다음 Cooler Backwashing을 진행한다.

*** 개인적으로 두가지 방법으로 Backwashing 시행 해본 결과 첫 번째 방법이 작업 시간도 훨씬 적게 걸리고 효과도 좋았으나 뻘들이 튀어나와 주변을 오손 시킨다는 점이 있다.**

하지만 우리가 생각해보면 Backwashing의 목적은 결국 뻘을 제거하는 것이다.

Cooler 상태 확인할 수 있고 소제 효과도 눈으로 직접 확인할 수 있기에 추천하는 방법이다.

*** 상기 언급한 Backwashing을 시행 했음에도 온도가 잘 떨어지지 않는다면, 다시 말해 Cooler 성능이 개선되지 않는다면 Backwashing 한번 더 한다.**

그럼에도 개선되지 않는다면 Chemical Cleaning을 시행한다.

Chemical Cleaning 방법은 기본적인 것이므로 추가적인 설명은 하지 않겠지만, Chemical은 친환경 Chemical인 G&E Bridge(S.W Side Cleaning)을 사용하는 것이 좋다.

간혹 SAF-ACID를 사용하는 경우가 있는데 용도가 완전 다른 것이다.

SAF-ACID : Scale 제거용 Chemical
S.W Side에 있는 '뻘'을 제거하기 위해서는 Scale 제거용 SAF-ACID가 아닌 G&E Bridge가 Chemical Cleaning에 있어 훨씬 더 효과적이다.

Backwashing은 사실 특별히 어려운 건 없으며 작업이 아주 간단하다.

대부분 선박들이 중국 항 기항하므로, 마지막 항구에서 Central Cooler S.W Strainer 소제 작업 및 Backwashing을 진행한다.

상기 설명한 바와 같이 저 RPM으로 운항 중에 Cooler 하나만 사용해도 될 경우 운항 중에 작업 가능하지만 급한 경우가 아니라면 가급적 정박 중에 작업을 진행한다.

2) 대체 해양 전력 컨테이너 (ALTERNATIVE MARITIME POWER)

AMP 장착 선박에는 Connection & Disconnection 절차 및 W.I(Working Instruction)이 있다.

무엇이든 한 번만 해보면 Image가 머리 속에 잡히므로 수월하게 할 수 있다.

Key 꼽고 A1 & A2 Panel Earthing, V.C.B(Vacuum Circuit Breaker) 취외, 장착 등 처음에는 무척 당황스럽고 혼동 될 수 차근차근 읽어보고 천천히 따라 하면 된다.

MC Panel에서 Connection & Disconnection에 들어가시면 하나씩 할 때마다 표시가 되니 천천히 순차적으로 하면 된다.

유의할 점은 Earthing 및 V.C.B 작동할 때 A1 Panel 및 A2 Panel 상부에 Earthing 및 V.C.B관련된 Button이 있다.

그것을 누르고 작동해야 하며 **무작정 그냥 힘으로 하면 부러진다.**

Key 또한 무리하게 작동시키면 부러질 수 있다.

입항 하루 전날에는 Ship's Cable을 미리 연결하도록 하고, Shore Cable은 입항 후 한 시간 안에 연결해야 한다.

따라서 Engine 정지 및 정리 마무리되면 Shore Cable 미리 조금 내려놓고 연결 준비를 한다.

반드시 작업하기 전에 여러 번 꼭 읽어보고 입항 전날 Ship's Cable 연결할 때 Remind를 한 후 기관실에서 조치하는 2기사에게도 다시 한번 교육 및 숙지시키도록 한다.

* 연결(CONNECTION)

Section	1) On sailing	2) Task before California port berthing		Remark
	Keep connect LV Cable(440V) always	Connect ship's HV Cable within 48HR before berthing		
1.Before berthing				Socket lever should be locked after the plug and socket are fully connected before connecting HV Cable 1/E, No.1 Oiler, Oilers

Section	1) Berthing	2) Stevedore is on board	Remark
	Request to remove lashing immediately after immigration	Remove AMP CNTR. Lashing	
2. After berthing			C/O

Section	1) Changeover panel position & AMP Shore receiving panel		2) BOW & AMP Panel	3) PIC for shore connection is on board & lower shore cable reel		Remark
	Set the changeover panel	Open earthing SW at AMP Receiving panel	Check the position of 2 Swithes	Shore power PIC is onboard. Lower AMP CNTR. Shore cable		
3.Lowering shore cable reel		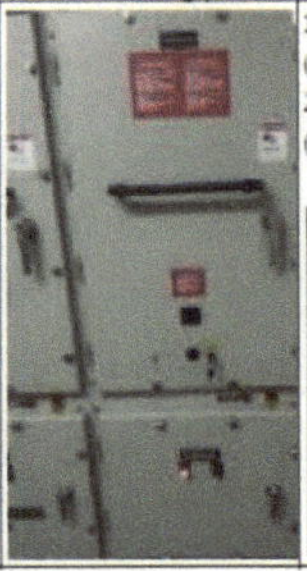	1) AMP CNTR Position (Port or STBD) 2) Mode Selection (BOW → AMP) 	The pilot loop system is applied by single type so that F16 CB turn off manually in order to process next sequence (only LAYTI port) 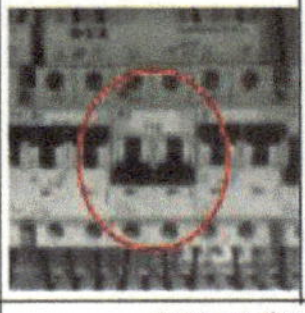	 	Changeover to manual mode and operate AMP CNTR. Cable reel can be operated by remote controller. 1/E, No.1, Oiler, Oilers

Section	1) Connect AMP CNTR. Shore plug~Shore socket box		2) Connection is completed	Remark
	Move the cable to the shore socket box	The HV Plug is connected	Fix the kellum grip ring	
4. Shore cable connection			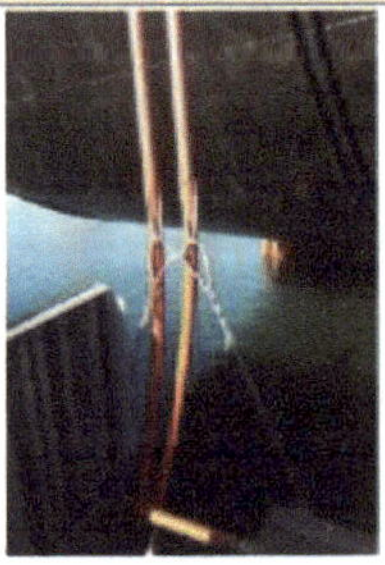	Check that connection is completed. Change over to cable reel auto mode from manual mode and check auto tension condition. 1/E

Section	1) Connect AMP CNTR. Shore plug~Shore socket box			2) Connection is completed		Remark
	1) A1 Panel	2) Shore cable reel operation mode	3) Safety area door	4) A2 Panel	5) A2 Panel	
	- Open the earthing S.W.D	- Change to the automatic mode	- Close the door & lock	- Open the earthing S.W	- Rack in the VCB	
5. AMP CNTR.	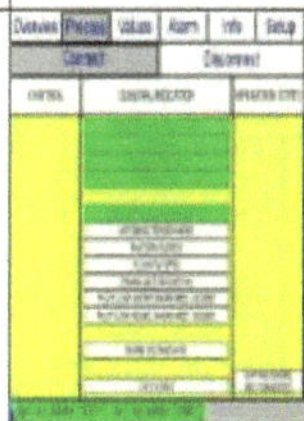	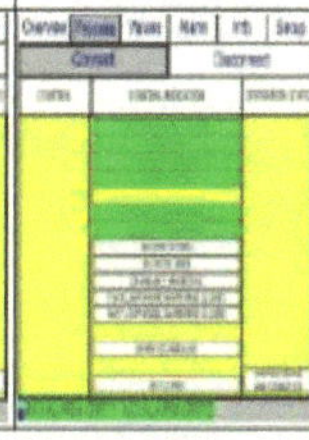	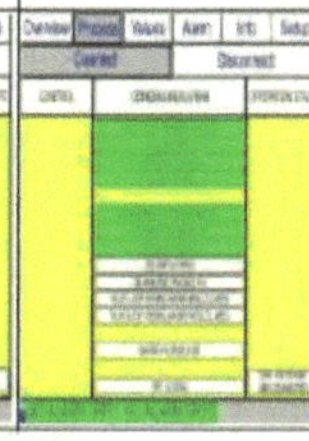	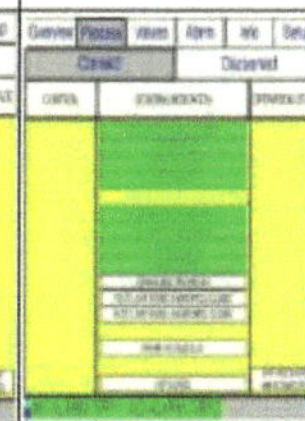	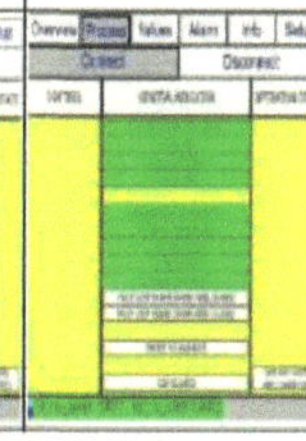	Check & record the power counter. 1/E
Section	3) Pilot Loop Check, Shore VCB Close			4) AMP CNTR. VCB Close		Remark
	6) Close pilot loop P2 at *MC Panel(Ship)	7) Close pilot loop P1 MC Panel(shore)	8) A1/A2 Panel	9) MC Panel	10) MC Panel /A1/A2 Panel	
	- Message to Harbor staff that we are going to close their pilot loop - If shore is Ok, Push S2 "Start Release"	- Message to harbor staff that system is ready to take shore power - Close pilot loop vessel (Shore) - **(Turn on F16 CB) - Close shore VCB	- Check phase, voltage and KW	- Push S3 "ON" - VCB of A2 Panel is closed	- Check current - Check Power consumption	
5. AMP CNTR.	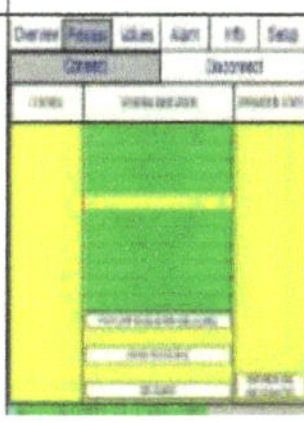	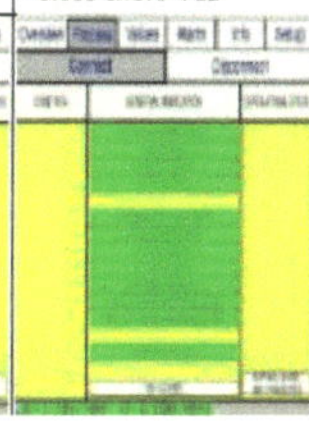		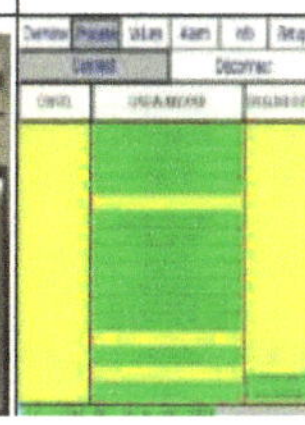		* MC Panel : Monitoring & Control Panel ** Only LAVTI 1/E

Section	1) AMP Shore receiving panel	2) Sync. Panel	3) AMP Control Panel - Changeover to shore power control from G/E control	4) AMP Control panel Arrival process (In case of operating more 2 G/E, it can be transferred, too)	5) AMP Control Panel Arrival Process	Remark
	- Check phase, voltage - Check that earthing SW is opened - Close VCB	① GEN. Control SW Auto → AMP ② Freq/Volt Meter SW → AMP Mode ③ Sync. Scope SW → AMP Mode	① Sync. Selection SW → AMP Mode ② Freq/Volt Meter SW → AMP Mode ③ AMP Selection SW → Auto mode	- Push Auto Arrival SW - Check closing BOW/AMP ACB	- Check shifting load G/E → Shore power -Check opening G/E ACB - Check stopping G/E (Stopping Priority : No.1,2,3,4)	
6. ECR Switch board		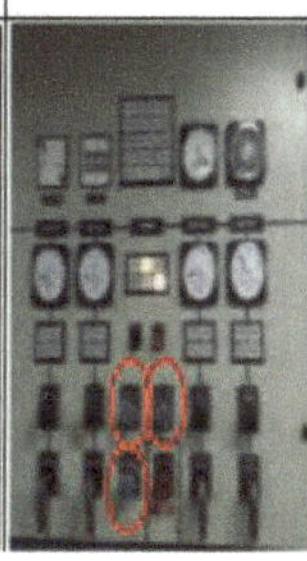		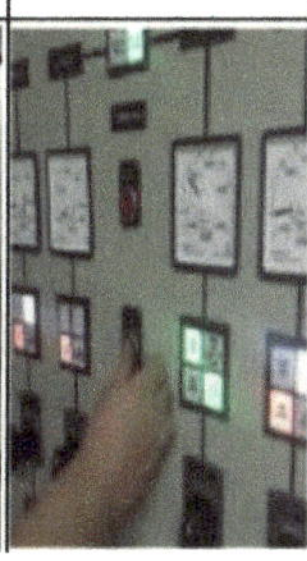		2/E

* 해제(DISCONNECTION)

Section	1) Start G/E at local area	2) AMP Control Panel	3) Sync. Panel	4) AMP Shore Receiving Panel	5) BOW & AMP Panel	Remark
	- Check AMP is ready - Check phase, voltage	- Push Auto Departure SW - Check Auto Sync. & G/E ACB Close - Check opening AMP ACB	- ① GEN. Control SW AMP → Auto ② Freq/Volt Meter SW → G/E Mode ③ Sync. Scope SW → G/E Mode	- Open the shore VCB	- Mode Selection SW AMP → BOW 6) *Message to harbor staff	
1. ECR Switch board						* Inform harbor staff that the system is shut down 2/E

Section	1) MC Panel	2) MC Panel/A1/A2 Panel	3) A2 Panel	4) A2 Panel	5) Safety area door	Remark
	- Push S1 Button - Open A2 Panel VCB	- Verify that voltage detectors at MV switchboard indicate no voltage	- Rack out VCB	- Close earthing S.W.	- Door unlock & Open□	
2. AMP CNTR	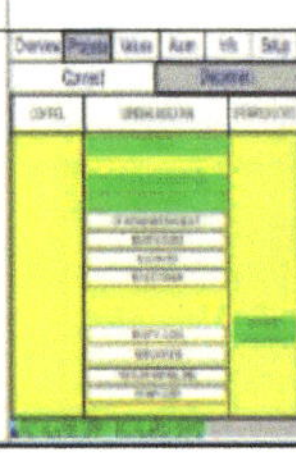		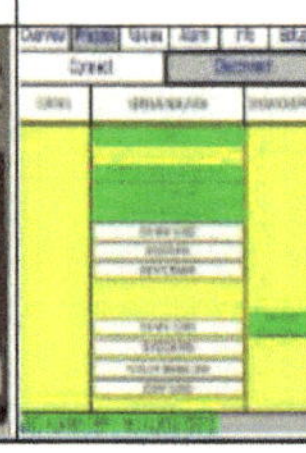	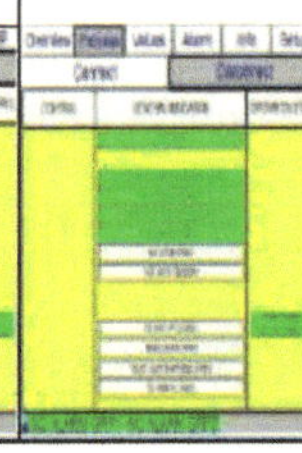	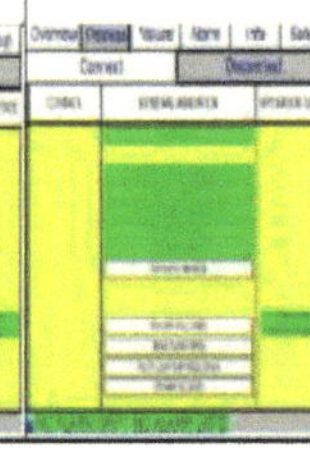	1/E

Section	6) Shore cable reel operation mode	7) A1 Panel □	8) Message to harbor staff	Remark
2. AMP CNTR	- Manual Mode	- Close earthing S.W.□	- Message to harbor staff that shore power system is switched off, earthed and ready for disconnection	1/E

Section	1) Remove AMP CNTR. Shore cable~Shore socket		2) Hoist AMP CNTR. Shore cable reel		Remark
	Disconnect the plug	Remove kellum grip ring	Hoist AMR CNTR. shore cable	Close the reel door	
3. Shore cable Disconnection					Message to crew that plugs are disconnected and cables are ready for lifting.
Section	1) Before departure	2) Task after departure			Remark
	Check AMP CNTR. Lashing	Disconnect ship's HV Cable within 24HR after departure			
4. Before departure & Task after departure					* For dismantling plug - Use the kirk key. - Lift up the socket lever * After dismantling plug - Close fully socket lever.

3) 발전기 진공 차단기(VACUUM CIRCUIT BREAKER)

본선에서 V.C.B 문제 관련 History가 많지 않고, V.C.B 취외하는 방법을 잘 모르는 사관들이 많다보니 실제로 점검이 필요한 상황에서도 기피하는 현상이 있다.

A.M.P 장착된 선박을 이미 승선하신 분들은 어렵지 않는데 모르면 당황할 수 밖에 없다.

G/E V.C.B 취외 하여 내부 보조접점이나 다른 부품들을 점검하거나 V.C.B Panel Front Door Open할 경우, V.C.B를 Test Position(Disconnection)에 와서 E.S(Earthing Switch)를 투입해야 Front Door Open 된다.

① 먼저 Stop 중인 G/E의 V.C.B(V.C.B Open 상태) 인출 홀을 개방한다.

② 인출 바를 삽입하여 V.C.B를 Test Position(Disconnection)에 놓으면 상부 Lamp에 V.C.B Test Position Lamp가 점등이 된다.

③ 전압 소자 및 상부 V.C.B Test Position Lamp 점등 확인 후 E.S(Earthing Switch)를 투입하고, E.S Key를 취외한다.
**(V.C.B를 Test Position으로 해야 E.S가 작동한다.
하단부 Door를 Open하여 확인하면 E.S에 Pin이 걸려있는 것이 보이며, V.C.B를 Test Position으로 하면 그 Pin이 들어가서 ES를 작동시킬 수 있다.
즉, Interlock이 되어있다.)**

④ Front Door를 Open 한다.
(Front Door는 E.S를 투입해야 Open된다. 즉, Interlock이 되어 있다.)

⑤ Earthing 되면 V.C.B Cover Open 되며 내부 보조접점이나 이외의 것들을 점검할 수 있다.

만약에 V.C.B를 취외하고 싶으면 본선에 V.C.B Lifting Machine이 있다.

그것을 이용하여 V.C.B를 장착하고, V.C.B를 취외 할 때는 연결된 Power Cable 취외하고 아래의 Locking Pin을 가운데로 누른 다음 취외 하면 된다.

반대로 다시 조립 시에는

① Lifting Machine 이용하여 V.C.B를 장착 후 Test Position에 놓고 Power Cable 연결한다.

② Front Door Close 한다.

③ E.S(Earthing Switch)를 개방(OFF) 한다.

④ V.C.B 인출 바를 Front Door에 있는 인출 홀에 삽입한다.

⑤ V.C.B를 Tool을 이용하여 Service Position으로 집어넣는다.

⑥ Key로 홀을 Close 후 Key를 취외한다.

V.C.B Open 상태에서 작업하고, V.C.B TEST Position으로 취외 시 Lamp 점등 확인한다.

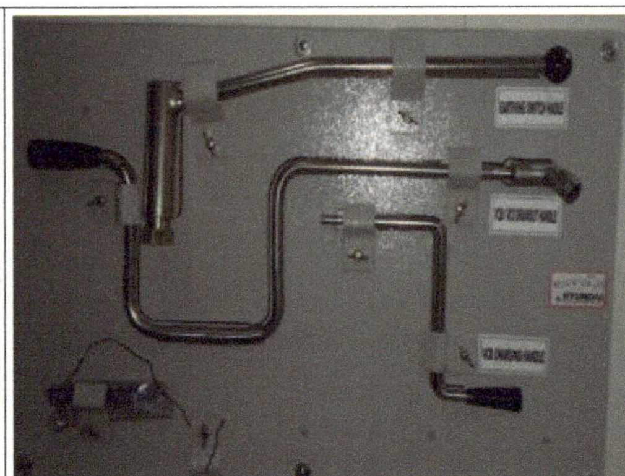

위에서부터 Earthing S/W Handle,
V.C.B Draw-out Handle
V.C.B Charging Handle

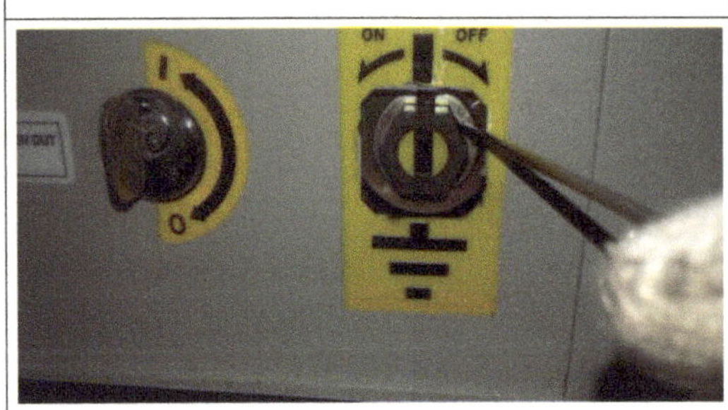

Earthing Hole
좌측 키로 Hole을 Open하고
우측과 같이 Earthing 을 한다.

하단부 Door를 Open하고
V.C.B가Test Position으로 되어 있다면
Locking Pin이 Earthing Switch 에서 빠진다.

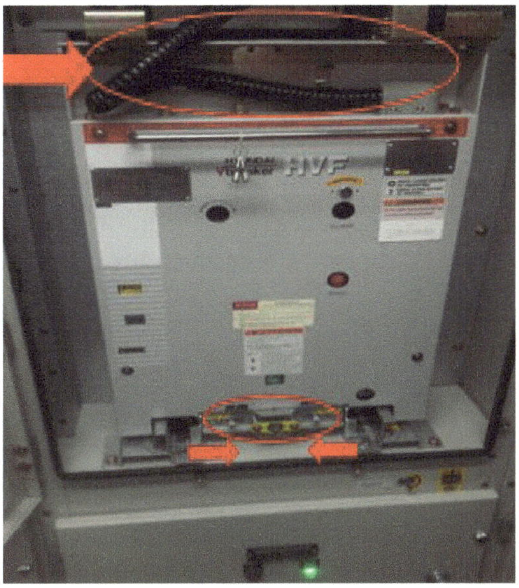

Earthing 투입 후 Front Door를
Open하면 V.C.B가 보인다.
상부 Power Cable을 취외 하고
아래쪽 두 개의 Locking Pin 가운데로 누르고
V.C.B 손잡이를 잡고 취외한다.

V.C.B Cover를 Open하면
AUX. Relay 확인 할 수 있다.

AUX. Relay를 취외한 모습

AUX. Relay 뒤 Linkage의 마모로 유격 발생하여
V.C.B Close후 AUX. Relay작동이 불량하여
V.C.B Close 오작동 발생시킨다.

*** 진공차단기 리프팅 머신**

Lifting Handle(화살표)/ 조정Valve(동그라미)

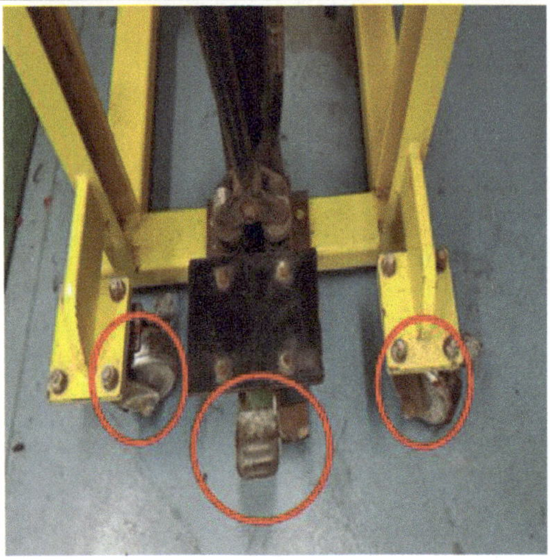

고정 발판

Lock Lever

실제 승선 중에 NO.3 G/E V.C.B가 문제가 발생하여 NO.2 REEFER TRANSFORMER V.C.B와 사양이 동일하여 교환 작업을 시행한 적이 있다.

Lifting Machine은 각 선박에 보유 중이다.

사용방법은 고정된 발판을 풀어내고 원하는 위치로 Machine을 움직인다.

V.C.B를 뺄 때는 원래의 위치보다 살짝 아래로 놓이게 조정한다.

조정은 두 번째 그림에서 보듯이 조정 Valve Close 상태에서 Lifting Handle을 좌우로 움직여서 올리면 된다.

내릴 때는 조정 Valve Open하여 손으로 살짝 아래로 누르면 아래로 움직이며, 위치를 조정했으면 네 번째 그림에서 Lock Lever 위를 향하도록 올린다.

V.C.B Power Cable 취외하고 중앙에 위치한 Locking Pin 중앙으로 누른 상태에서 V.C.B Handle을 당기면 빠진다.

조립 할 경우에는 V.C.B를 Lifting Machine에 올린 상태에서 조립하는 위치보다 살짝 위에 위치하도록 한 다음에 밀어 넣는다.

작업을 하기 전에 순서를 미리 충분히 숙지하고 작업을 하면 어려운 작업이 아니다.

4) 비상발전기 공기 차단기(AIR CIRCUIT BREAKER)

ACB 발출 가이드	선명	SM MUMBAI
	POSITION	EM'CY G/E ACB
	MAKER	TERASAKI
	TYPE	AR208S

0. 작업에 앞서
1) EM'CY G/E ACB OPEN상태 확인 (ACB 전면부 INDICATOR)
2) EM'CY G/E [AUTO / MANUAL] SWITCH MANUAL로 전환 (비상발전기 자동운전 방지)
3) 작업장소 주변정리, 내부 활선 접촉 주의

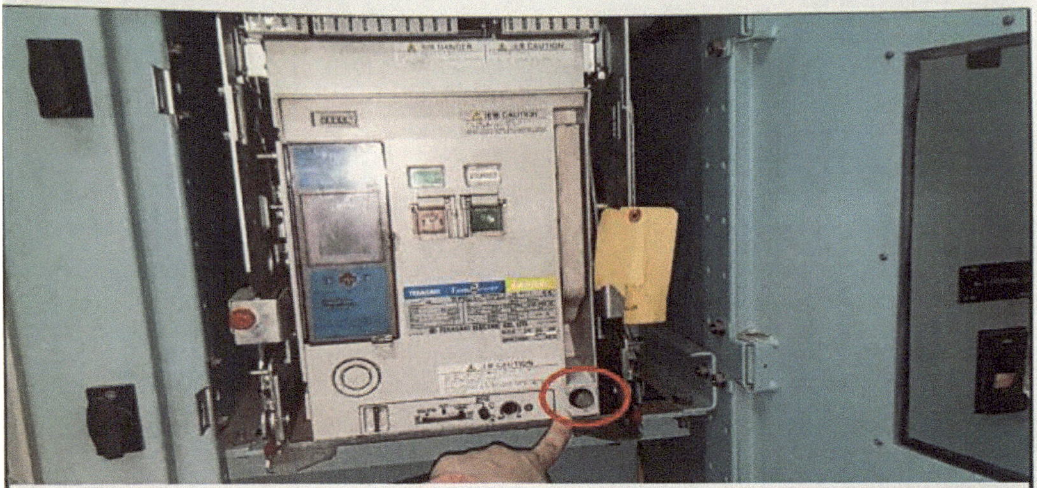

1. ACB DRAW OUT BAR 취외

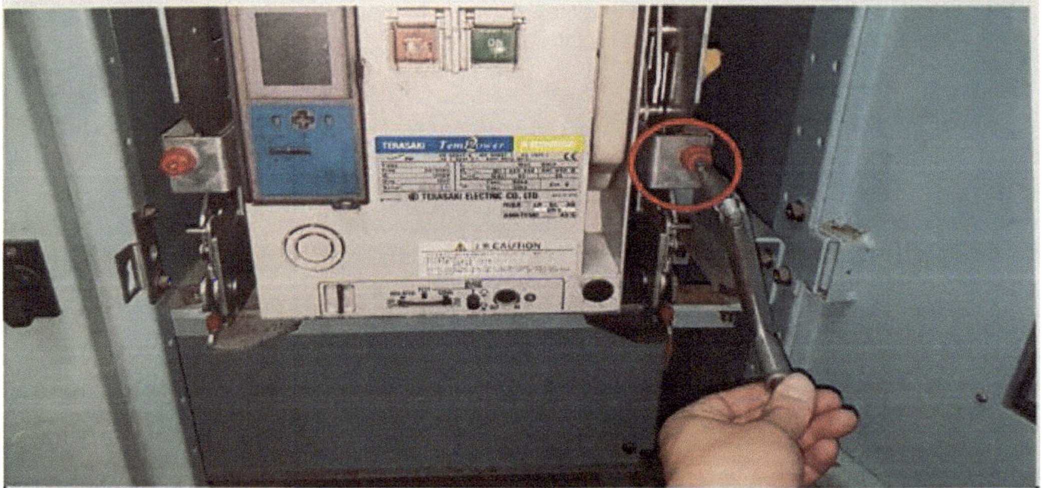

2. ACB FIX BOLT 취외

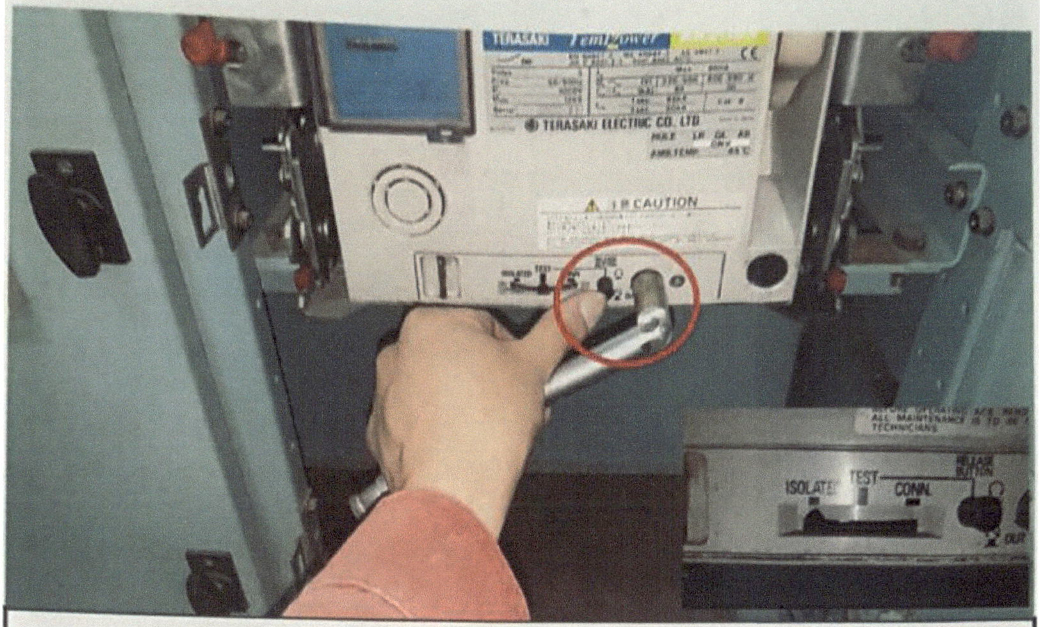

3. 반시계로 회전하여 ACB 발출. TEST MODE에서 걸림현상 확인 후 RELEASE BUTTON 누르고 계속해서 취외 진행. ISOLATED INDICATING 확인

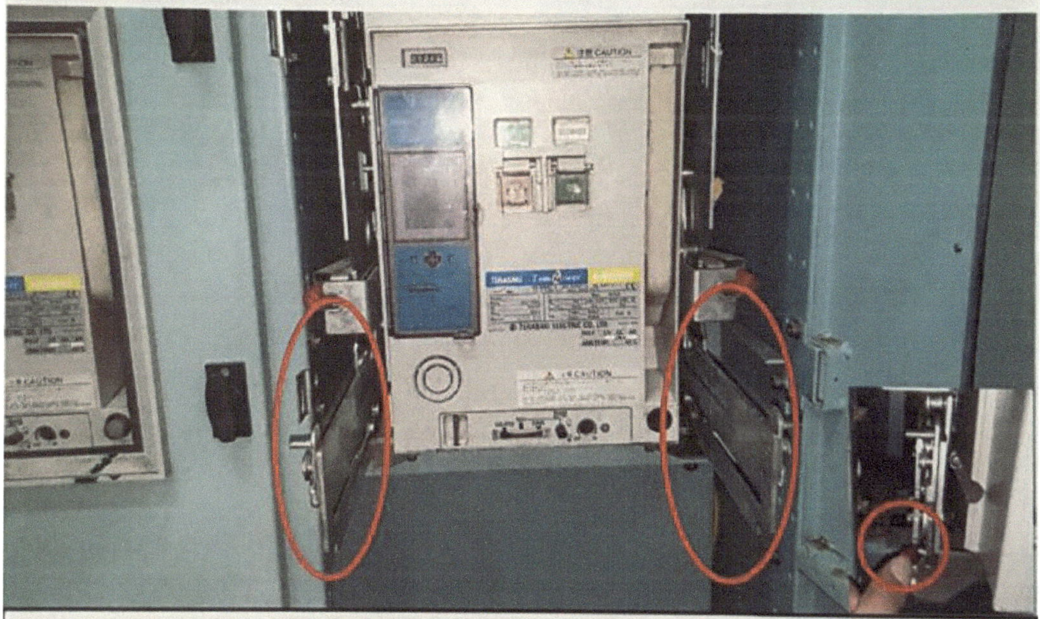

4. ACB DRAW - OUT RAIL 발출. (빨간색 힌지 누르고 RAIL 발출)

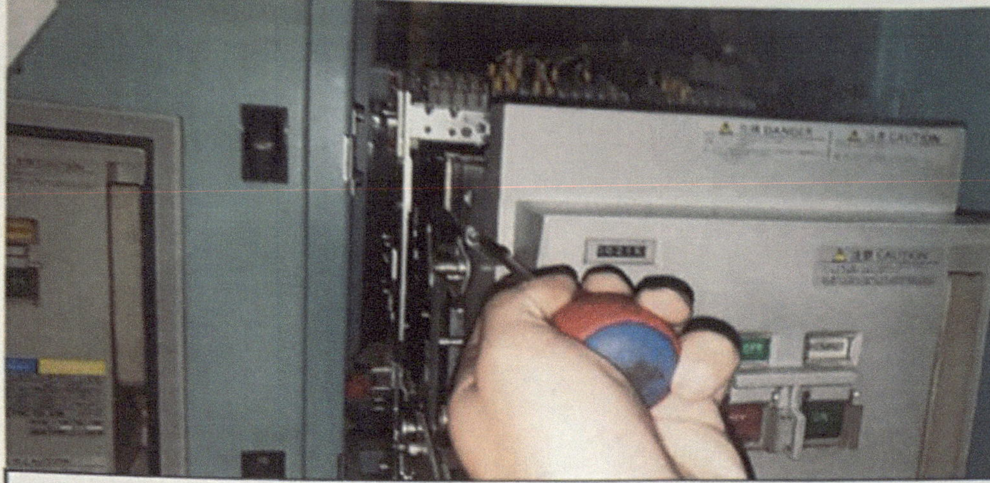

5. ACB를 잡아당겨 발출후 전면부 BOLT 취외 및 FRONT COVER 제거

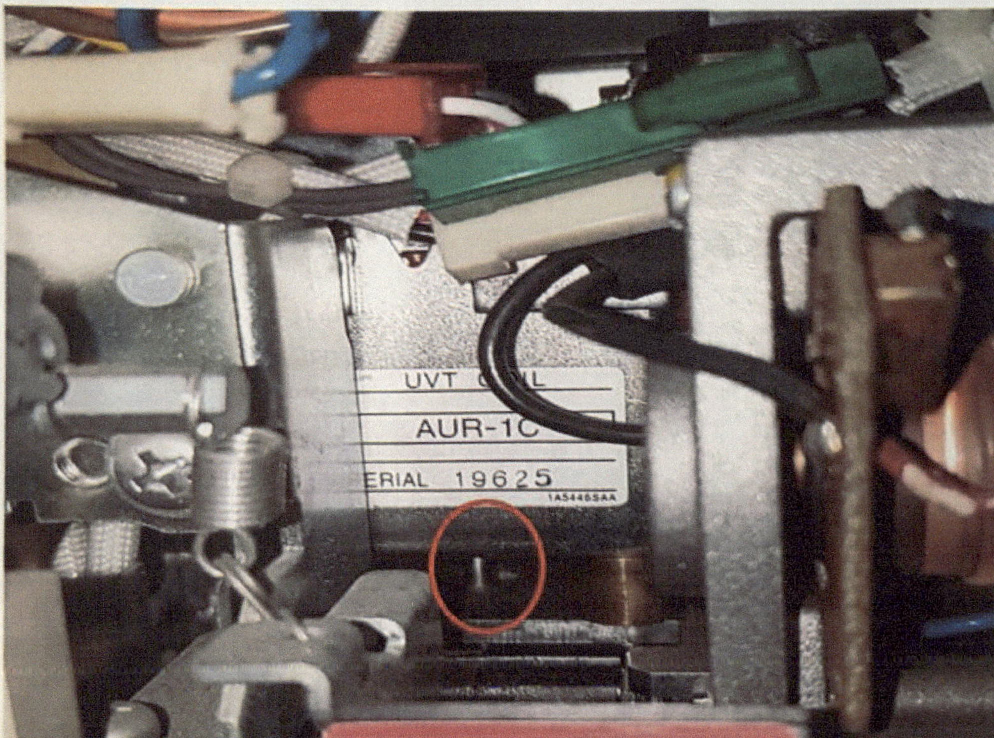

6. UVT COIL PULL-UP 상태 확인. [전원인가시(비상발전기 가동시) - ROD 상승]
7. 비상발전기 미가동상태로 TEST를 하기 위하여 UVT COIL 상부를 롱노우즈등으로 잡아 위쪽으로 잡아당겨 놓은 상태로 [OPEN / CLOSE] TEST 가능.

8. TEST 후 역순으로 복구 (ACB INDICATOR CONN. 확인)

5) 비상발전기 RTS 테스트

Main Switch Board & Emergency Switch Board의 Air Circuit Brake Close되어 선내 전원 및 비상전원이 공급되고 있다.

Emergency Switch Board의 Air Circuit Brake 수동으로 Open하여 Emergency Generator 자동으로 기동 되고, Emergency Generator의 Air Circuit Brake 자동으로 Close되어 비상 전원 공급되기까지 **45초 이내**에 완료되어야 한다.

상기 관련된 결과물을 사진을 찍어서 Emergency Generator Room에 게시한다.

① RTS TEST 전 비상 기기 및 중요 항해 기기 'Power Off' 한다.

② Emergency Switch Board Panel 내부에 Sequence Test Switch가 있다.

'Normal' Position → 'ENG & ACB' Position 변경하면 비상 전원이 꺼지고, Emergency Generator 자동으로 기동한 다음 A.C.B(Air Circuit Brake) Close되어 다시 비상 전원이 공급된다.

Switch Position 변경하여 비상 전원 다시 공급되기까지 45초 이내에 완성되는지 확인한다.

③ 'ENG & ACB' Position → 'Normal' Position으로 변경 시 비상 전원 순간적으로 꺼졌다가 다시 복구된다. 즉, 총 2번 전원이 떨어진다.

Emergency Generator Stop Timer Setting 값에 따라 자동으로 Stop 된다.

④ Sequence Test Switch에서 'ENG' Position으로 하면 Emergency Generator 자동으로 기동되나 A.C.B는 Close되지 않는다.

⑤ RTS TEST관련 결과물(Sequence Test Switch Position, 소요시간, Emergency Generator 부하 등) 현장에 게시한다.

'Normal' Position → 'ENG & ACB' Position

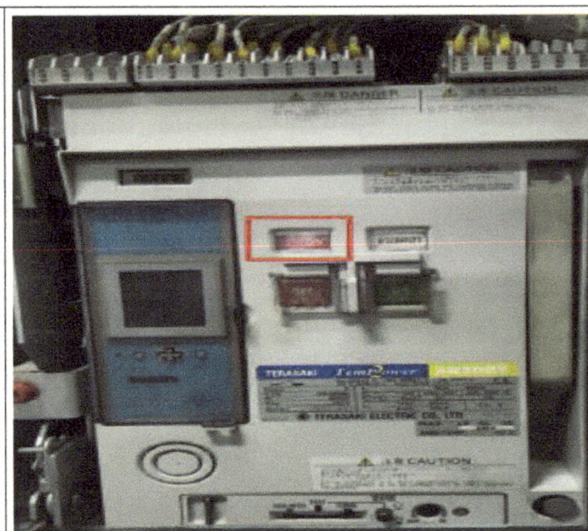

ACB 'Close'

Emergency Generator 부하

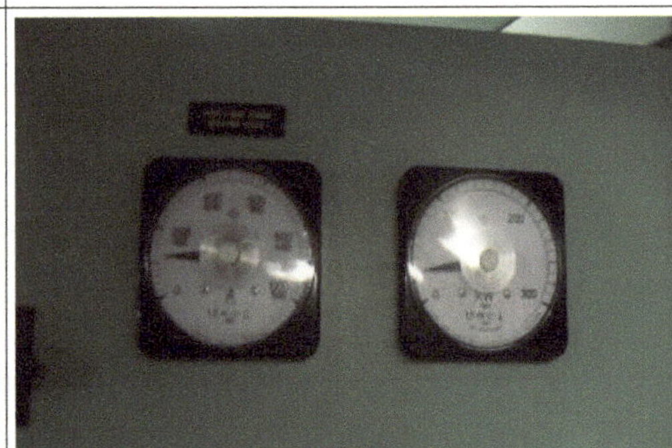

Emergency Generator 부하

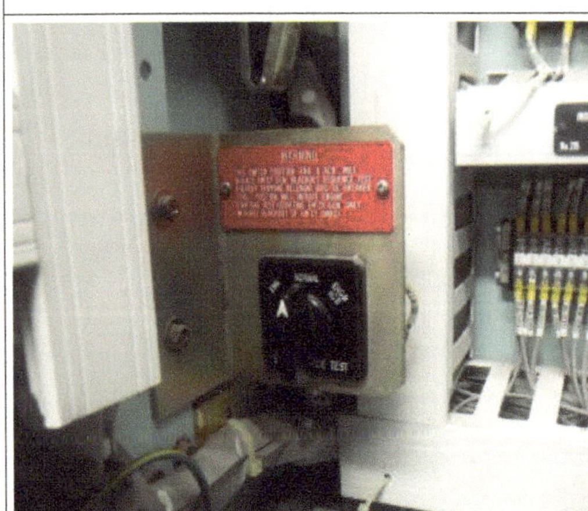

'ENG & ACB' Position → 'Normal' Position

게시물 사진

6) 기관실 오버헤드 크레인

Overhead Crane Controller는 반드시 기관장 또는 1기사가 조작한다.

Fuse 끊어짐, Limit Switch작동 불량, Remote Controller 내부 Cable 단선, Micro Switch작동 불량 및 Motor Brake Pad 교환 및 Air Gap 점검, Wire 교환 등의 상황들이 발생한다.

Controller Button을 촌동으로 조작하지 말고 만약에 작동이 잘 안되거나, Lower & Hoist 등 특정 위치 조작 불량할 경우 Controller 내부, Overhead Crane Controller Panel 내부 Fuse 확인한다.

Limit Switch, Brake Pad, Fuse등은 쉽게 점검 가능하다.

Hoist & Lower 관련 Micro Switch 위치는 아래의 그림을 참고한다.

반드시 유념할 사항은 E/R Overhead Crane 작업은 고소 작업이므로 안전을 위해 안전벨트를 꼭 장착하고 정비해야 한다.

Micro Switch 작동 불량 시 좌측에 표시된 Panel에서 Cable Jumping시 사용 가능하다. 우측 표시된 부분 Panel에 Micro Switch 있다.

우측 Panel Cover Open하면 Micro Switch가 있다.

	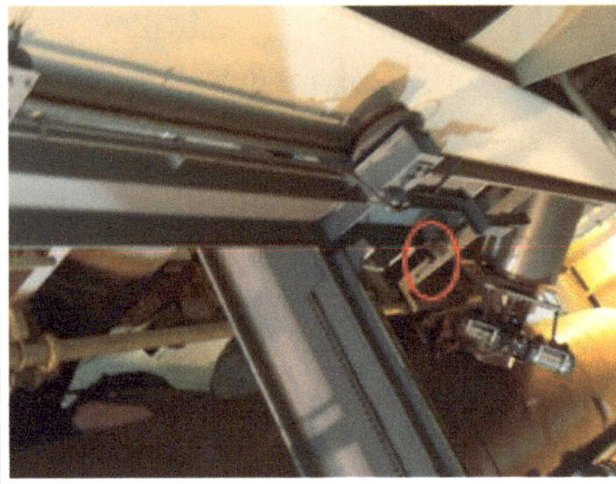
Limit Switch	Limit Switch

Limit Switch는 각 Overhead Crane(FWD, AFT)마다 앞,뒤로 1개씩 있다.

Limit Switch가 없거나 작동이 불량하면 Overhead Crane 조작 시 부딪히거나 파손 위험이 있으니 조작 시 항상 주의한다.

* 모터 에어 갭 점검

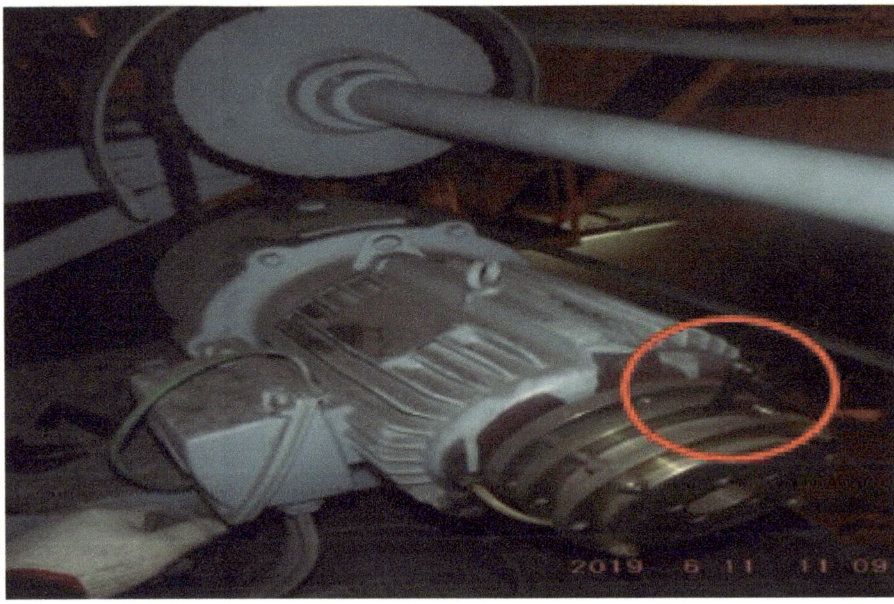

상기 사진은 Travelling(FWD′ & AFT′) Motor Air Gap점검 사진

Hoisting은 Gap Table 있으나 Full Tight후 한 바퀴 Open하여 점검하지만 보통 0.5mm 이내로 생각하고 측정하면 된다.(Instruction Book 참조)

7) 공기 압축기

본선 정비 주기 참조하여 작업을 진행한다.

→ 1ST & 2ND Valve 교환, L.O 교환, Suction Filter 교환, L.O Filter 소제, Oil & Air Check Valve 점검, Lubricator 점검, Cooler 소제, Compressor OV'HL 등

아래 그림과 같이 Pressure S/W(Main Air Tank 측에 위치)를 이용하여 Start / Stop 을 설정한다.

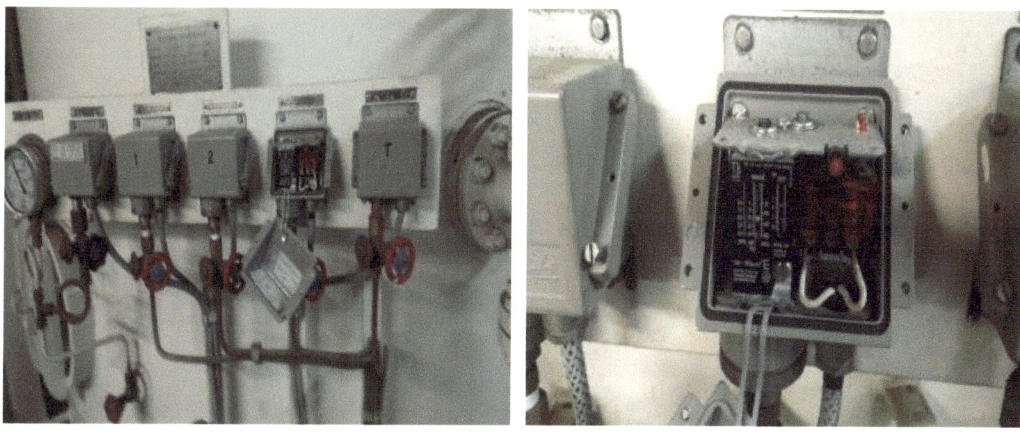

< Start & Stop Pressure Switch >

게다가 아래 그림처럼 Instruction Book을 참조하여 Torque를 확인하고 정비를 한다.

⚠ CAUTION

Order of tightening

Fig. 26

* ***To assemble the cooler cover and cylinder head, first tentatively fix the whole, then tighten diagonally in the order from inner to outer side with the tightening torque shown in Table 7.***

(Unit : N · m)

Type	H(LV)-63.64.264		H(LV)-73.74.273.274.373.374	
Number of stages	1st stage	2nd stage	1st stage	2nd stage
Cylinder head bolt	M10 : 70 M12 : 120	——	M10 : 70 M12 : 120	——
Valve clamping nut	25	——	45	——
Valve set nut	——	12	——	25
Cooler cover mounting bolt	——	M8 : 40	——	M8 : 40
Valve flange mounting bolt	——	M10 : 70	——	M10 : 70
Conn. rod bolt	110		180	

Table. 7

(1) 언로딩 밸브

→ Unloading 운전을 할 경우 아래에 표시된 두 Valve의 Pipe 풀어내고 Valve Open하여 운전을 하면 Air Compressor가 Ampere가 걸리지 않고 Unloading을 하게 된다.

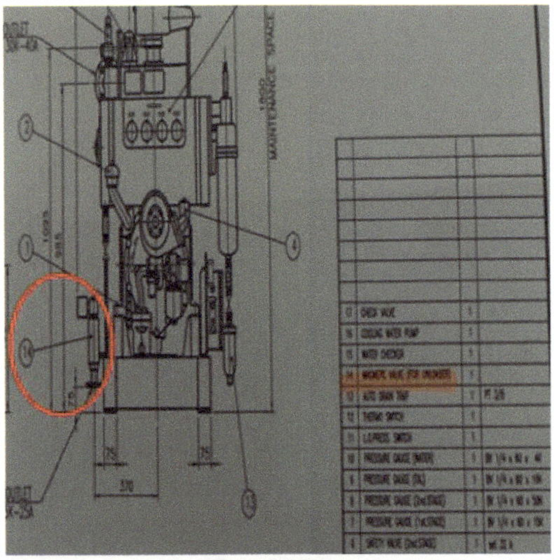

Unloading Valve

표시된 Valve의 Pipe를 분해하고
Valve Open 한다.

표시된 Valve의 Pipe를 분해하고
Valve Open 한다. (Air Drain Valve)

Compressor Crank Chamber를 확인하고 싶으면

① 먼저 Air Inlet & Outlet Valve를 잠그고, Air Drain Valve Open하여 Air Drain 한다.

② Fresh Water Inlet & Outlet Valve Close 한다.

③ Crank Chamber Open 한다.

(Crank Chamber Gasket 손상 부 발견하면 교환한다.)

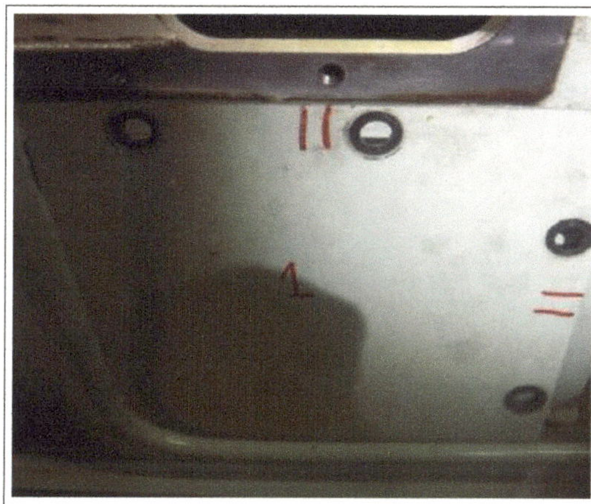

Crank Chamber Cover

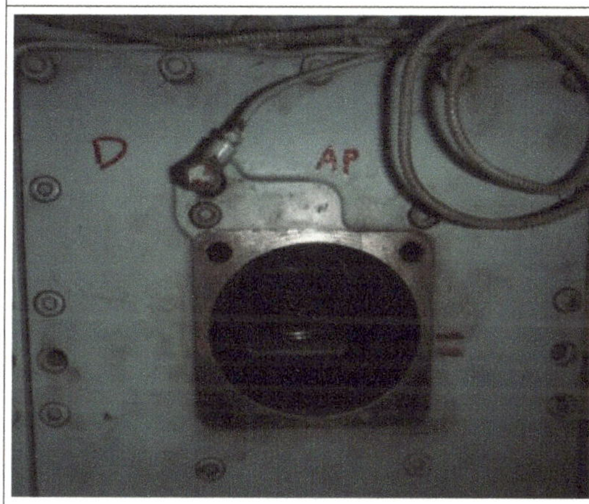

2nd Stage Valve

2nd Stage Valve 내부

Crank Chamber Cover Open 한 상태에서 2nd Stage Valve를 취외하여 Hand Turning Tool을 이용하여 Hand Turning을 하면서 내부 Piston Skirt, Piston Ring을 확인 할 수 있다.

2nd Stage Valve 조립 시에 Copper Gasket 조립 잊지 말고 불량 시 교환한다.

FWD

A
A

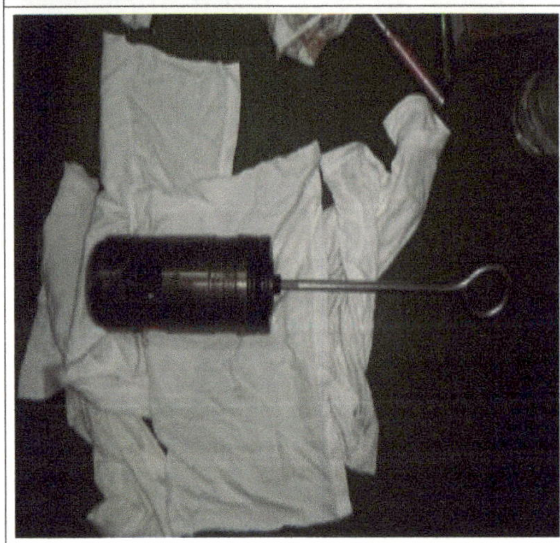

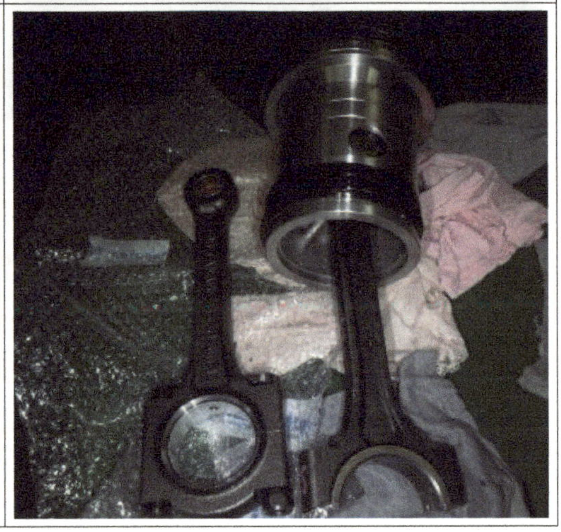

Air Compressor의 Air가 누설하여 LT Chamber로 누설하는 경우, Sight Glass가 있는 Type이면 Air 누설로 인하여 기포가 보일 것이고, LT 온도가 올라간다.

Air Compressor가 Sight Glass Type이 아니라, Plate Type 경우는 Air 누설하여 LT Chamber로 들어가게 되어 Plate가 터진다.
(일종의 Safety역할)

보통 Air가 누설하는 경우는, Head Gasket가 찢어지는 경우에 주로 발생한다.
(Air Hole → L.T Hole로 Air 누설)

따라서 Head Gasket을 먼저 점검하여 상태를 확인하고, Head를 취외하는 경우에는 반드시 Head Gasket을 교환하도록 한다.

Cooler 점검 시에도 Instruction Book을 참조하여 Gasket 조립에 유의하도록 하고,

체결 Torque도 앞서 언급한 대로 Instruction Book을 참조하고,

1st Stage Valve 조립 시 Head Gasket 교환,

2nd Stage Valve 조립 시에도 Copper Gasket 조립,

주기적으로 Suction Filter 교환,

Crank Chamber Open하여 Metal Piece 또는 L.O 오손 여부 확인한다.

(2) 공기 압축기 오일 실 교환

Compressor 운전 시 Flywheel 안쪽에서 누설하면 Oil-Seal에서 누설하는 것이다.

Oil-Seal 교환은 Motor 분해하고 Flywheel 분해하면 Oil-Seal Cover를 분해할 수 있다.

Oil-Seal Cover 내에 장착된 Oil-Seal을 교환하고 Gear pulley 체결하여 Flywheel을 분해한다. **(분해가 잘 안될 경우 열을 가한 다음 Gear pully에 Hammering한다.)**

상기 방법으로 시도했음에도 분해가 잘 되지 않는다면, 아래의 그림처럼 Flywheel 안쪽에 볼트 & 너트 4SET를 준비하여 위, 아래, 좌, 우 방향으로 설치하고 볼트와 너트를 펴주면 분해된다.

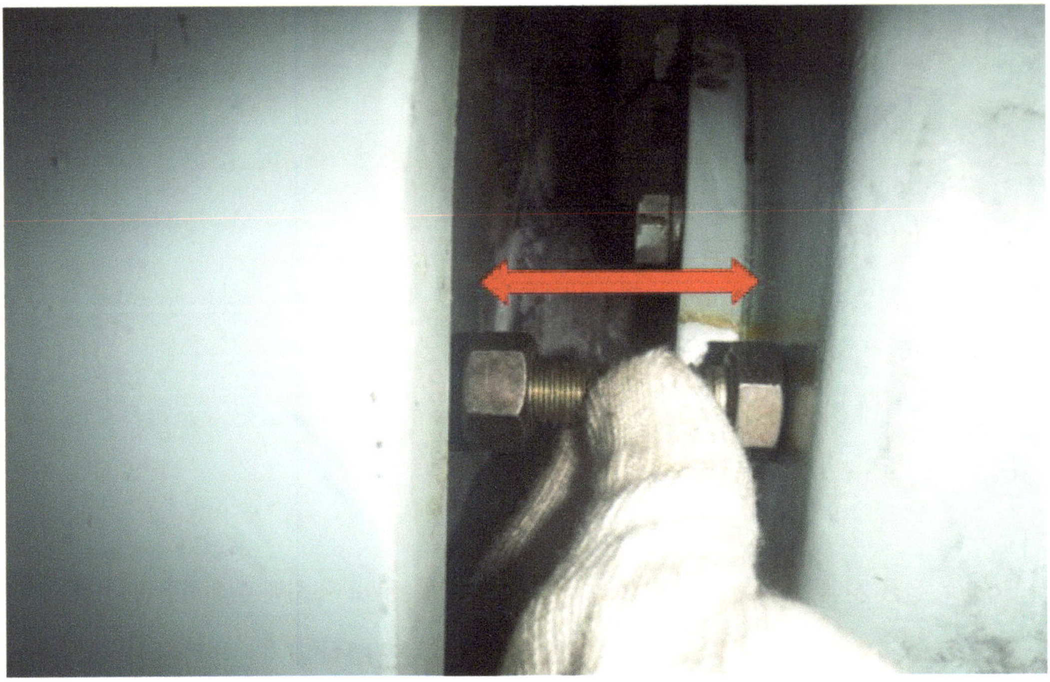

다음은 실린더와 쿨러 일체형 타입인 HATLAPA Maker 압축기에 대한 설명이다.

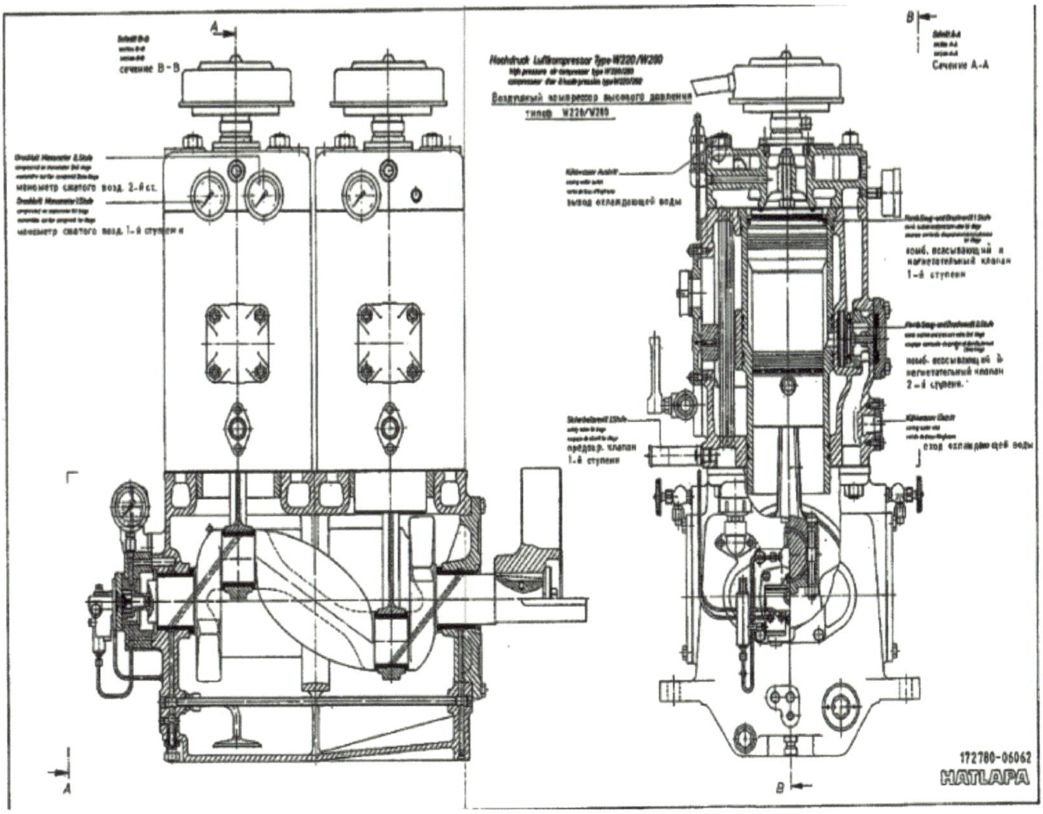

HATLAPA	Anzugsdrehmomente Tightening Torques Couples de serrage	Typ W110 - W420
Zylinderkopf cylinderhead culasse de cylindre		370 Nm
Zylinder cylinder cylindre		370 Nm
Pleuel connecting rod bielle		83 Nm 1)
Ventilzentralschraube valve fitting bolt goujon de soupape	1. Stufe / 1st Stage / 1ère étage 2. Stufe / 2nd stage / 2ème étage	100 Nm 22 Nm
Lagerdeckel Kurbelgeh. crankshaft bearing cover couvercle palier/carter		78 Nm
Ventildeckel valve cover couvercle de soupape	2. Stufe 2nd stage 2ème étage	78 Nm
1)	Kurbelzapfen und Gewinde einölen Oil crank pin and bolts threads Huilage tourillon de vilebr. et filetage	

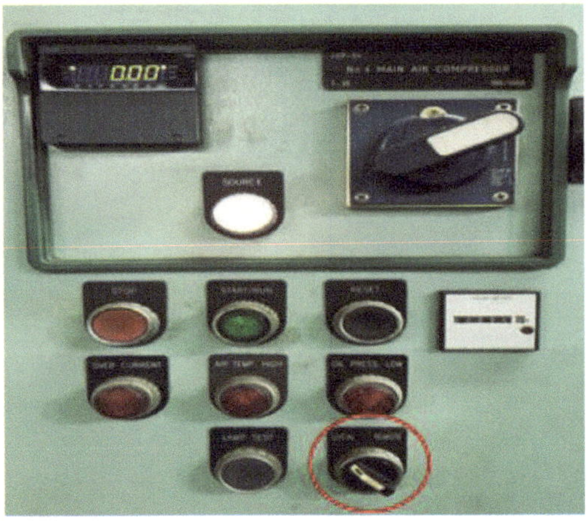

Control Panel 'Remote' → 'Local'

C.F.W Inlet Valve (왼) / Drain Valve (오른)

Drain Valve (위) / C.F.W Outlet Valve (아래)

C.F.W Inlet Line(위) / Drain Valve(아래)

C.F.W Outlet Line

Air Supply Valve (왼) / Check Valve (오른)

Air drain

Air Gauge

Compressor air temp control switch & Gauge

L.O pump & L.O Gauge

Drip indicator

Oil Pressure Switch

Oil and water separator

L.O Filling

Safety Valve

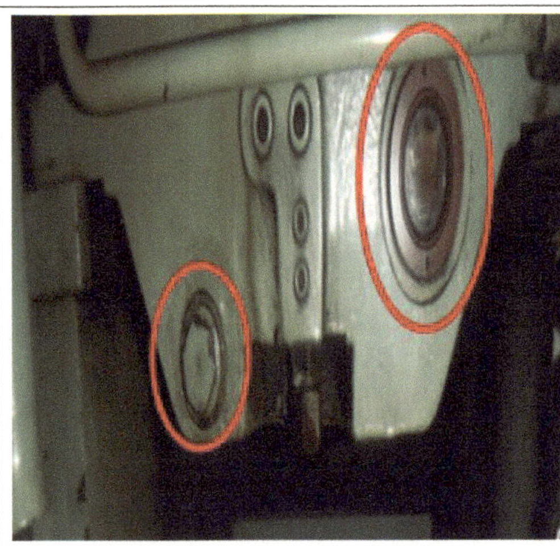

Sight Glass & L.O DRAIN COCK

(3) 드립 인디케이터

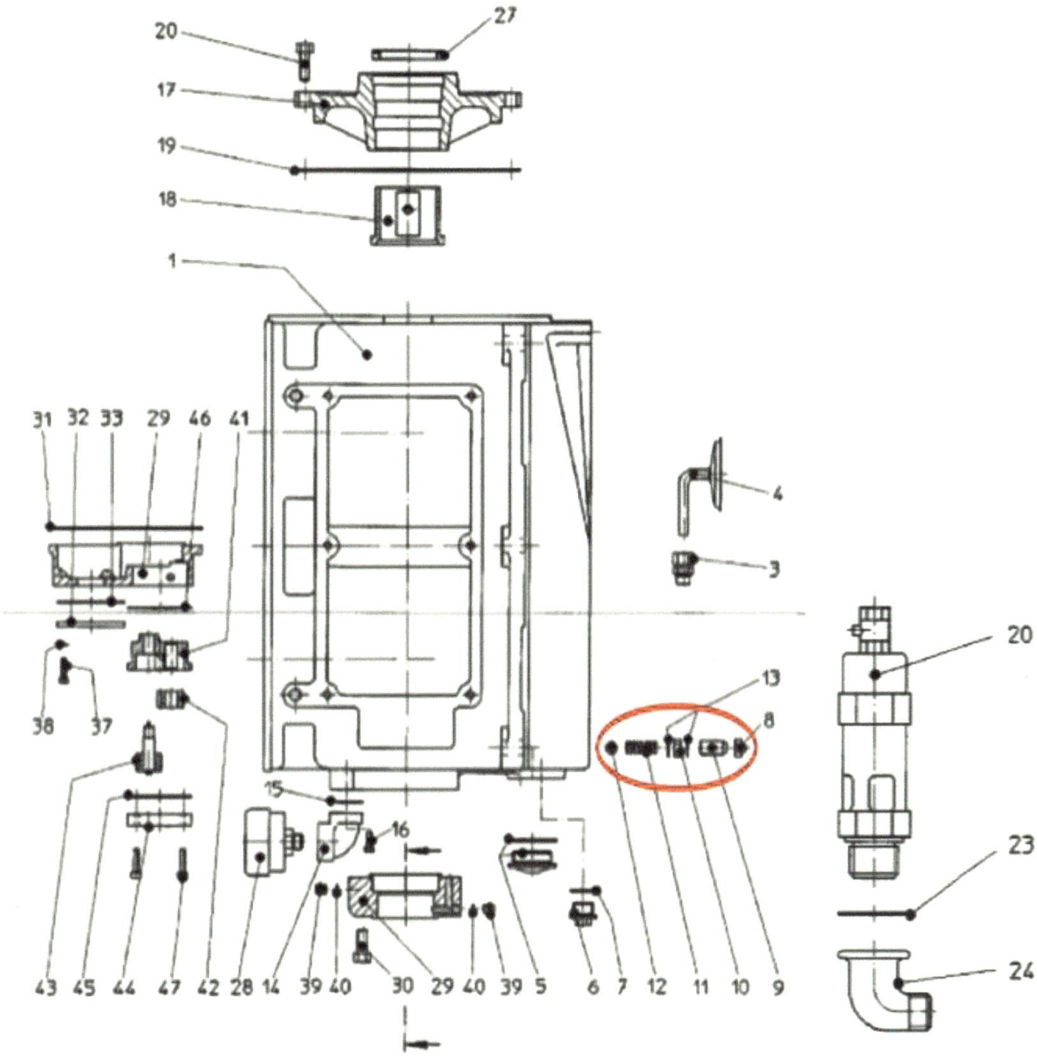

Air Compressor 운전 중에 Drip Indicator Sight Glass에서 L.O Drip 되는지 반드시 확인한다.

L.O Drip 되지 않는다면 1ST Stage 측 Piston 상부로 L.O가 공급되지 않아 Piston과 Liner 윤활이 제대로 되지 않는다.

L.O Drip 되지 않으면 Control Panel에서 'Remote' → 'Local' Position으로 변경하고 점검한다.

Drip Indicator 내부 부품이 매우 작으므로 신중하게 분해하여 내부 관통 소제하며, 내부 부품이 매우 작으므로 신중하게 분해, 정비한다.

Oil Pump 압력이 낮을 경우 Drip Indicator 의 내부 아래쪽 Ball을 빼고 조립하면 된다.

당연히 Oil Pump 압력이 낮은 것에 대해서는 L.O Filter 소제 & Oil Pump OV'HL 한다.

* 드립 인디케이터 분해 및 관통 소제

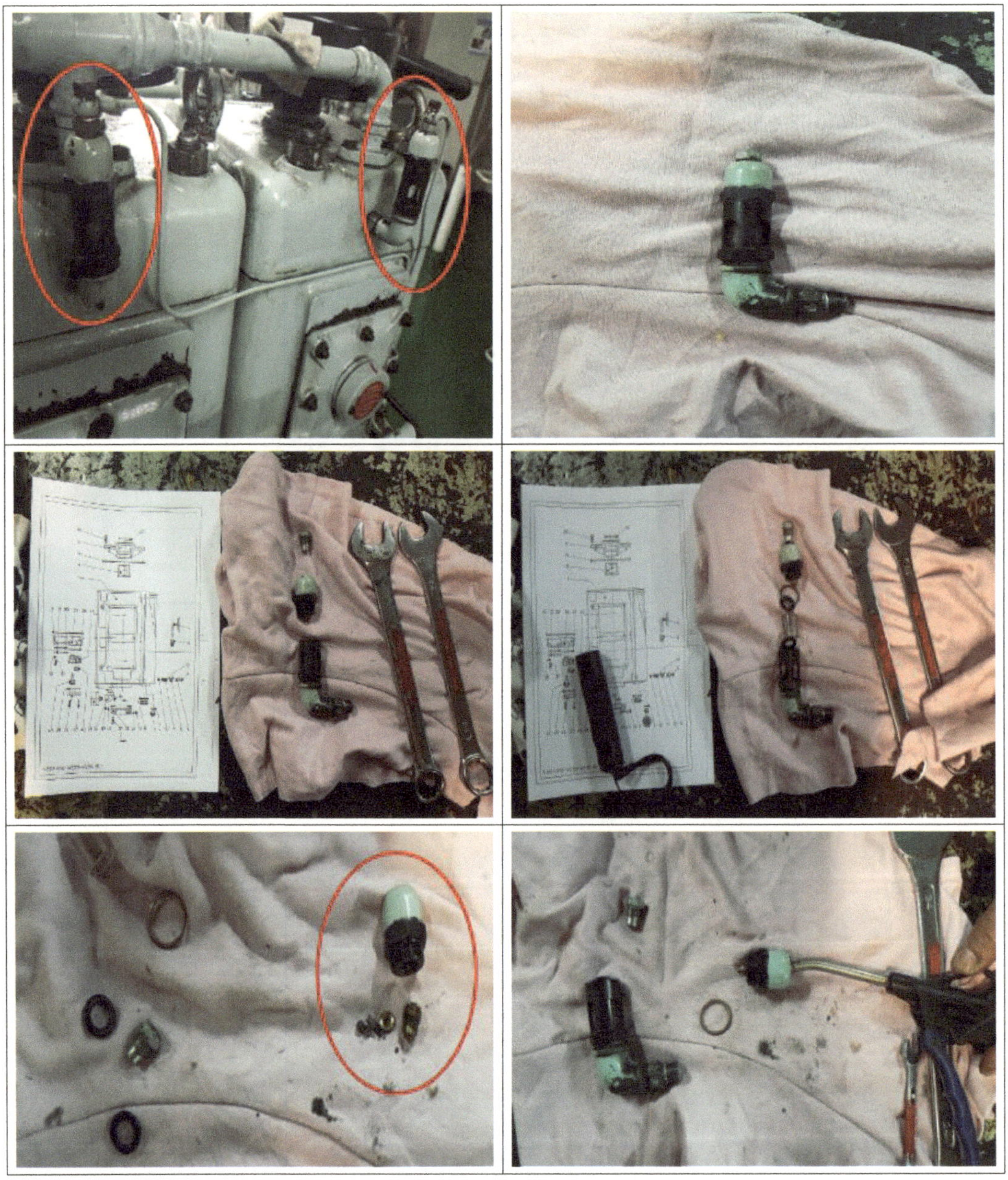

(4) 1ST & 2ND STAGE 밸브

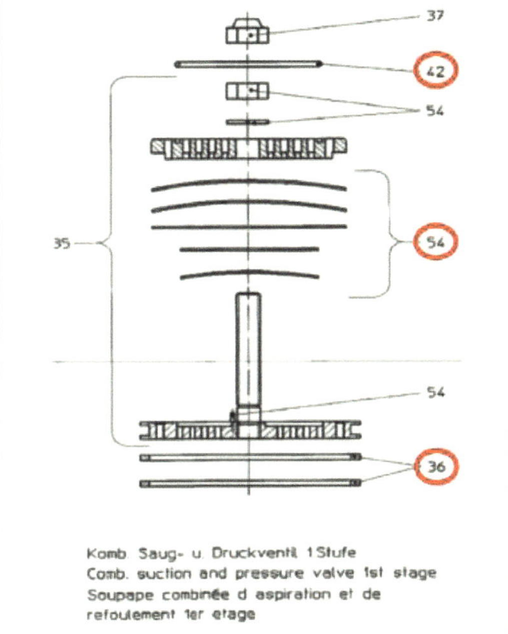

1ST Stage Valve

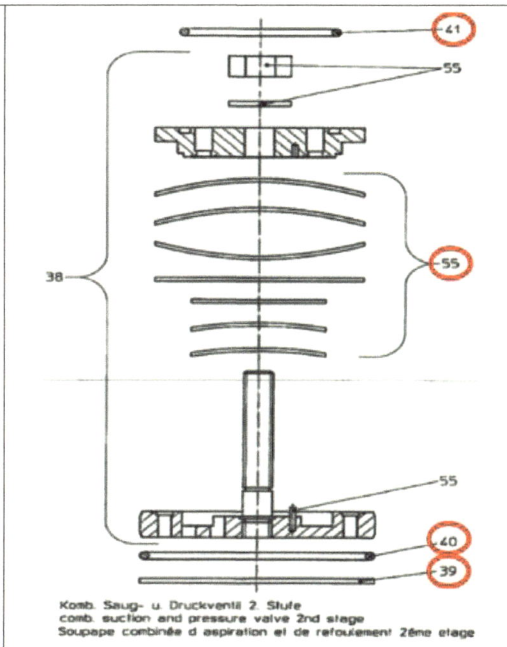

2ND Stage Valve

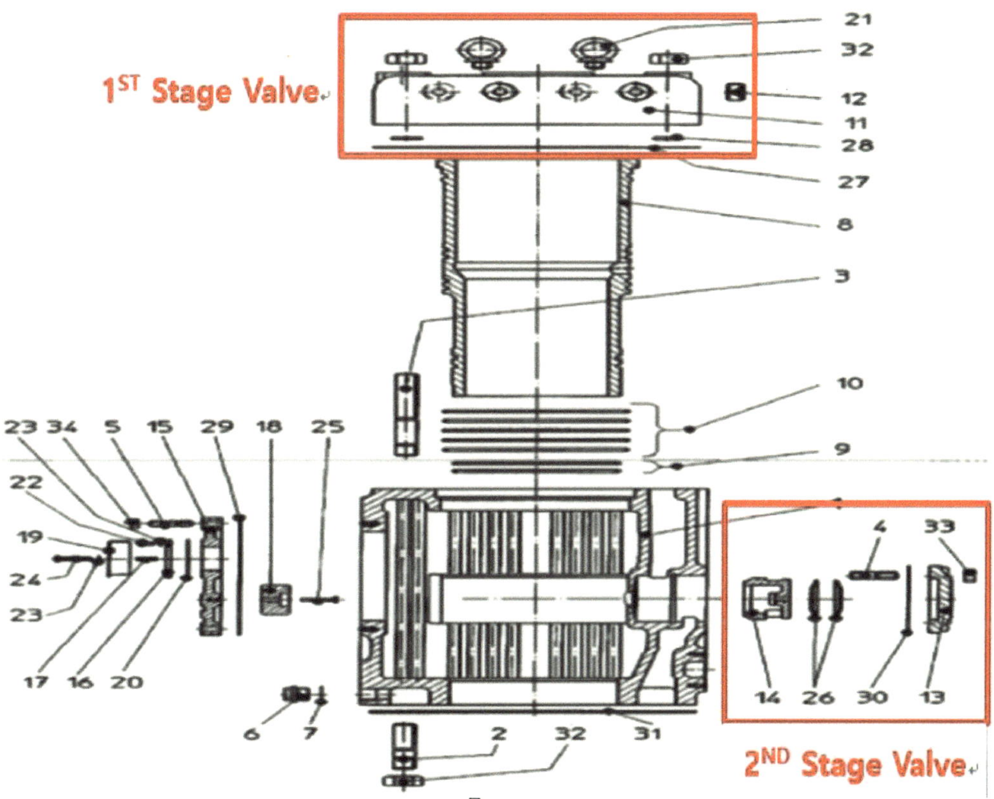

① **교환 방법**

Compressor 운전 중에 1ST & 2ND Stage Valve의 Pressure Gauge 지시치가 다른 Compressor 보다 유독 낮게 지시된다면 Pressure Gauge 교환해보고, 상태 동일하면 Valve 교환한다.

㉠ Control Panel 'Remote' → 'Local' 변경하거나 Power Off 한다.

㉡ Air Inlet & Outlet Valve Close 후 Air Drain 한다.
Water Inlet & Outlet Valve Close 후 Water Drain 한다.

㉢ Air Filter Silencer를 포함하여 Cylinder Head에 연결된 액세서리 및 Pipe들 분리한 다음 Cylinder Head Nut 분해한다.
Cylinder Head에 Eye Bolt, Shackle, Rope를 연결하고 Chain Block으로 분해한다.

㉣ 취외한 Cylinder Head에서 1ST Stage Valve 분해하고, Cylinder 상부와 Cylinder Head Gasket 조립 부분 Brush 소제한다.

㉤ 1ST Stage Valve 정비품 또는 신품에 Packing Ring(**No.36**), O-ring(**No.42**) 조립한다.
1ST Stage Valve 조립 시 Torque(**100 Nm**) 참조하여 Cylinder Head에 조립한다.
(1ST Stage Valve 하부의 Packing Ring 조립 시 방향 확인)

㉥ Cylinder Head 조립 전 Piston 상부 Air Compressor L.O를 도포하여 초기 기동 시 Liner와 Piston에 Metal Touch 되지 않도록 한다.
Cylinder Head Gasket(**No.27**) 방향에 맞게 놓고 O-ring(**No.28**) 조립한다.
Cylinder Head 조립 시 1ST Stage Valve Packing Ring 제대로 조립될 수 있도록 '-' 드라이버를 이용하여 3~4방향에서 눌러주면서 Chain Block을 내려 Cylinder Head 조립한다.
(Cylinder Head Gasket 반드시 제대로 된 방향으로 조립한다.)

㉦ Cylinder Head Nut를 Torque(**370Nm**) 확인하여 조립하고, 액세서리 및 Pipe들 연결한다.
Air Filter Silencer를 조립하고 필요 시 내부 Air Filter 교환한다.

㉧ 2ND Stage Valve Cover Nut(**No.33**) 분해한 다음Cover(**No.13**), Spring Plate (**No.26**), 2ND Stage Valve 취외한다.
(Valve 취외하면 Shaft Hand turning하여 Piston Ring, Piston 상태 확인할 수 있다.)
2ND Stage Valve 취외한 Housing 내부 Packing Ring(**No.39**), O-ring(**No.40**)을

교환한다.

취외한 2ND Stage Valve에서 Thrust Member(**No.14**) 분리하여 정비품 또는 신품에 조립한다.

2ND Stage Valve를 집어넣고 Spring Plate 방향을 확인하여 Cover와 같이 조립한다.

Cover Bolt Torque(**78Nm**)확인하여 Tight 한다.

ⓩ Water Inlet & Drain Valve Open하여 Line Air Venting 후 Water Outlet Valve Open 한다.

Air Inlet & Outlet Valve Open 한다.

ⓒ Compressor 기동하여 1ST & 2ND Stage Pressure Gauge 확인한다.

Air Filter Silencer 분해

액세서리 및 Pipe들 분해

Cylinder Head Nut 분해

Cylinder Head 취외

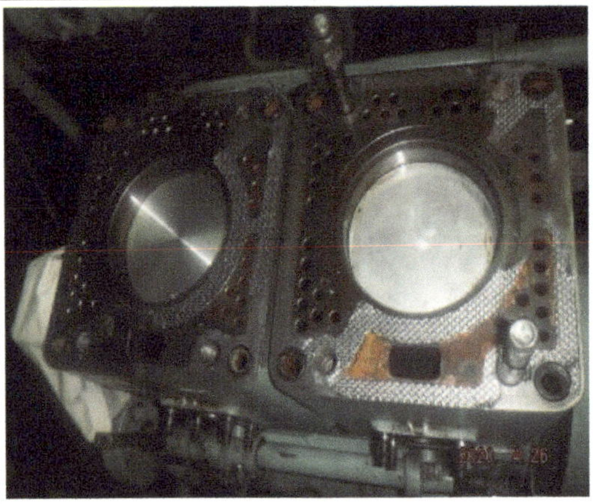

Cylinder Head Gasket (Brush 소제)

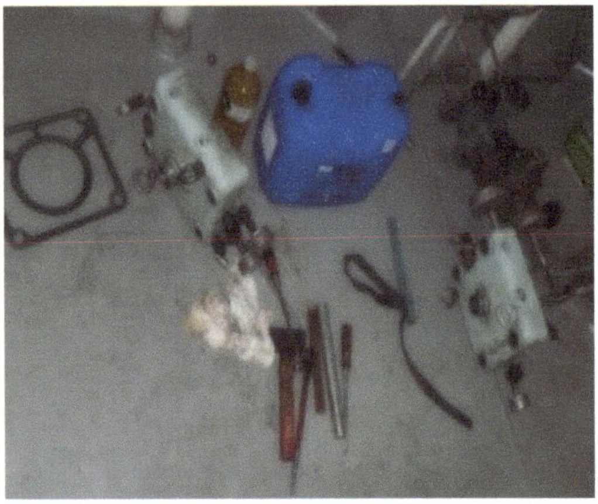

Cylinder Head & 1ST Stage Valve

Cylinder Head & 1ST Stage Valve
Tight Torque(100Nm)

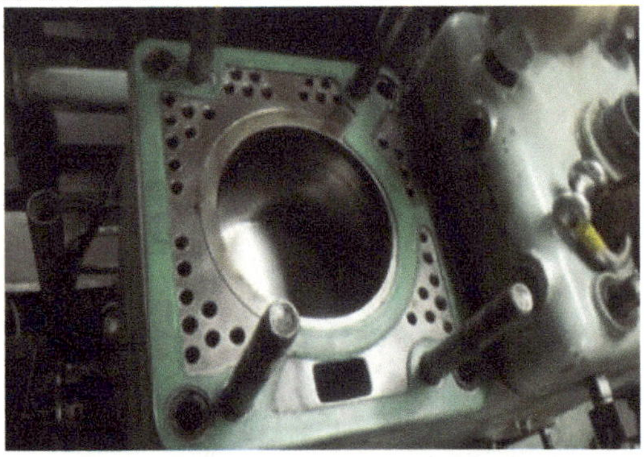

Cylinder Head Gasket 방향에 맞게 넣고
(예비품 Gasket보관 상태 불량하여 임시 제작)
Piston 상부에 Compressor L.O 뿌려준다.

Cylinder Head Nut Torque(370Nm)

주변 액세서리, Pipe, Air Filter Silencer 조립

2ND Stage Valve

2ND Stage Valve, Spring Plate & Cover

Hand Turning하여 Piston Ring & Piston 확인

Packing & O-ring 교환

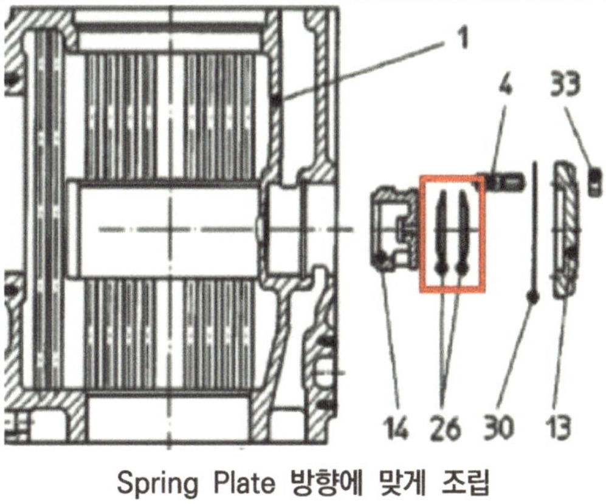

Spring Plate 방향에 맞게 조립

2ND Stage Valve Cover Nut Torque 조립

② **정비 방법**

㉠ 취외품 1ST & 2ND Stage Valve 도면을 참조하여 하나씩 분해한다.

㉡ 아래 표시한 부분 Valve Lapping 한다.

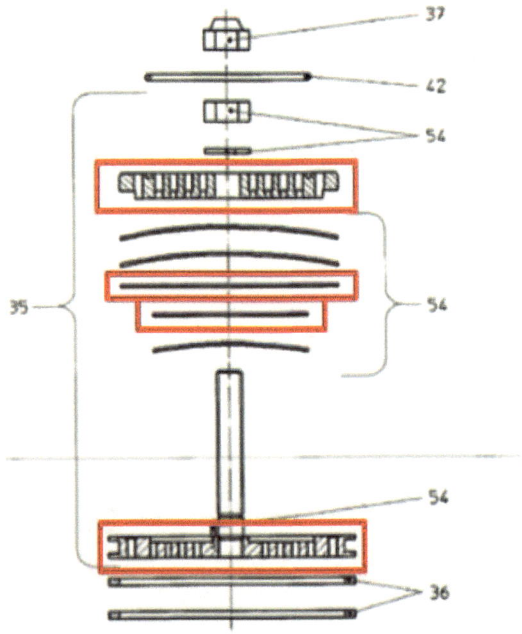

Komb. Saug- u. Druckventil 1.Stufe
Comb. suction and pressure valve 1st stage
Soupape combinée d aspiration et de refoulement 1er etage

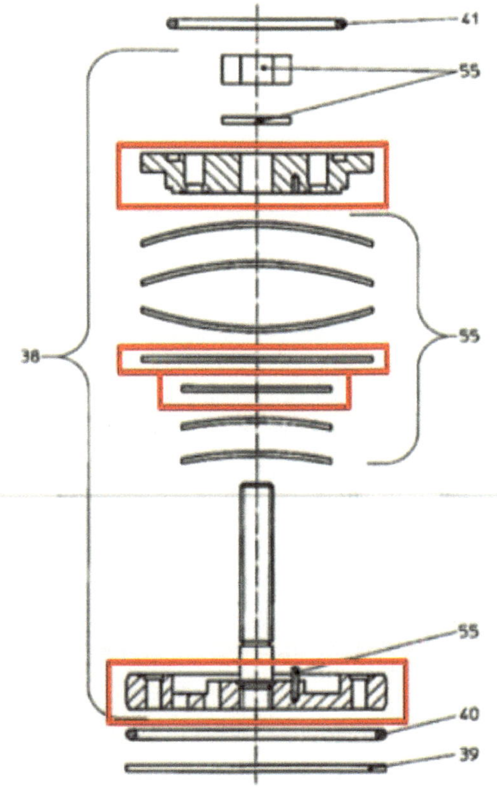

Komb. Saug- u. Druckventil 2. Stufe
comb. suction and pressure valve 2nd stage
Soupape combinée d aspiration et de refoulement 2ème etage

㉢ 상기 그림 Valve Plates & Springs(**No.54 & 55**) 예비품이 있다면 사용한다. 예비품이 없다면 상기 표시한 부분 Plate & Spring을 Lapping하여 재사용한다.

㉣ 1ST & 2ND Stage Valve Nut(**No.37 & 55**) Torque 확인 후 Tight 한다.

㉤ Valve를 뒤집어서 Body 뒷부분에 물을 집어넣었을 때 물이 빠지지 않고 유지되면 조립이 잘된 것이고, 만약에 물이 빠진다면 Spring, Plate & Valve를 다시 Lapping 한다.

물이 빠지지 않아야 제대로 Seal이 되는 것이다.

몇 번의 Lapping을 하여도 똑같다면 예비품 사용하고 해당 Plate & Spring 폐기한다.

㉥ 1ST & 2ND Stage Valve O-ring (**No.42, 41**) 조립 후 Wrap에 씌워 보관한다.

(5) 에어 필터 & 오일 SIEVE 점검

점검 전에 반드시 Control Panel에서 'Remote' → 'Local' Position으로 변경한다.

Air Filter(**No.6**) & Oil Sieve(**No.4**) 주기적으로 정비 시간에 맞게 확인한다.

Air Filter Silencer의 나비 볼트를 풀어 Silencer 상부를 들면 Filter가 있다.

Air Compressor Chamber 점검 시 Oil Sieve를 풀어서 내부 Air로 불어 관통 소제한다.

Oil Pump 압력이 낮다면 Oil Sieve 확인한다.

Crankcase L.O Drain Cock Open하여 L.O Drain 후 Crankcase Cover Open 한다.

Oil Sieve 분해 시 Burst Plate Cover 아래의 Crankcase Cover Open해야 분해할 수 있다.

Crankcase 점검 시 Hand Turning하면서 Crankcase L.O상태(수분, Metal Piece 등) Crankshaft, Con-rod, Con-rod Bearing, Piston, Liner 등을 전반적으로 점검한다.

L.O에 거품이나 흰색의 기포들이 보이면 Water 누설하는 것이므로, L.O Drain 후 Hand Turning하여 Liner 점검하고, 금속 조각이 식별된다면 Piston Ring, Liner, Con-rod Bearing 등 확인한다.

L.O 보충하는 곳에도 Filter가 있으며 주기적으로 확인하여 필요 시 교환한다.

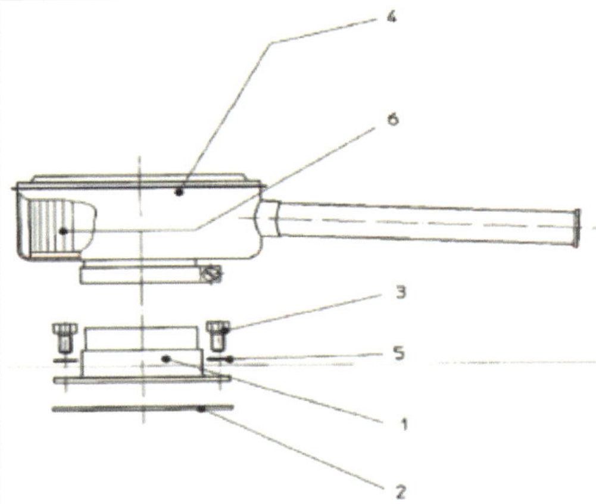

Air Filter(No.6)

Air Filter

L.O Drain Cock & Sight Glass

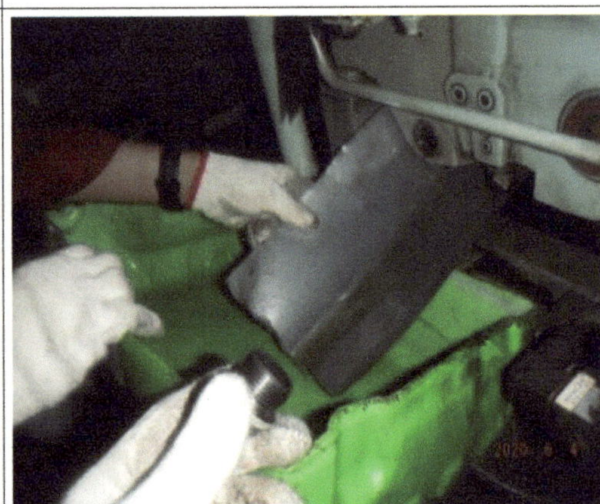

L.O Drain Cock & L.O Drain

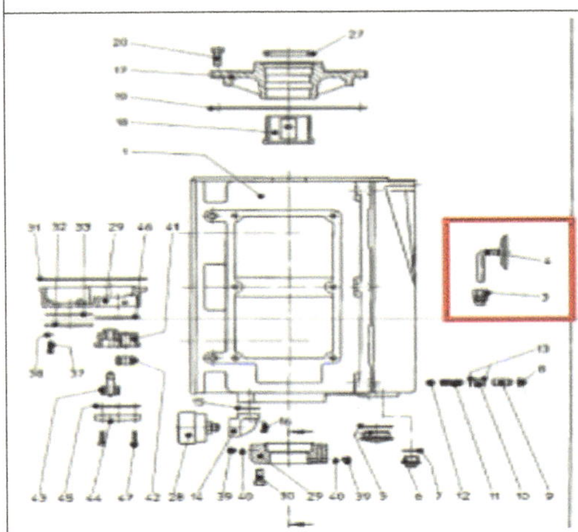

Oil Sieve(No.4)

Oil Sieve

(6) BURST PLATE

Cylinder Head Gasket 손상되어 Air가 LT라인으로 공급되거나 Cooler의 Air Line 파공될 경우, Burst Plate(**No.20**)가 터지며 일종의 Safety라고 볼 수 있다.

Burst Plate 장착된 부분이 Cooler와 연결되므로 분해 시 Water Inlet & Outlet Valve Close하고 Water Drain 한다.

분해 시 Cooler 상태를 육안으로 확인할 수 있으며 Cooler 안으로 Air Line지나간다.

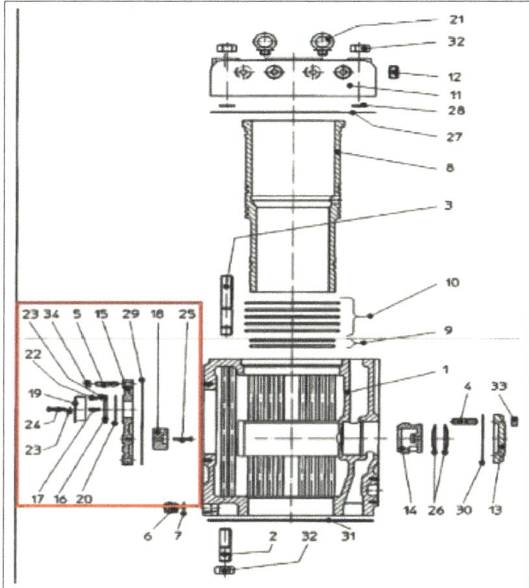

Burst Plate(No.20)

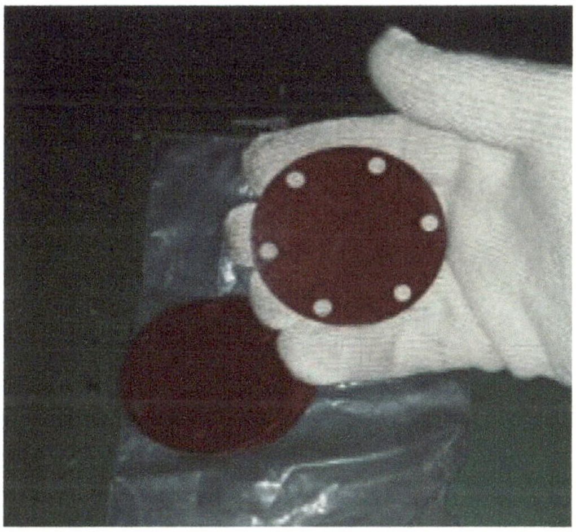

〈 Burst Plate 〉

(7) 피스톤 분해 및 정비

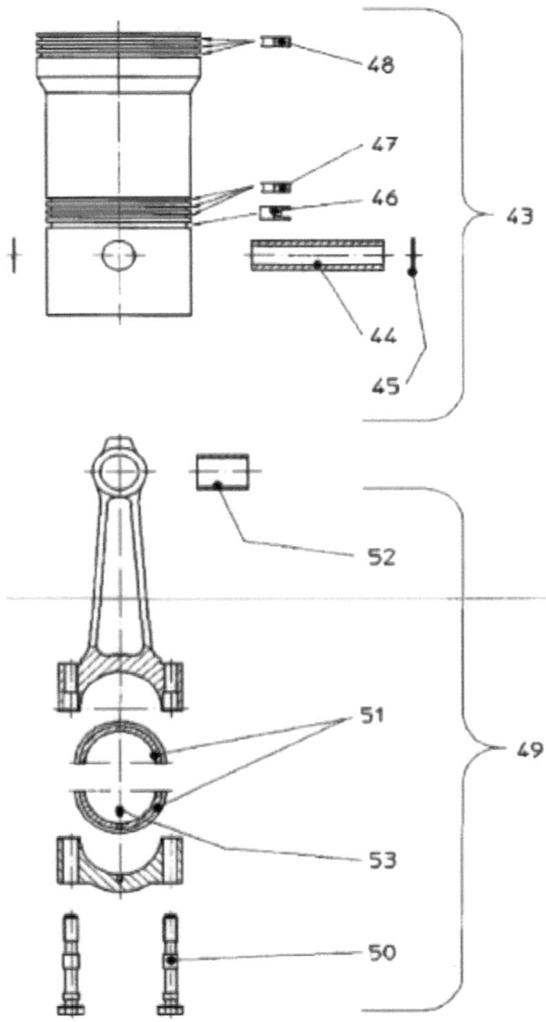

Piston OV'HL의 경우 앞서 설명한 부분의 연장선이다.

점검 전에 반드시 Control Panel에서 'Remote' → 'Local' Position으로 변경한다.

Crankcase L.O Drain 후 Crankcase Cover Open 한다.

㉠ Crankcase Cover Open 및 Shaft Hand Turning하여 Con-rod Bolt(**No.50**), Bearing(**No.51**)

Con-rod 아래측 Cover 분해한다.
Cylinder 하부의 Base Bolt 분해한다.

㉡ Cylinder 상부에 Eye Bolt, Shackle, Rope를 연결하고 Chain Block으로 Cylinder

분해한다.

분해한 Cylinder를 눕히거나 세로 방향으로 고정한 다음 Rubber Hammer, 나무 또는 Zinc로 Cylinder 하부 Con-rod(**No.49**)를 Piston(**No.43**) 방향으로 쳐서 Piston과 함께 분해한다.

ⓒ 분해한 Cylinder 내부를 깨끗한 천으로 닦아서 Liner 상태 및 분해한 Piston 상태를 확인하고, Cylinder 상부 Head Gasket조립 부분을 Brush 소제하고 Compress Air Line 관통 소제한다.

Cooler Cover(**No.15**) 분해하여 Cooler 상태 확인하고 필요 시 Chemical Cleaning 한다.

Cooler Cover Gasket(**No.29**) 불량 시 교환한다.

Piston, Con-rod, Con-rod Bearing & Cylinder Scratch, Scuffing 확인 후 필요 시 교환한다.

ⓔ 1ST & 2ND Piston Ring(**No.48,47**), Oil Scraper Ring(**No.46**) 교환하되, Top Mark 있는 Piston Ring이라면 조립 시 Top Mark가 상부를 향하도록 조립한다.

Piston, Piston Ring & Liner에 조립이 원활하도록 Crankcase chamber에 주입되는 L.O를 도포한다.

(Ring 조립할 때 천 또는 노끈으로 Piston Ring 양단을 고정하고 당기면서 조립하면 된다.)

ⓜ Piston을 Cylinder에 조립할 때 2ND Stage Valve가 조립되는 구멍으로 Piston Ring 제대로 조립되면서 들어가는지 확인한다.

Cylinder에 Piston 올리고 Con-rod를 돌리면서 2ND Stage Valve 구멍으로 Piston Ring 조립상태를 확인하고, '-' 드라이브로 눌러주면서 Piston Ring이 제대로 조립될 수 있도록 한다.

Piston Ring 제대로 조립되었다면 Rubber Hammer로 Piston 상부를 살살 치면서 조립한다.

Piston Ring 조립 상태 확인하지 않고 망치로 계속 치면서 조립하면 Piston과 Ring 깨진다.

Piston의 경우 알루미늄 재질이므로 타격에 약하며 Piston Ring도 마찬가지다.

ⓗ Piston조립 된 Cylinder 조립 전에 Cylinder 하부의 Gasket(**No.31**)을 방향에 맞게 놓는다.

(Cylinder 하부 Gasket 반드시 제대로 된 방향으로 조립한다.
방향이 다르면 조립 후 Compressor 기동 시 Safety Valve 작동한다.)

Cylinder 상부에 Eye Bolt, Shackle, Rope 연결하고, Chain Block으로 조립 시 Hand Turning을 하여 Crankcase의 Crankshaft와 Piston Con-rod 사이 Con-rod Bearing 및 Bearing Cover를 조립하여 Bolt를 Torque(**78Nm**)에 맞게 체결한다.

*** Con-rod Bearing에 Compressor L.O를 도포하여 조립한다.**

ⓢ Cylinder Base Nut 조립하고 Hand Turning하여 Piston 및 Crankshaft 작동 확인한다.

나머지 부분은 앞서 이미 설명했으니 생략한다.

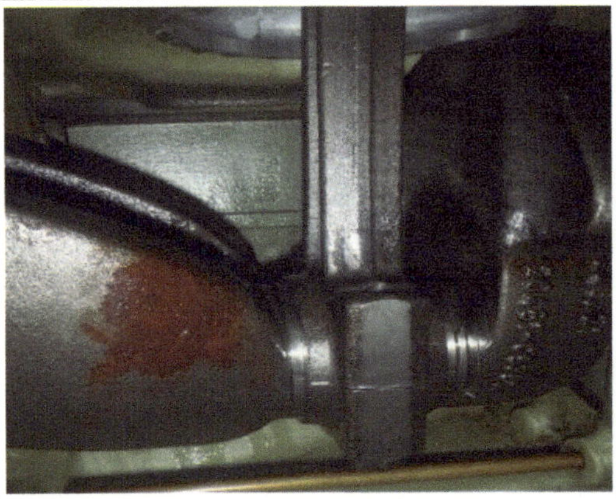

Con-rod & Con-rod Bearing

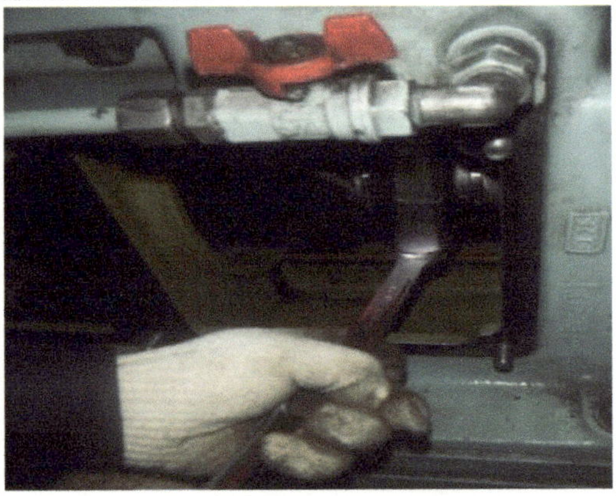

Con-rod Bearing Cover Bolt

Shaft (Hand Turning)

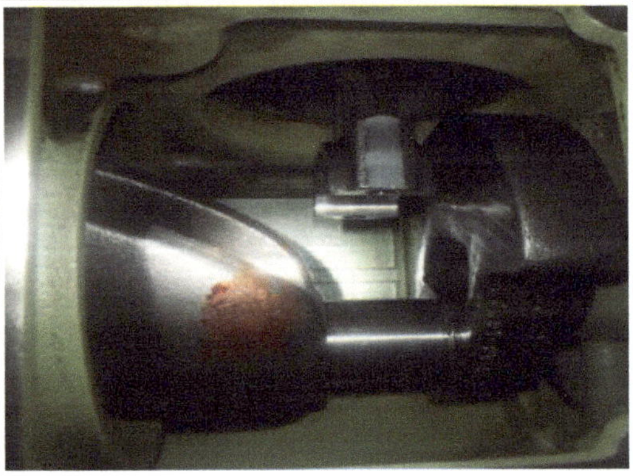

Crankshaft에서 Con-rod분해

Con-rod Bearing, Cover & Bolt

Cylinder Base Nut

Liner OV′HL

분해한 Piston Scratch & Scuffing

Piston Ring & 노끈

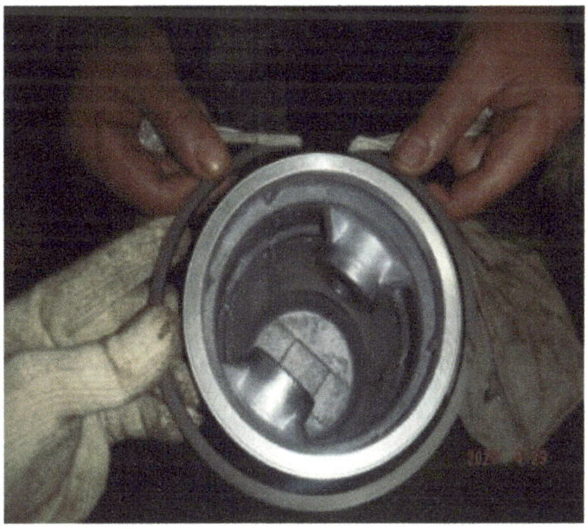

Piston Ring 조립

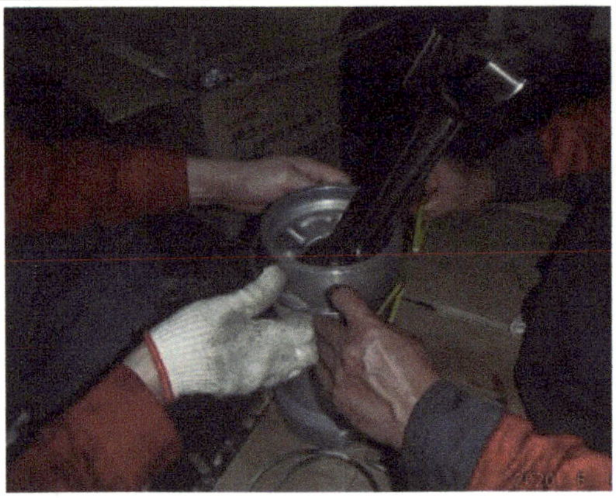

Piston Ring 조립

Piston Ring 조립된 Piston

Cylinder에 Piston 장착

상기와 같이 Piston을 Cylinder에 조립

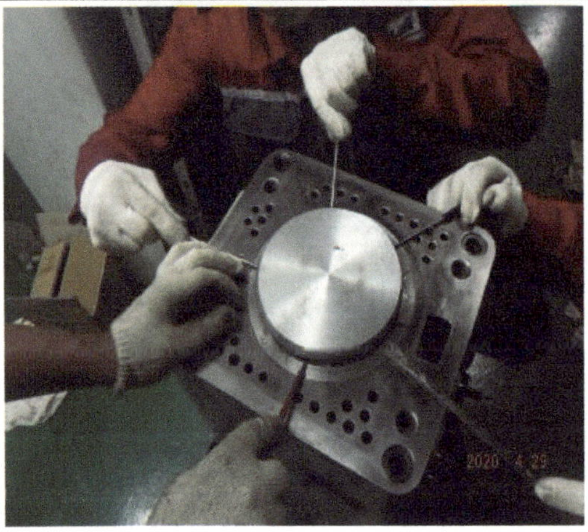

상기와 같이 Piston을 Cylinder에 조립

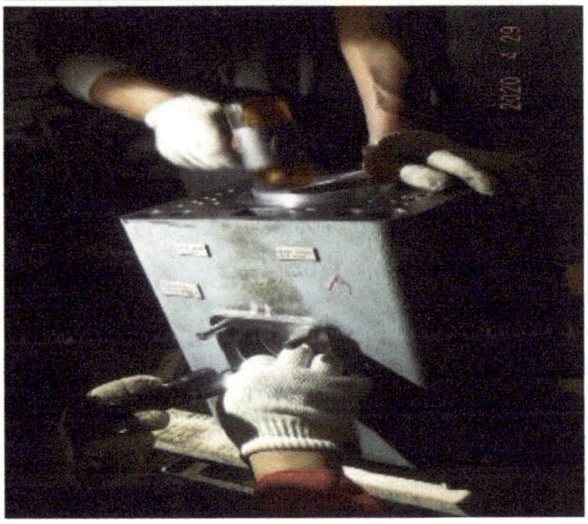

2ND Stage Valve 측으로 Piston Ring 조립 확인

Piston 조립

Cylinder 하부 Gasket

Cylinder 하부 Gasket

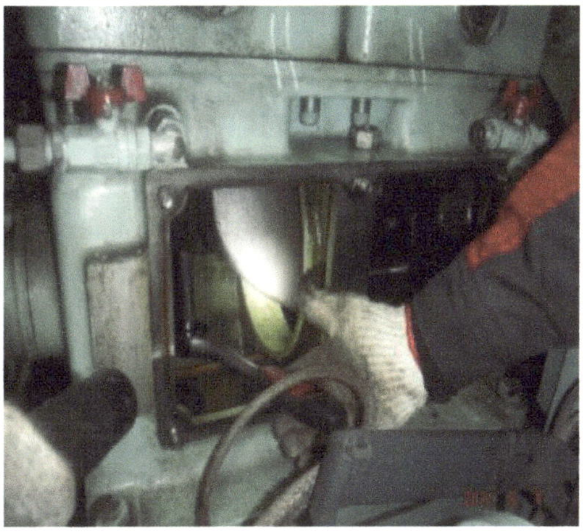

Con-rod Bolt Tight (Torque : 78Nm)

Cylinder Base Nut

L.O Filling

L.O 보충 완료하였으면 C.F.W(Cooling Fresh Water) Inlet Valve Open하여 Line Venting 한다.

Venting 완료되었으면 C.F.W Outlet Valve Open, Air Valve Open 한다.

Piston Ring교환하면 초기 기동 시 이전보다 Ampere가 조금 올라간다.

L.O는 Piston Ring 교환 후에는 수시로 L.O 색깔을 확인하여 오손되면 바로 교환한다.

만약에 기동하였는데 Safety Valve 작동한다면 Cylinder 하부 Gasket(**No.31**) 방향 재확인한다.

Gasket 제대로 조립되었다면 정비된 1ST & 2ND Stage Valve를 재확인한다.

(8) 라이너 분해 및 정비

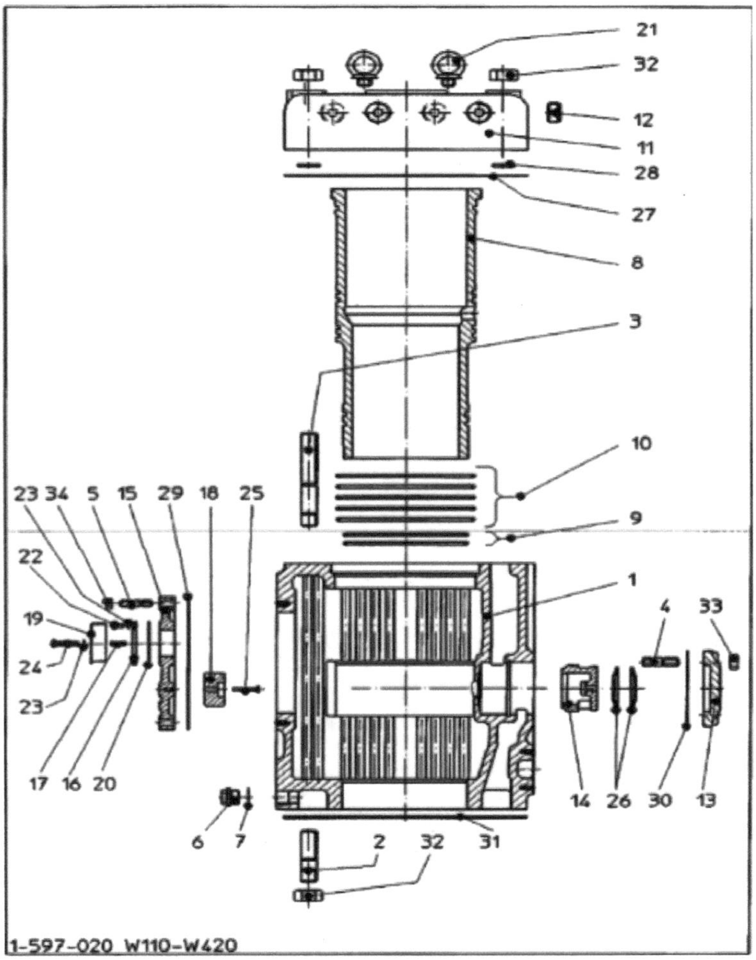

Liner OV'HL의 경우 앞서 설명한 부분의 연장선이다.

㉠ Liner(**No.8**) OV'HL 시에는 Cylinder를 다시 장착한다.

Cylinder Base Nut를 체결한 상태에서 Tool설치하여 Liner를 취외한다.

실제 승선한 선박에서, 이전 Liner OV'HL중에 Tool 설치하여 분해가 잘 되지 않았는지 Tool의 Thread와 Nut 부분을 용접하여 Cylinder 눕힌 다음 Tool을 설치하고 Hammer로 타격하여 취외한 이력이 있었다.

본선에 보유 중인 Tool이 하나밖에 없어 똑같이 분해하였고, 신품 Tool 을 청구하였다.

㉡ 분해한 Liner Scratch & Scuffing 확인 후 필요 시 교환한다.

Liner O-ring(**No.9** & **10**)들 전부 교환한다.

㉢ Liner 취외하고 Cylinder 내부 깨끗한 천으로 소제한다.

또한 Liner 취외하면 Cylinder 내부 Cooler 상태를 확인할 수 있다.

㉣ Liner를 Cylinder에 조립 시 기존에 장착된 방향에 맞게 조립한다.

조립 시 나무를 대고 Zinc로 쳐서 조립한다.

Piston과 마찬가지로 2ND Stage Valve 조립되는 곳으로 Liner 제대로 조립되는지 확인한다.

Liner 분해 Tool

Liner 분해 Tool(Liner 하부)

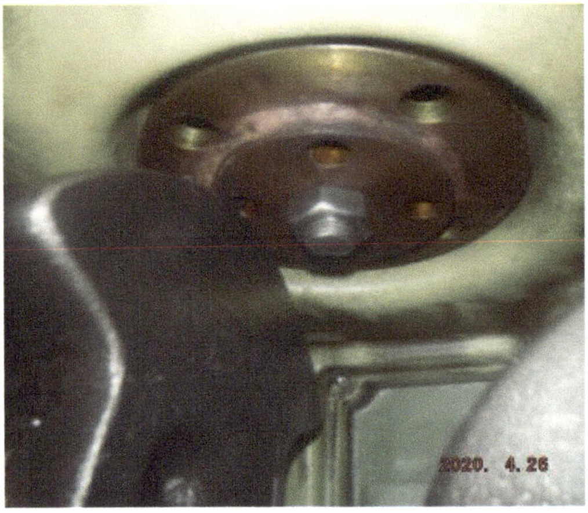

Liner 분해 Tool(Liner 상부)

이전 작업자들 분해 Tool Thread와 Nut 용접

Liner 분해

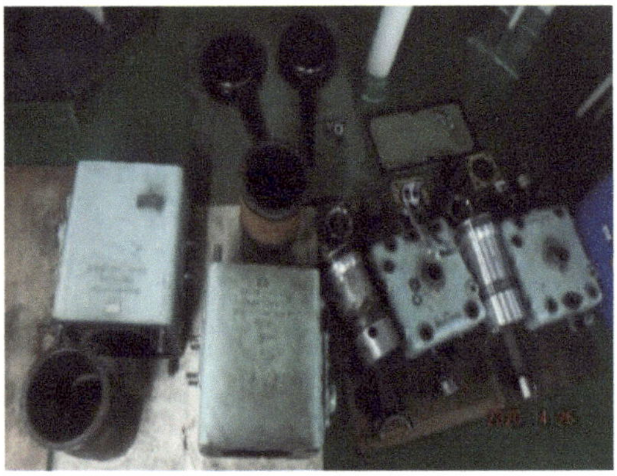

Liner OV′HL

취외품 Liner

신품 Liner O-ring(No.9 & 10) 조립

	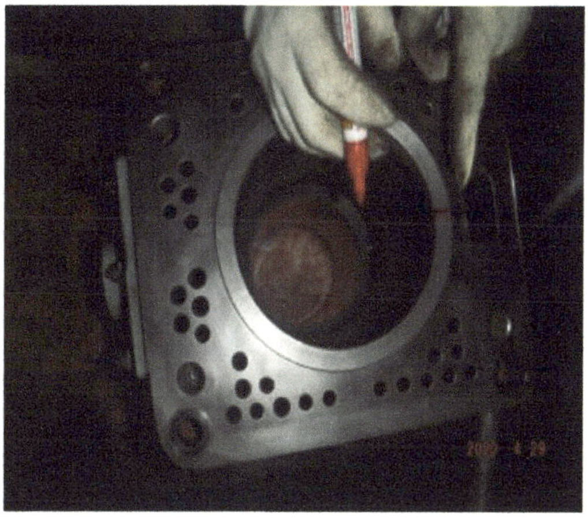
나무와 Zinc로 Liner 조립	Cylinder에 Liner조립 완료

* 순찰 중 기동 중인 Main Air Compressor Sight Glass에 흰색 거품 및 기포가 상당히 많이 형성되어 있다면, Local Position으로 변경 후 Crankcase Cover Open하여 L.O에서 Water가 혼입되었는지 반드시 확인한다.

Water 혼입되었다면, L.O 전량 Drain 한 다음 Shaft Hand Turning하여 Liner 누설 확인한다.

Liner에서 Water 누설한다면 Liner 분해, 정비를 진행한다.

따라서 항상 순찰 중에는 항상 Sight Glass를 확인하는 습관을 가져야 한다.

Compressor Shaft측의 Oil Seal은 앞에서 설명한 것과 동일하다.

맺음말

선박에서 일등 기관사는 결국 현장 리더가 되는 것입니다.

우리 모두 리더에 대해서 인간관계에 대해서 다시 한번 생각해 볼 필요가 있다고 생각합니다. 전문적인 지식과 기술을 습득하는데는 우리 모두가 많은 노력을 기울이지만, 정작 인간에 대해 인간관계에 대해서는 상대적으로 적은 노력을 기울이는 편인 것 같습니다.

인간을 이해하지 못 하고서 단순히 명령을 내리고, 조직원이 순순히 복종하는 것을 기대하기를 바라는 시니어 사관이 되어서는 안됩니다.

나 자신부터 분명한 가치관과 비전을 갖고 모범을 보여야 하며,

마음을 열고서 진정으로 다가서야 하며,

같이 승선하고 있는 주니어 사관들에게 마음을 열고 열렬한 협력을 얻어내야 하며,

갈등들을 효과적으로 해소해야 한다고 생각합니다.

오늘날 진정으로 필요한 리더에 대해, 미국의 저명한 경영 컨설턴트 피터 드러커는 이렇게 말했습니다.

"이제는 지위로 조직원 위에 군림하는 시대는 지나갔다.

인간적인 매력과 영향력으로 추종자를 만들어내야 한다."

조직원들의 사기를 북돋고 나아갈 길을 지도하고 조언하면서, 그들과 성공적인 인간관계를 유지해야 합니다.

우리가 살고 있는 이 21세기는 하이테크(High Tech)의 시대입니다. 치열한 경쟁 속에서 수많은 분야에서 많은 전문가가 탄생하고 있습니다. 그러나 조직원의 열렬한 협력을 얻어내는 하이 터치(High Touch) 리더는 주변에 그리 많지 않은 것 같습니다.

인간의 마음은 낙하산과 같아서 펼쳐지지 않으면 쓸 수가 없다고 했듯이, 우선 자기 마음부터 열고 모범을 보이면서 상대방의 마음을 열어 열정과 풍부한 정감을 불어넣는 리더, 창의력을 바탕으로 성과를 내는 리더, 인생의 의미와 가치를 전달해서 직업을 취미로 바꿔나가는 리더, 그가 바로 진정한 의미에서의 리더라고 생각합니다.

여러분의 가장 넓은 바다는 아직 항해되지 않았고, 가장 먼 여행은 아직 끝나지 않았습니다. 가보지 않은 그 바다와 여행은 그 끝이 창대할지, 그 마지막이 미약할지 알 수는 없고, 가보지 않은 길에 대한 두려움도 크지만 기대감은 여러분들의 발걸음을 재촉할 것입니다.

지금 여러분들이 걸어가는 그 길은 가고자 했던 길의 어디쯤인가요?

이 책을 읽는 여러분의 앞날에 항상 행운이 깃들고, 가족과 같은 분위기로 모든 선박에서 인정받는 사람이 되길 바라며 항상 다치지 말고 건강하길 바랍니다.

"남다른 삶이 선사하는 정신적인 풍요를 소중히 하며,
많은 사람들과 그 가치를 나눌 수 있는 그런 사람이 되길 바라며…"

편저자 **정 현 우**

Slow and Steady Wins the Race!

저자 정 현 우

목포해양대학교 졸업
한진해운 실습 기관사 및 산학 장학생
한진해운 3등 기관사, 2등 기관사
SM 상선 2등 기관사, 1등 기관사
SM 상선 (견)기관장
현] 한국 선급(KR) 선임 검사원

저서

3등 기관사 실무기초
선박용 전자제어엔진 실무
선박 내연기관 운용 실무

해광출판기
해광출판기

선박 내연기관 운용 실무

- MAN ES 캠 구동 엔진 (MC-C) -

2025년 5월 21일 인쇄　　　　정가 39,000원
2025년 5월 31일 발행

저　　자　정현우
편　　집　송영미 ⓣ
발 행 인　조준형　　http://www.hgbookpub.com
발 행 처　해광출판사　　E-mail : munbookcokr@naver.com
주　　소　부산광역시 중구 해관로 41-1 (중앙동2가)
전화번호　(051) 253-0001　　팩스 (051) 245-1187
등　　록　제 2012-000005호

해광출판사에서는 여러분의 소중한 원고와 함께 할 기회를 기다리고 있습니다. 책으로 엮을 원고나 아이디어가 있으신 분들은 이메일 munbookcokr@naver.com로 책에 대한 간단한 개요와 원고 전체 또는 일부를 연락처와 함께 보내주십시오.

ISBN 979-11-6881-815-6